CLYMER™

TECUMSEH

L-HEAD ENGINES

The world's finest publisher of mechanical how-to manuals

INTERTEC PUBLISHING CORPORATION
P.O. Box 12901, Overland Park, Kansas 66282-2901

FIRST EDITION
First Printing June, 1994

Printed in U.S.A.

ISBN: 0-89287-617-4

Library of Congress: 93-61211

Technical photography by Mike Morlan.

Technical illustrations by Steve Amos.

COVER: Photo courtesy of Tecumseh Products Company.

Chapter One
General Information 1

Chapter Two
Engine Fundamentals 2

Chapter Three
General Engine Information 3

Chapter Four
Troubleshooting 4

Chapter Five
Lubrication and Maintenance 5

Chapter Six
Fuel and Governor Systems 6

Chapter Seven
Ignition System and Flywheel Brake 7

Chapter Eight
Rewind Starters 8

Chapter Nine
Electrical System 9

Chapter Ten
General Inspection and Repair Techniques 10

Chapter Eleven
Engine Overhaul 11

Chapter Twelve
Tecumseh Tools 12

Chapter Thirteen
Craftsman/Tecumseh Model Numbers 13

Index 14

CONTENTS

QUICK REFERENCE DATA IX

CHAPTER ONE
GENERAL INFORMATION 1

Safety
Notes, cautions and warnings
Service hints
Special tools
Basic hand tools
Test equipment
Precision measuring tools
Fasteners
Sealants, cements and cleaners
Lubricants

CHAPTER TWO
ENGINE FUNDAMENTALS 21

Operating principles
Major engine components
Fuel system
Ignition system
Electrical system

CHAPTER THREE
GENERAL ENGINE INFORMATION 42

Fuel
Oil
Engine identification
Basic engine specifications
Purchasing parts

CHAPTER FOUR
TROUBLESHOOTING 49

Fuel
Controls
Starter
Troubleshooting

CHAPTER FIVE
LUBRICATION AND MAINTENANCE 57

Lubrication
Maintenance
Safety devices
Oil change
Air cleaner
Cooling system
Hoses and wires
Operating cable
Muffler
Compression test
Spark plug
Ignition breaker points and condenser
Ignition coil armature air gap (models with external ignition coil)
Ignition timing
Carburetor adjustment (except Vector VLV50, VLXL50, VLV55 and VLXL55)
Carburetor adjustment (Vector VLV50, VLXL50, VLV55 and VLXL55)
Governor linkage
Combustion chamber

CHAPTER SIX
FUEL AND GOVERNOR SYSTEMS 76

Operating linkage
Diaphragm carburetors
Tecumseh float carburetors
Walbro LME carburetor
Governor system

CHAPTER SEVEN
IGNITION SYSTEM AND FLYWHEEL BRAKE 106

Ignition coil (breaker-point ignition)
Ignition module/coil
Flywheel and key
Breaker point cam
Stop switch
Flywheel brake

CHAPTER EIGHT
REWIND STARTERS 115

Rewind starters mounted on blower housing
Early "teardrop" shaped starter
Starter used on HM and VM series engines
Stylized rewind starter
Vertical-pull rewind starters

CHAPTER NINE
ELECTRICAL SYSTEM 135

Electrical precautions
Testing
Battery
Charging/lighting system
Electric starter system
Troubleshooting
Electric starter motors

CHAPTER TEN
GENERAL INSPECTION AND REPAIR TECHNIQUES 151

Failure analysis
Precision measurements
Repair techniques

CHAPTER ELEVEN
ENGINE OVERHAUL 160

Muffler
Crankcase breather
Flywheel
Cylinder head
Valve system
Oil seals (service with crankshaft installed)
Internal engine components
Engine disassembly
Piston rings
Piston and piston pin
Connecting rod
Camshaft and tappets
Oil pump
Governor
Crankshaft and main bearings
Cylinder
Oil seals
Balancer shaft
Auxiliary drive shaft
Engine assembly

CHAPTER TWELVE
TECUMSEH TOOLS 213

CHAPTER THIRTEEN
CRAFTSMAN/TECUMSEH MODEL NUMBERS 215

INDEX 222

QUICK REFERENCE DATA

ENGINE OIL

Type	Viscosity	Ambient operating temperature
Regular grade	SAE 10W	Below 32° F (0° C)
	SAE 30	Above 32° F (0° C)
Multigrade	SAE 5W-30	Below 32° F (0° C)
	SAE 10W-30	Above 32° F (0° C)

TUNE-UP SPECIFICATIONS

Spark plug gap	0.030 in. (0.76 mm)
Breaker point gap	0.020 in. (0.50 mm)
Ignition coil armature air gap	0.0125 in. (0.32 mm)
Valve clearance (both valves)	
H50 (ignition under flywheel), H60, H70, HM70, H80, HM80, HM100, TVM125 (ignition under flywheel), TVM140, TVM170, TVM195, TVXL195, TVM220, TVXL220, V50, V60, V70, VM70, V80, VM100	0.010 in. (0.25 mm)
All other engines	0.008 in. (0.20 mm)
Tecumseh float carburetor	
Float level setting	0.162-0.215 in. (4.11-5.46 mm)
Walbro LME carburetor	
Float level setting	
8 hp (6 kW) engines	0.070-0.110 in. (1.78-2.79 mm)
All other engines	0.110-0.130 in. (2.79-3.30 mm)

SPECIAL TIGHTENING TORQUES

	ft.-lb.	in.-lb.	N•m
Carburetor to intake pipe		48-72	5.4-8.1
Cylinder head bolts			
Light frame & VLV50, VLXL50, VLV55 & VLXL55		180-200	20.3-24.9
All other engines		160-200	18.1-22.6
Flywheel nut			
Light frame & VLV50, VLXL50, VLV55 & VLXL55	33-36		45-49
Medium frame			
Ignition coil behind flywheel	35-42		48-57
Ignition coil outside flywheel	50-55		68-75
Intake pipe to cylinder		72-96	8.2-10.8
Muffler (small frame)		20-35	2.3-3.9
Muffler (medium frame)		90-150	11.3-14.7
Spark plug	15		20
Starter, electric		50-80	5.7-9.0
Starter, rewind			
Top mount		40-60	4.5-6.7
Side mount		50-70	5.7-7.9

CLYMER

TECUMSEH

L-HEAD ENGINES

CHAPTER ONE

GENERAL INFORMATION

SAFETY

Safety must be a constant concern for anyone working on or around machinery. Accidents can cause disabling injury to humans as well as damaging equipment. Although anticipating all manner of accidents possible is impossible, adhering to the following rules will reduce the possibility of accidents:

1. Never use gasoline as a cleaning solvent.
2. Never smoke or use a torch near flammable liquids such as cleaning solvent. Remember that open flames are present in some heaters, including water heaters and stoves.

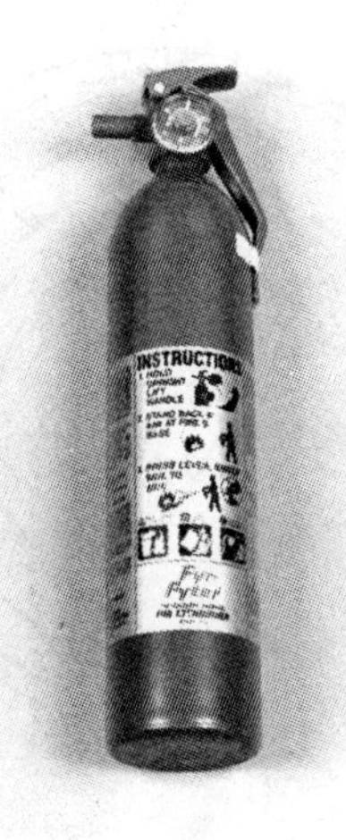

3. Never smoke or use a torch in an area where batteries are being charged. Highly explosive hydrogen gas is emitted during the charging process.
4. Place oil-soaked or solvent-soaked rags in a suitable closed metal container.
5. Disconnect the ground cable from the battery terminal when working on the electrical system. Never connect the posts on a battery either with wire or other metal objects, such as tools. The sparks may ignite hydrogen emitted from the battery, causing an explosion.
6. If welding or brazing, follow all recommended safety precautions prescribed by the American Welding Society. If in doubt as to proper procedure, take the work to an experienced welding shop.
7. Use the right tool for the job. Worn, improper or modified tools may cause injury and/or damage.
8. Be sure the parts meet the standards specified by the engine or equipment manufacturer. Installing incorrect or substandard parts can cause failure which may injure the technician or operator.
9. Keep the work area clean, uncluttered and well-lighted.
10. Wear appropriate safety equipment and clothing. Be sure safety equipment is designed to provide maximum protection and is properly used.
11. Be sure the shop is equipped with fire, safety and first aid equipment. An approved fire extinguisher (**Figure 1**) rated for gasoline (Class B) and electrical (Class C) fires should be nearby. A telephone should

be nearby with phone numbers of fire and medical service agencies highly visible on or near the phone.

12. Exercise extreme caution when using compressed air equipment, particularly blow guns. Always wear safety eyewear (**Figure 2**) when using compressed air. Do not use compressed air around other people and pets. Compressed air can hurl objects with sufficient force to cause injury. Never direct compressed air into skin or body openings, such as a cut, as this can cause severe injury or death.

13. When drying bearings or other rotating parts with compressed air, never allow the air jet to rotate the bearing or part. The air jet is capable of rotating them at speeds far in excess of those for which they were designed. The bearing or rotating part is very likely to disintegrate and cause serious injury and damage. To prevent bearing damage when using compressed air, hold the inner bearing race (**Figure 3**) by hand.

14. Small children, bystanders and pets should be kept out of the work area and away from equipment that is in an unsafe condition, i.e., equipment with safety guards removed while undergoing service.

15. Observe all safety notations on equipment. If safety systems must be defeated for servicing, be sure safety systems are functional before operating equipment.

16. If the equipment or engine must be lifted or supported, be sure proper equipment is used and in a safe manner. Be sure the equipment or engine is secure before applying great force to tools, such as when loosening or tightening cylinder head screws. Do not leave the equipment or engine in a position that someone else may inadvertently knock or bump over.

17. Be sure all electrical tools are in safe working order and properly grounded. The work area must be dry.

18. Do not use liquids from unmarked containers.

19. Follow directions and note safety precautions specified by manufacturers of fluids, cleaning solvents and adhesives.

20. Think through all procedures before-hand to anticipate possible problems.

21. Do not run an engine in an enclosed space. The area must be well-ventilated.

22. Wear rubber gloves and safety eyewear when handling a battery or battery acid.

NOTES, CAUTIONS AND WARNINGS

The terms NOTE, CAUTION and WARNING have specific meanings in this manual. A NOTE provides additional information to make a step or procedure easier or clearer. Disregarding a NOTE could cause inconvenience, but would not cause damage or personal injury.

A CAUTION emphasizes areas where equipment damage could occur. Disregarding a CAUTION could cause permanent mechanical damage; however, personal injury is unlikely.

A WARNING emphasizes areas where personal injury or even death could result from negligence. Mechanical damage may also occur. WARNINGS *are to be taken seriously.* In some cases, serious injury and death have resulted from disregarding similar warnings.

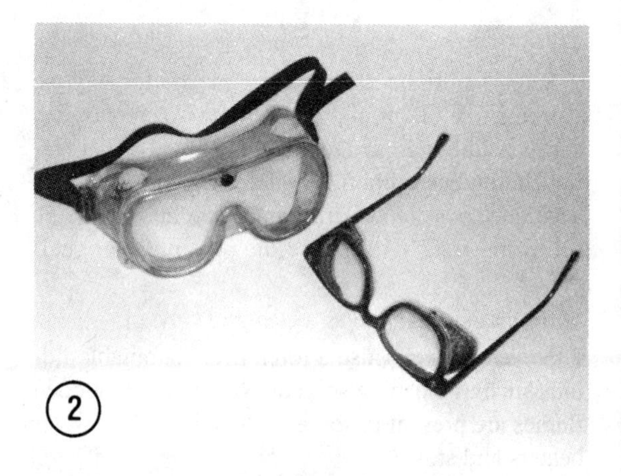

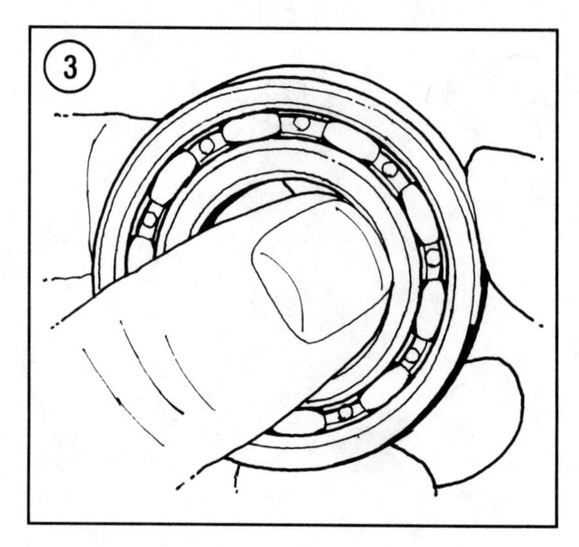

SERVICE HINTS

Most of the service procedures outlined in this manual are straightforward and can be performed by anyone reasonably competent with tools. It is suggested, however, that you consider your own capabilities carefully before attempting any operation involving major disassembly.

1. Safety first must be the overriding concern of any mechanic, regardless of experience. Read and follow the safety rules in the preceding *Safety* section.
2. With regard to procedural terms used in this manual, the term "replace" means to discard a defective part and install a new or exchange unit. "Overhaul" means to remove, disassemble, inspect, measure, repair or replace defective parts, reassemble and install major systems or parts.
3. Before undertaking any work on an engine or piece of equipment, be sure the unit is secured in a safe manner. Poorly supported or secured equipment can cause damage to the unit, as well as possible personal harm. Two hands are often needed to perform a task, which is difficult if one hand is required to hold the unit in place.
4. Repairs go much faster and easier if the engine or equipment is clean before work starts. There are several special cleaners for washing the engine and related parts. Spray or brush on the cleaning solution, following the manufacturer's directions. Rinse parts with a garden hose. Be sure the solution can be safely disposed of according to local regulations. Clean all oily or greasy parts with cleaning solvent when removed.
5. Much of the labor cost charged by mechanics is to remove and disassemble other parts to reach the defective unit. It is usually possible to perform the preliminary operations yourself and then take the defective unit to the dealer for repair.

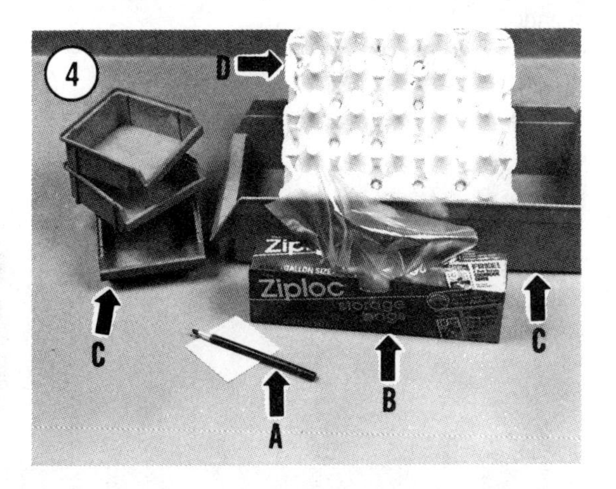

6. Read the complete service procedure in this manual while looking at the actual parts before performing any work. Be sure the proper procedure is being used (the procedure may be different for various models). Study the illustrations and text until you have a good idea of what is involved in completing the job satisfactorily.
7. If special tools or replacement parts are required, make arrangements to get them before starting work. If special tools must be obtained, either by purchase or renting, determine whether it is better to have the job performed by a professional shop, keeping in mind that if the tool is purchased it will be available for future jobs. Also keep in mind that some jobs may require experience that is available only at a professional shop. This experience can often be acquired through adult-education courses, as well as experimentation, if the consequences of a mistake are understood.
8. Other than simple tests, electrical testing may require sophisticated test equipment and a knowledge of electronics.

CAUTION

Improper electrical testing can sometimes damage electrical components.

9. During disassembly, keep a few general cautions in mind. Excessive hand force is rarely needed to get things apart. If parts are a tight fit, such as a bearing in a case, there is usually a tool designed to separate them. Never use a screwdriver to pry parts with machined surfaces. You will mar the surfaces, which will promote leaks.
10. Make diagrams (or take a Polaroid picture) wherever similar-appearing parts are found. For instance, retaining screws may be of different lengths. Trying to remember where everything was originally located may prove costly. If the job must be stopped for a long period of time, which can occur when ordering parts, remembering details may be impossible.
11. Tag all similar internal parts for location and mark all mating parts for position. Record the number and thickness of any shims as they are removed (A, **Figure 4**); measure with a Vernier caliper or micrometer. Small parts such as screws can be identified by placing them in plastic sandwich bags (B, **Figure 4**). Seal and label them with masking tape.

12. Place parts from a specific area of the engine (e.g., valves, camshaft, balancer, etc.) in boxes (C, **Figure 4**) to keep them separated.
13. When disassembling shaft assemblies, use an egg carton (D, **Figure 4**) and set the parts in the depressions in the order they are removed.
14. Wiring should be tagged with masking tape and marked as each wire is removed. Again, do not rely on memory alone, especially if the wiring has been altered.
15. Finished surfaces should be protected from physical damage and corrosion. Keep gasoline off painted surfaces.
16. Use penetrating oil on frozen or tight screws or bolts, then strike the screw or bolt head a few times with a hammer and punch (use a screwdriver on screws). Avoid the use of heat where possible, as it can warp, melt or affect the temper of parts. Heat also ruins paint and plastics.
17. No parts removed or installed (other than bushings and bearings) in the procedures found in this manual should require unusual force during disassembly or assembly. If a part is difficult to remove or install, find out why before proceeding.
18. Cover all openings after removing parts or components to prevent entrance of dirt or other foreign material.
19. When assembling parts, be sure all shims and washers are installed exactly as they were removed.
20. Whenever a rotating part butts against a stationary part, look for a shim or washer.
21. Installing new gaskets rather than reusing old gaskets is a good practice. Although using old gaskets may be the only possibility in some instances, the probability of a leak occurring is high, resulting in another teardown. Gaskets can be cut from sheets or rolls of gasket material of the same thickness as the old gaskets. Use of gasket forming compounds may be a better alternative to reusing old gaskets.
22. Heavy grease can be used to hold parts in place if they tend to fall out during assembly.
23. Never use wire to clean carburetor jets and air passages. The wire may disfigure the orifices, thereby affecting fuel or air flow.
24. Compressed air is helpful when drying parts and to dislodge debris from passages. Be sure compressed air is not directed against diaphragms or other delicate parts that may be damaged by the sudden force of the air stream. Do not spin bearings or rotating parts with compressed air; the part can be damaged by rotating at high speed.
25. A baby bottle makes a good measuring device for liquids. Get one that is graduated in fluid ounces and cubic centimeters. *Do not* allow a baby to drink out of it after used in the workshop as residue from the oil or solvent will always remain.
26. Treat a rebuilt engine as a new engine and follow the proper break in procedure.
27. Rushing to complete a job usually creates mistakes that require repeating the job. Take your time and do it right the first time.

SPECIAL TOOLS

Tecumseh offers tools that are designed to accomplish specific jobs on Tecumseh engines. Where the use of a Tecumseh tool is necessary to satisfactorily accomplish a particular job, the Tecumseh tool number is specified in the text. Tecumseh tools can be obtained through a dealer or distributor. See Chapter Three for a list of Tecumseh distributors who can provide the name of a local dealer. In some cases, the tool is also available from a tool manufacturer or from an aftermarket parts supplier. A small-engine dealer can usually obtain the tools or provide the name of a supplier.

Be careful when substituting a "home-made" tool for a recommended tool. Although time and money may be saved if an existing or fabricated tool can be made to work, consider the possibilities if the tool does not work properly. If the substitute tool damages the engine, the cost to repair the damage, as well as losing time, may exceed the cost of the recommended tool.

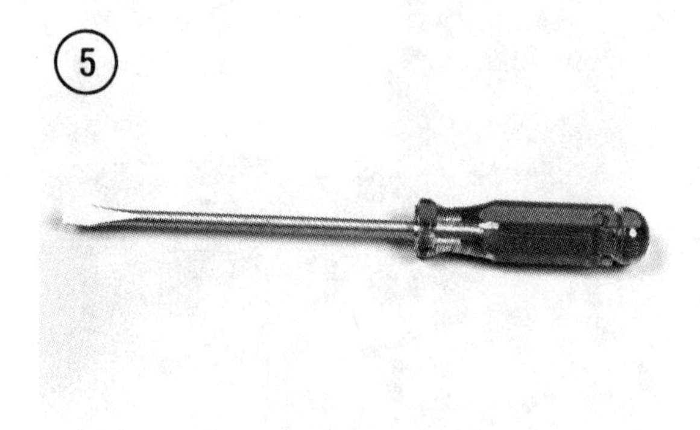

BASIC HAND TOOLS

A good set of tools is essential to undertaking engine repair in a safe, efficient and satisfactory manner. The quality and extent of a mechanic's tool collection should be determined by the work to be performed and the frequency of use. Cheap, low-quality tools often break and do not perform well, often damaging the equipment as well as exposing the mechanic to injury if the tool should fail during a high-risk job. Buying a tool for use one time may be a waste of money if the tool could be rented instead, or the job requiring the special tool could be performed by a repair or machine shop for a nominal fee.

There is a collection of basic tools that is present in the tool set of any engine mechanic. A journeyman mechanic may purchase a complete "A to Z" set of tools at the outset, since the tools are essential for quality, productive work, but most experienced do-it-yourselfers build their tool collections based on what is needed. As experience is gained, more complex jobs are tackled and the tools needed are added to the collection.

The following points should be considered when purchasing tools. Quality tools may be expensive at the outset, but they last longer, fit better and may have features not found in cheaper tools. Tools can be purchased by mail order, at parts stores, at tool supply outlets, and from general merchandisers. All of the former offer high quality as well as cheap tools. A quality tool manufacturer will replace a tool if it breaks or wears out when used for jobs it was designed. Before purchasing tools, ask the supplier if free replacement is guaranteed by the tool manufacturer and what the procedure is; free replacement may be a waste of time if not easy and prompt.

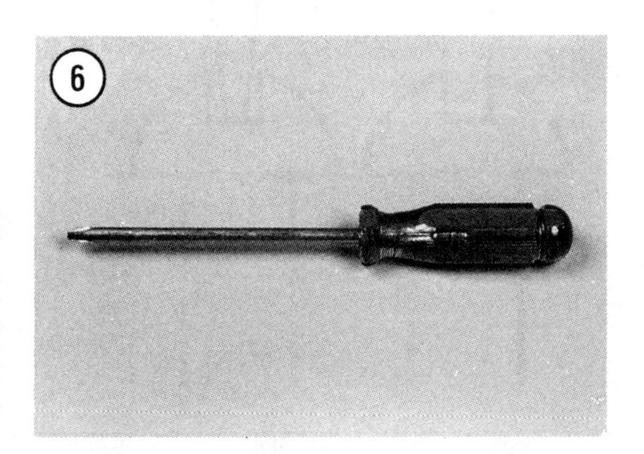

Before purchasing tools, be sure of the type fasteners used on the engine, whether U.S. standard or metric. Be aware that both U.S. and metric tools may be needed as both types of fasteners may be present; the engine block may have U.S. fasteners while the engine components, such as the carburetor, may have metric fasteners.

Some of the more common hand tools that are frequently needed during engine service and repair are outlined in the following paragraphs.

Screwdrivers

The screwdriver is a very basic tool, but if used improperly it will do more damage than good. The slot on a screw has a definite dimension and shape. Through improper use or selection, a screwdriver can damage the screw head, making removal of the screw difficult. A screwdriver must be selected to conform to the shape of the screw head used. Two basic types of screwdrivers are required: standard (flat-blade or slot-blade) screwdrivers (**Figure 5**) and Phillips screwdrivers (**Figure 6**).

Note the following when selecting and using screwdrivers.

1. The screwdriver must always fit the screw head. If the screwdriver blade is too small for the screw slot, damage may occur to the screw slot and screwdriver. If the blade is too large, it cannot engage the slot properly and will result in damage to the screw head.
2. Standard screwdrivers are identified by the length of their blade. A 6 in. screwdriver has a blade 6 in. long. The width of the screwdriver blade will vary, so make sure that the blade engages the screw slot the complete width of the screw.
3. Phillips screwdrivers are sized according to their point size and numbered one through four. The degree of taper determines the point size with a number one being the most pointed. The points become more blunt as their number increases.

NOTE

You should also be aware of another screwdriver similar to the Phillips, and that is the Reed and Prince tip. Like the Phillips, the Reed and Prince screwdriver tip forms an "X" but with one major exception, the Reed and Prince screwdriver has a much more pointed tip. The Reed and Prince screwdriver should never be used on Phillips screws and vice versa. Intermixing these screwdrivers will cause damage to the screw and screwdriver. If you have both types in your tool box and they are similar in appearance, you may want to identify them by painting the screwdriver shank underneath the handle.

4. When selecting screwdrivers, note that you can apply more power with less effort with a longer screwdriver than with a short one. Of course, there will be situations where only a short handled screwdriver can be used. Keep this in mind though, when removing tight screws.

5. Because the working end of a screwdriver receives quite a bit of abuse, you should purchase screwdrivers with hardened tips. The extra money will be well spent.

Screwdrivers are available in sets which often include an assortment of standard and Phillips blades. If you buy them individually, buy at least the following:

a. Standard screwdriver—5/16 × 6 in. blade.
b. Standard screwdriver—3/8 × 12 in. blade.
c. Phillips screwdriver—size 2 tip, 6 in. blade.
d. Phillips screwdriver—size 3 tip, 6 and 8 in. blade.

Use screwdrivers only for driving screws. Never use a screwdriver for prying or chiseling metal. Do not try to remove a Phillips, Torx or Allen head screw with a standard screwdriver (unless the screw has a combination head that will accept either type); you can damage the head so that the proper tool will be unable to remove it.

Keep screwdrivers in the proper condition and they will last longer and perform better. Always keep the tip of a standard screwdriver in good condition. **Figure 7** shows how to grind the tip to the proper shape if it becomes damaged. Note the symmetrical sides of the tip.

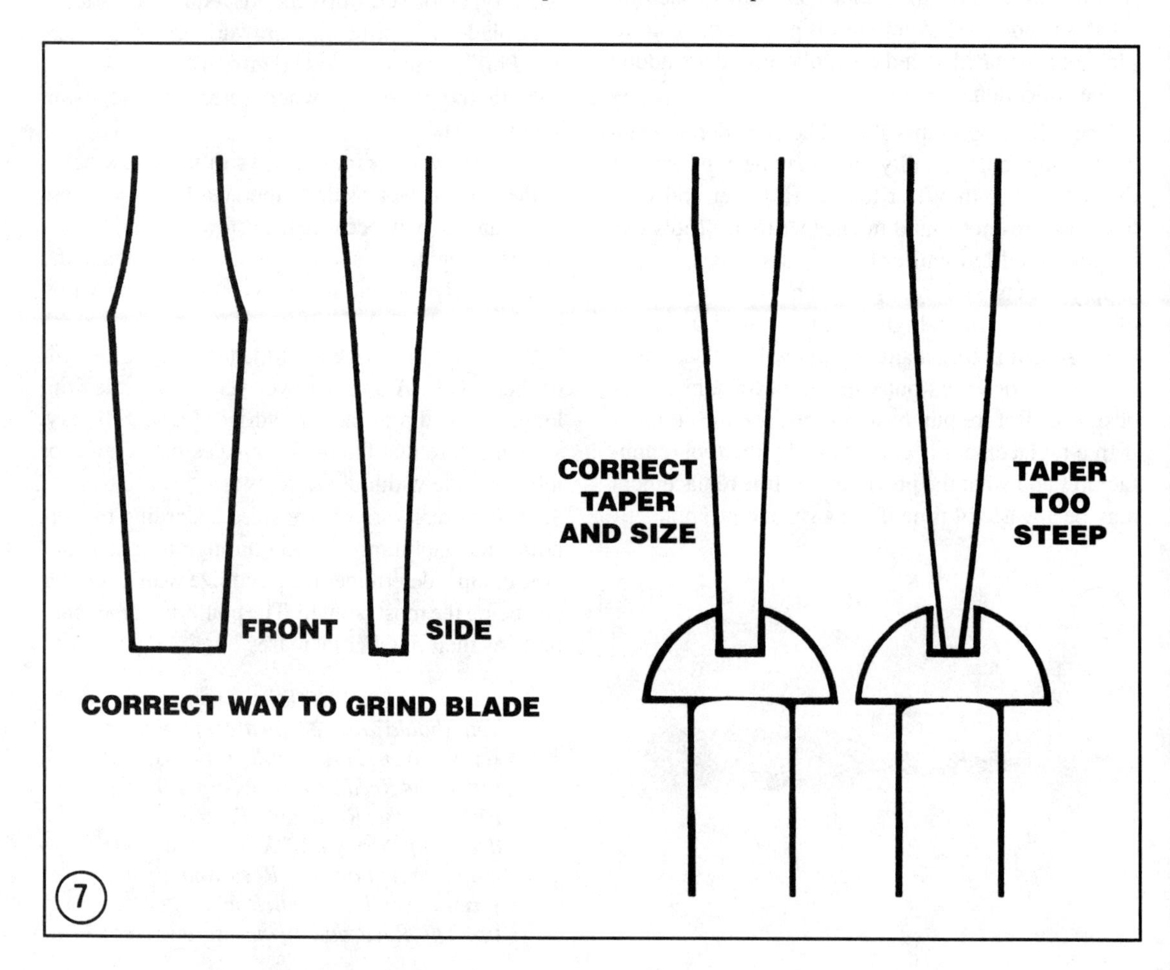

Pliers

Pliers come in a wide range of types and sizes. Pliers are useful for cutting, bending and crimping. They should never be used to cut hardened objects or to turn bolts or nuts. **Figure 8** illustrates several types of pliers useful for engine service and repair.

Each type of pliers has a specialized function. Slip-joint pliers are general purpose pliers and are used mainly for holding things and for bending.

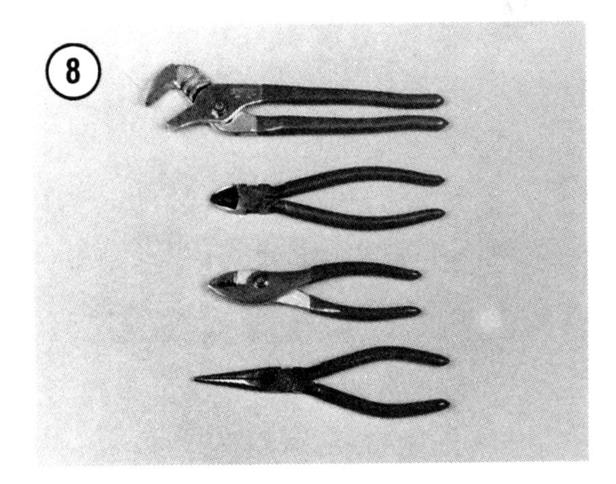

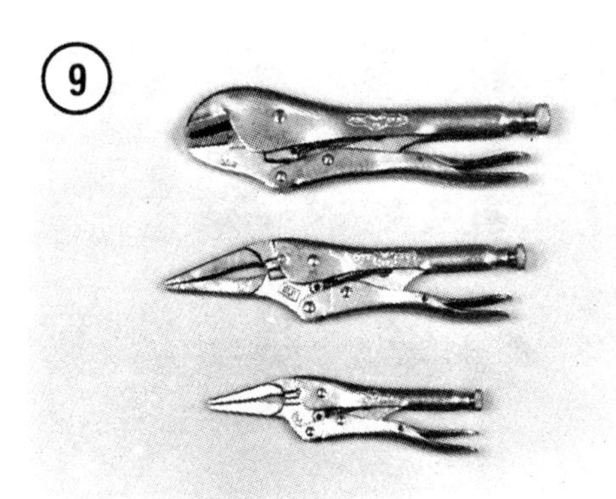

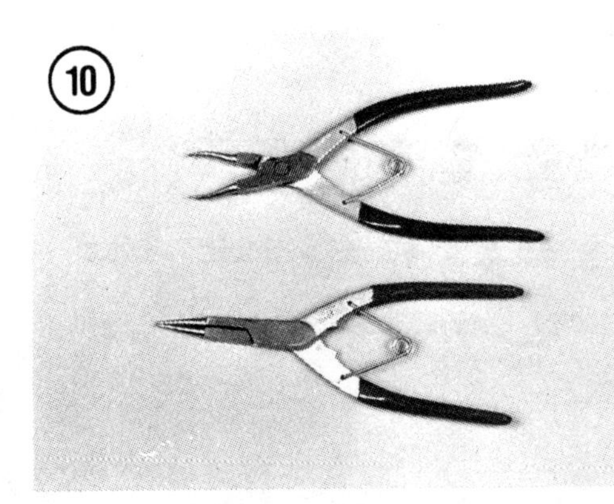

Needlenose pliers are used to hold or bend small objects. Water pump pliers can be adjusted to hold various sizes of objects; the jaws remain parallel to grip around objects such as pipe or tubing. There are many more types of pliers.

CAUTION
Pliers should not be used for loosening or tightening nuts or bolts. The pliers' sharp teeth will grind off the nut or bolt corners and damage it.

CAUTION
If slip-joint or water pump pliers are going to be used to hold an object with a finished surface, wrap the object with heavy tape or rubber for protection.

Locking Pliers

Locking pliers (**Figure 9**) are used to hold objects very tightly while another task is performed on the object. While locking pliers work well, caution should be followed with their use. Because locking pliers exert more force than regular pliers, their sharp jaws can permanently scar the object. In addition, when locking pliers are locked into position, they can crush or deform thin-walled material.

Locking pliers are available in many types for more specific tasks.

Snap Ring Pliers

Snap ring pliers (**Figure 10**) are special in that they are used to remove or install snap rings. When purchasing snap ring pliers, there are two kinds from which to choose. External pliers (spreading) are used to remove snap rings that fit on the outside of a shaft. Internal pliers (squeezing) are used to remove snap rings that fit inside a housing.

WARNING
Because snap rings can sometimes slip and "fly off" during removal and installation, always wear safety glasses when servicing them.

Box-end, Open-end and Combination Wrenches

Box-end and open-end wrenches (**Figure 11**) are available in sets or separately in a variety of sizes. The size number stamped near the end refers to the distance between two parallel flats on the hex head bolt or nut.

Box-end wrenches are usually superior to open-end wrenches. Open-end wrenches grip the nut on only two flats. Unless a wrench fits well, it may slip and round off the points on the nut. The box-end wrench grips on all six flats. Both 6-point and 12-point openings on box-end wrenches are available. The 6-point gives superior holding power; the 12-point allows a shorter swing.

Combination wrenches which are open on one side and boxed on the other are also available. Both ends are the same size.

No matter which style of wrench is used, proper use is important to prevent personal injury. When using a wrench, get into the habit of pulling the wrench toward you. This technique will reduce the risk of injuring your hand if the wrench should slip. If you have to push the wrench away from you to loosen or tighten a fastener, open and push with the palm of your hand; your fingers and knuckles will be out of the way if the wrench slips. Before using a wrench, always think ahead as to what could happen if the wrench should slip or if the fastener strips or breaks.

Adjustable Wrenches

An adjustable wrench can be adjusted to fit nearly any nut or bolt head which has clear access around its entire perimeter. Adjustable wrenches are best used as a backup wrench to keep a large nut or bolt from turning while the other end is being loosened or tightened with a proper wrench. See **Figure 12**.

Adjustable wrenches have only two gripping surfaces which makes them more subject to slipping off the fastener and damaging the part and possibly your hand. See *Box-end, Open-end and Combination Wrenches* in this chapter.

These wrenches are directional; the solid jaw must be the one transmitting the force. If you use the adjustable jaw to transmit the force, it will loosen and possibly slip off.

Adjustable wrenches come in a variety of sizes, but 6 in. and 8 in. wrenches are generally most useful.

Socket Wrenches

This type is undoubtedly the fastest, safest and most convenient to use. Sockets which attach to a ratchet handle (**Figure 13**) are available with 6-point or 12-point openings and 1/4, 3/8, 1/2 and 3/4 in. drives. The drive size indicates the size of the square hole which mates with the ratchet handle.

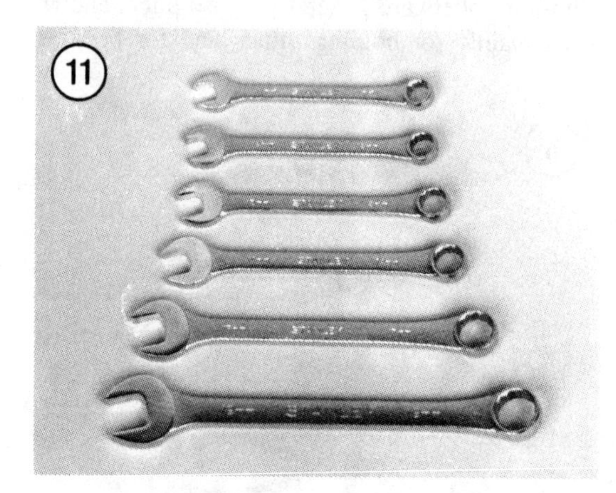
11

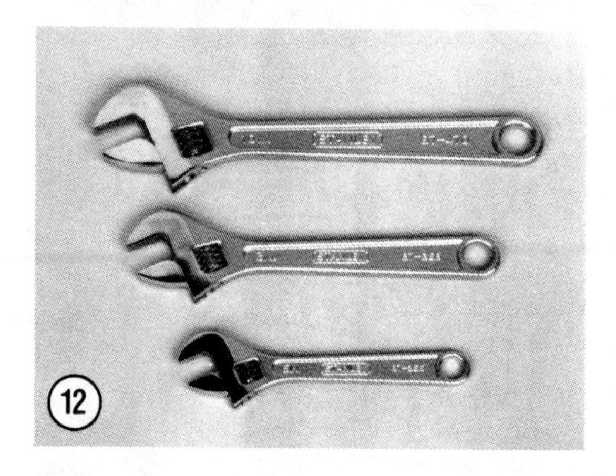
12

13

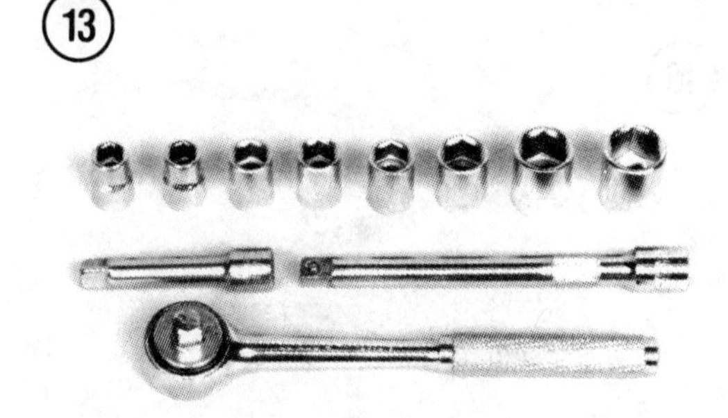

Torque Wrench

A torque wrench (**Figure 14**) is used with a socket to measure how tightly a nut or bolt is installed. They come in a wide price range and with either 3/8 or 1/2 in. square drives. The drive size indicates the size of the square drive which mates with the socket.

For general small engine repair, a torque wrench that measures 0-200 in.-lb. (0-23 N•m) and one that measures 0-150 ft.-lb. (0-200 N•m) will be most useful.

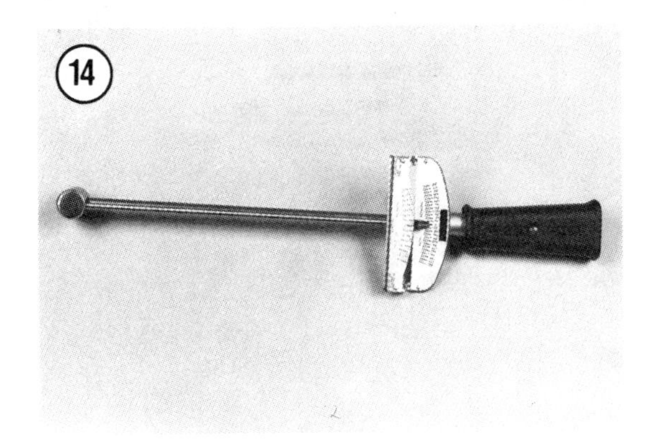

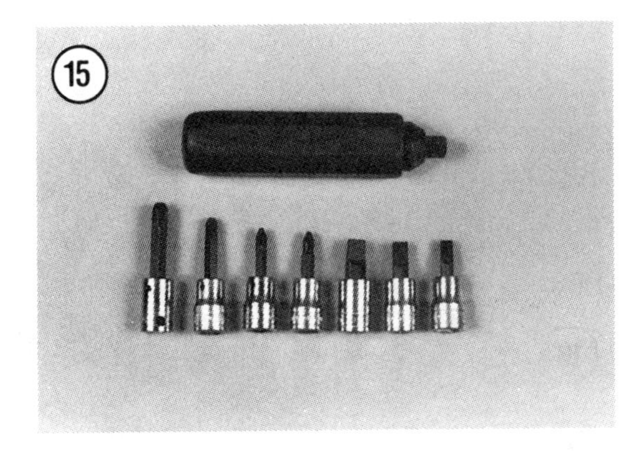

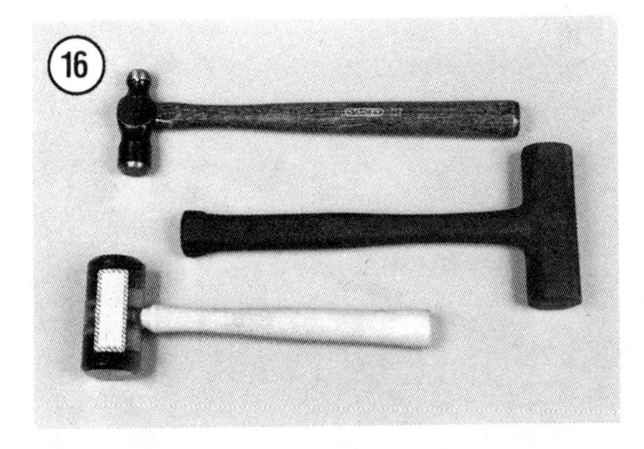

Impact Driver

This tool makes removal of tight fasteners easy and eliminates damage to bolts and screw slots. Impact drivers and interchangeable bits (**Figure 15**) are available at most large hardware and tool stores. Don't purchase a cheap impact driver as it won't operate as well as a moderately priced impact driver. Sockets can also be used with a hand impact driver. However, make sure the socket is designed for use with an impact driver or air tool. Do not use regular hand-type sockets, as they may shatter during use.

Hammers

The correct hammer (**Figure 16**) is necessary for repairs. Use only a hammer with a face (or head) of rubber or plastic or the soft-faced type that is filled with lead shot. These are sometimes necessary in engine teardowns. *Never* use a metal-faced hammer on engine components as severe damage will result in most cases. Ball-peen or machinist's hammers will be required when striking another tool, such as a punch or impact driver. When striking a hammer against a punch, cold chisel or similar tool, the face of the hammer should be at least 1/2 in. larger than the head of the tool. When it is necessary to strike hard against a steel part without damaging it, a brass hammer should be used. A brass hammer can be used because brass will give when striking a harder object.

When using hammers, note the following.

1. *Always* wear safety glasses when using a hammer.
2. Inspect hammers for damaged or broken parts. Repair or replace the hammer as required. *Do not* use a hammer with a taped handle.
3. Always wipe oil or grease off of the hammer before using it.
4. The head of the hammer should always strike the object squarely. Do not use the side of the hammer or the handle to strike an object.
5. Always use the correct hammer for the job.

Allen Wrenches

Allen wrenches (**Figure 17**) are available in sets or separately in a variety of sizes. These sets come in SAE and metric size. Allen screws are sometimes called socket screws. Note that a variety of tools are available as shown in **Figure 17** to fit Allen screws, which may be located in confined areas.

Chisels and Punches

Chisels and punches (**Figure 18**) are made of tool steel and configured in a variety of shapes and sizes. Punches can be used to create a locating dimple for drilling, aligning holes in mating parts, driving out pins and applying force in tight areas. Chisels are useful when metal must be chipped, gouged or cut, usually in a last-resort situation.

CAUTION
Never use a screwdriver as a punch or chisel. The handle and metal in the blade are not capable of withstanding sharp hammer blows. The screwdriver will be ruined and a chisel or punch would probably have done the job better.

Files

A selection of files (**Figure 19**) is needed to cut metal, such as smoothing large flat areas or removing burrs and irregularities. Files are available in various lengths, shapes and cutting patterns. Double-cut files are generally preferred for general engine work. A file handle should be attached to the file. A file card should be used to clean metal chips from the file teeth.

Tap and Die Set

A complete tap and die set (**Figure 20**) is a relatively expensive tool. But when you need a tap or die to restore a damaged thread, a tap and die set recoup the price over a period of time both in parts costs and lost time. Tap and die sets are available for both U.S. standard and metric threads.

Screw Extractors

When a screw or bolt is broken off in a hole, a screw extractor (**Figure 21**) can sometimes be used to remove the screw or bolt. After a hole is drilled in the screw or bolt, the extractor is inserted in the hole and the left-hand flutes on the extractor grip the screw or bolt so it can be unscrewed. Use of screw extractors is discussed in Chapter Ten under *Removing Broken Screws or Bolts.*

Drivers and Pullers

These tools are used to remove and install oil seals, bushings, bearings, gears and flywheels. These will be called out as needed in the service sections of this manual.

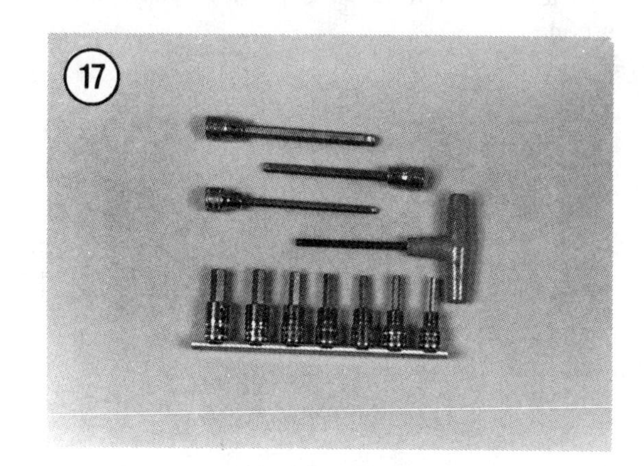
17

18

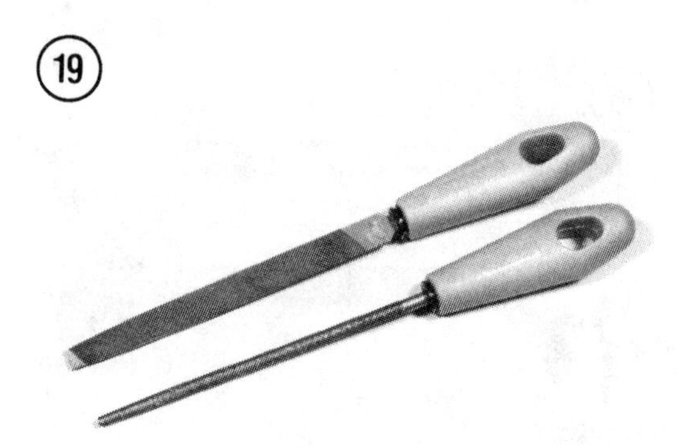
19

Engine Overhaul Tools

Tools such as valve spring compressors, cylinder ridge reamers, cylinder hones, piston ring compressors and valve service tools are discussed in Chapter Eleven.

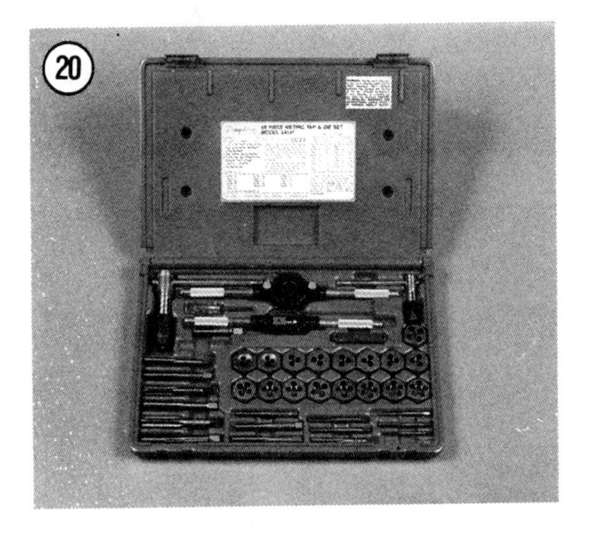

TEST EQUIPMENT

Spark Tester

A quick way to check the ignition system is to connect a spark tester (**Figure 22**) to the end of the spark plug wire and operate the engine's starter. A visible spark should jump the gap on the tester. A variety of spark testers is available from engine and aftermarket manufacturers. The gap distance is adjustable on some testers, while more sophisticated testers have a pressure chamber that simulates compression pressure in the cylinder.

Multimeter or Volt-Ohmmeter (VOM)

This instrument (**Figure 23**) is invaluable for electrical system troubleshooting and service. A few of its functions may be duplicated by homemade test equipment, but for the serious mechanic it is a must. Its uses are described in the applicable section of the book.

Compression Gauge

An engine with low compression cannot be properly tuned and will not develop full power. A compression gauge measures engine compression. The one shown in **Figure 24** has a flexible stem with an extension that permits more convenient reading of the gauge. The fitting on the hose end screws into the spark plug hole. Some types of compression gauges have a rubber tip (**Figure 25**) that is held manually in the spark plug hole.

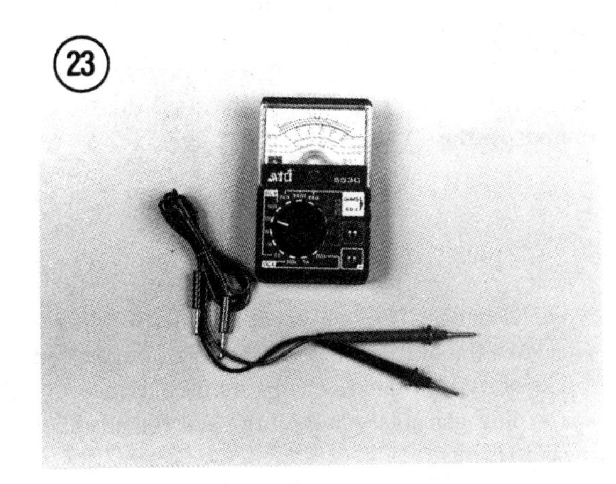

Battery Hydrometer

A hydrometer (**Figure 26**) is the best way to check a battery's state of charge. A hydrometer measures the weight or density (specific gravity) of the battery acid.

Portable Tachometer

A portable tachometer (**Figure 27**) is necessary for tuning the engine. Carburetor and governor adjustments must be performed at specific engine speeds. Two types are available, a mechanical type that uses a reed to sense engine rpm or an electrical type that senses engine rpm inductively at the spark plug wire.

PRECISION MEASURING TOOLS

Measurement is an important part of servicing any engine. When performing many of the service procedures in this manual, it will be necessary to make a number of measurements. These include basic checks such as engine compression and spark plug gap. Engine overhauling will require measurements to determine the condition of the piston, cylinder bore, crankshaft and other engine components. When making these measurements, the degree of accuracy will dictate which tool is required. Precision measuring tools are expensive. To avoid purchase of an expensive tool for one-time use, it may be wise to have the measurement performed by a professional shop. Some measuring tools, such as micrometers, require experience to be used accurately. This "feel" should be attained before making critical measurements. Following is a descriptive list of measuring tools that might be used during an engine overhaul.

Feeler Gauges

Feeler gauges (**Figure 28**) are made of either a piece of a flat or round hardened steel of a specified thickness. Wire (round) gauges are used to measure spark plug gap. Flat gauges are used for all other measurements.

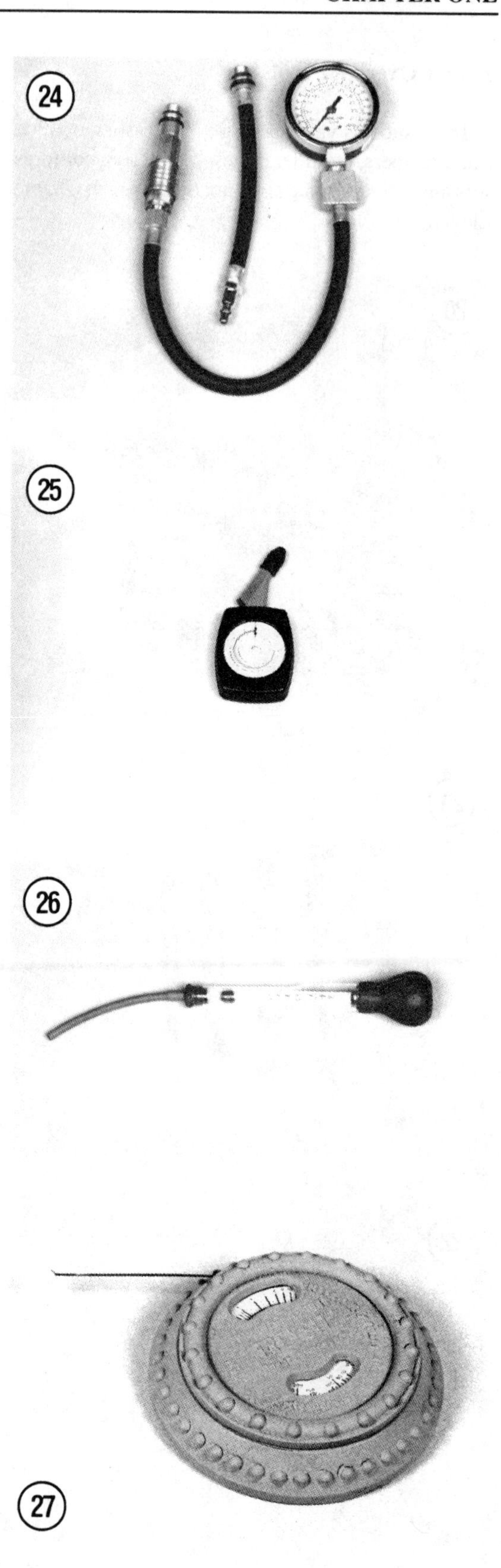

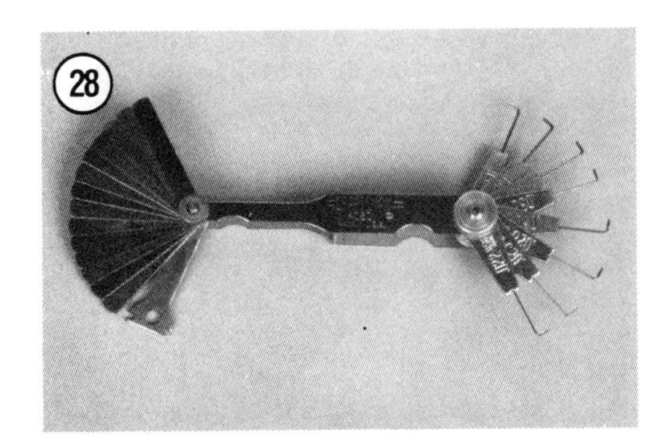

28

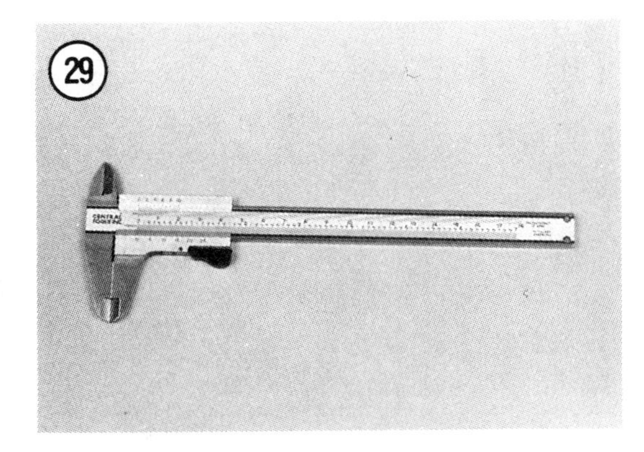

29

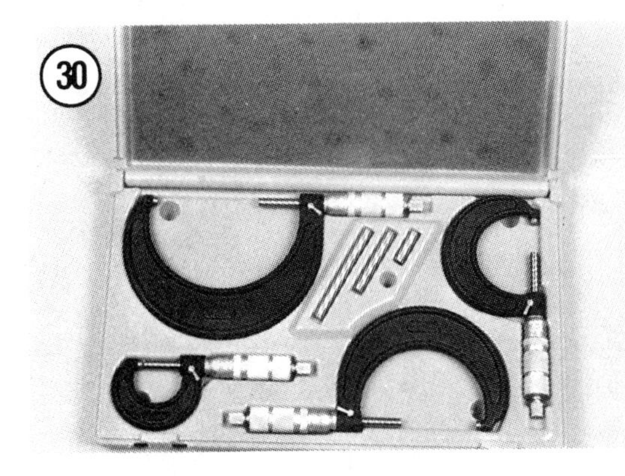

30

Vernier Caliper

A Vernier caliper (**Figure 29**) is invaluable when it is necessary to measure inside, outside and depth measurements with close precision. It can be used to measure the thickness of shims and thrust washers. Vernier calipers are available in a wide assortment of styles and price ranges.

Outside Micrometers

The outside micrometer (**Figure 30**) is used for very exact measurements of close-tolerance components. It can be used to measure the outside diameter of a piston as well as for shims and thrust washers. Outside micrometers will be required to transfer measurements from bore, snap and small hole gauges. Micrometers can be purchased individually or in a set.

Dial Indicator

Dial indicators (**Figure 31**) are precision tools used to check crankshaft runout and end play limits. Dial indicators may be purchased individually or as a set with adapters that facilitate mounting the indicator in a position for accurate measurement.

Cylinder Bore Gauge

The cylinder bore gauge is a very specialized precision tool. The gauge set shown in **Figure 32** is comprised of a dial indicator, handle and a number of length adapters to adapt the gauge to different bore sizes. The bore gauge can be used to make cylinder bore measurements such as bore size, taper and out-of-round. An outside micrometer must be

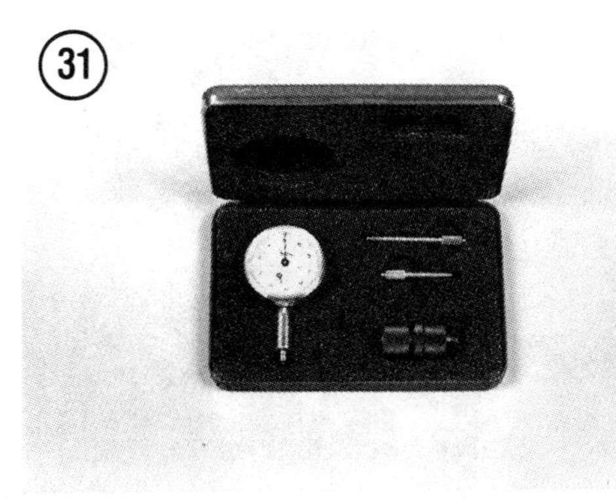

31

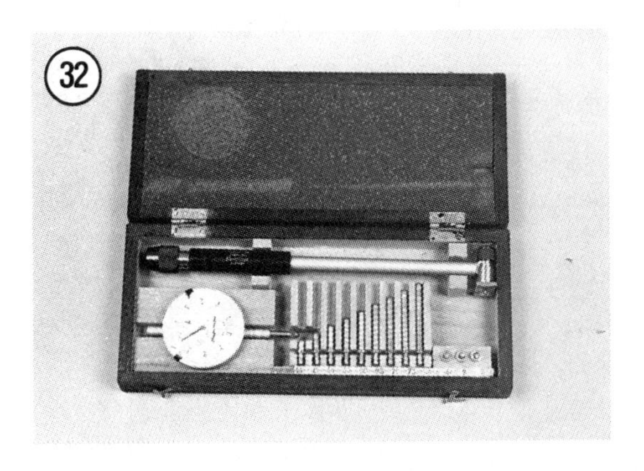

32

used together with the bore gauge to determine bore dimensions.

Telescoping Gauges

Telescoping gauges (**Figure 33**) can be used to measure hole diameters from 5/16 in. to 6 in. The telescoping gauge does not have a scale gauge for direct reading. Thus an outside micrometer must be used in conjunction with the telescoping gauges to determine bore dimensions.

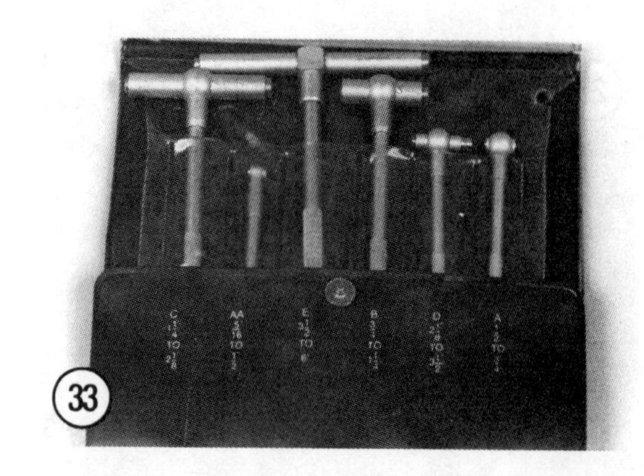

33

Small Hole Gauges

A set of small hole gauges (**Figure 34**) allows measurement of a hole, groove or slot ranging in size up to 1/2 in. An outside micrometer must be used together with the small hole gauge to determine bore dimensions.

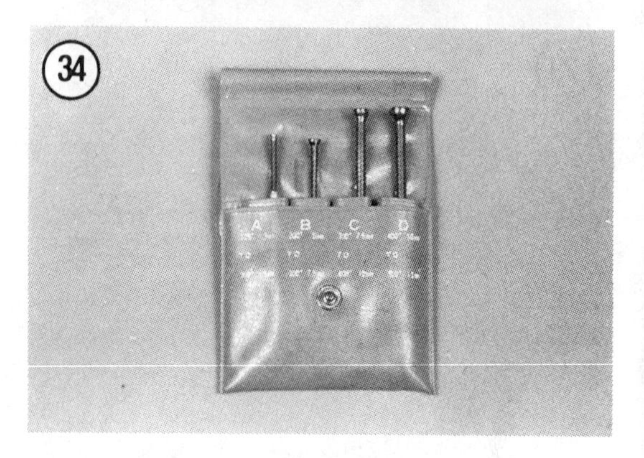

34

Screw Pitch Gauge

A screw pitch gauge (**Figure 35**) determines the thread pitch of bolts, screws, and other threaded fasteners. The gauge is made up of a number of thin plates. Each plate has a thread shape cut on one edge to match one thread pitch. When using a screw pitch gauge to determine a thread pitch size, try to fit different blade sizes onto the bolt thread until both threads match.

35

Surface Plate

A surface plate can be used to check the flatness of parts or to provide a perfectly flat surface for minor resurfacing of cylinder head or other critical gasket surfaces. While industrial quality surface plates are very expensive, a suitable substitute can be improvised using a thick metal plate. The metal plate shown in **Figure 36** has a piece of sandpaper glued to its surface that is used for cleaning and smoothing cylinder head and crankcase mating surfaces.

NOTE

Check with a local machine shop on the availability and cost of having a metal plate machined for use as a surface plate.

36

FASTENERS

Fasteners (screws, bolts, nuts, studs, pins, clips, etc.) are used to secure the various pieces of the engine together. Proper selection and installation of fasteners is important to ensure that the engine operates satisfactorily, otherwise, engine failure is possible.

Threaded Fasteners

Most of the components of an engine are held together by threaded fasteners, i.e., screws, bolts, nuts and studs. Most fasteners are tightened by turning clockwise (right-hand threads), although some fasteners may have left-hand threads if rotating parts can cause loosening.

Two dimensions are needed to match threaded fasteners: the number of threads in a given distance and the nominal outside diameter of the threads. Two standards are currently used in the United States to specify the dimensions of threaded fasteners, the U.S. common system and the metric system. Particular attention must be paid when working on a later model engine as both U.S. and metric fasteners may be used, causing damage to threads if fasteners are mismatched during assembly.

NOTE
Threaded fasteners should be hand tightened during initial assembly to be sure mismatched fasteners are not being used and crossthreading is not occurring. If fasteners are hard to turn, determine cause before applying tool for final tightening.

Screws and bolts built to U.S. common system standard are classified by length (L, **Figure 37**), nominal diameter (D) and threads per inch (TPI). A typical bolt might be identified by the numbers 7/16—14 × 1 1/2, which would indicate that the bolt has a nominal diameter of 7/16 in., 14 threads per inch and a length of 1 1/2 in.

U.S. screws and bolts are graded according to Society of Automotive Engineers (SAE) specifications to indicate their strength. Slash marks are located on the top of the screw or bolt as shown in **Figure 37** to indicate the strength grade with a greater number of slashes indicating greater strength. Ungraded screws and bolts (no slash marks on head) are the weakest.

Metric screws and bolts are classified by length (L, **Figure 38**), nominal diameter (D) and distance between thread crests (T). A typical bolt might be identified by the numbers 12 × 1.25—130, which would indicate that the bolt has a nominal diameter of 12 mm, the distance between threads crests is 1.25 mm and bolt length is 130 mm.

The strength of metric screws and bolts is indicated by numbers located on the top of the screw or bolt as shown in **Figure 38**. The higher the number the stronger the screw or bolt. Unnumbered screws or bolts are the weakest.

CAUTION
***Do not** install screws or bolts with a lower strength grade classification than installed originally by engine or equipment manufacturer. Doing so may cause engine or equipment failure and possible injury.*

Tightening a screw or bolt increases the clamping force it exerts, the stronger the screw or bolt, the greater the clamping force. In most cases, the

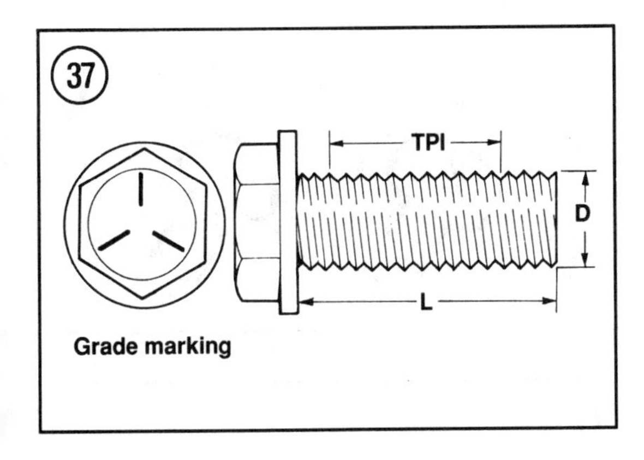

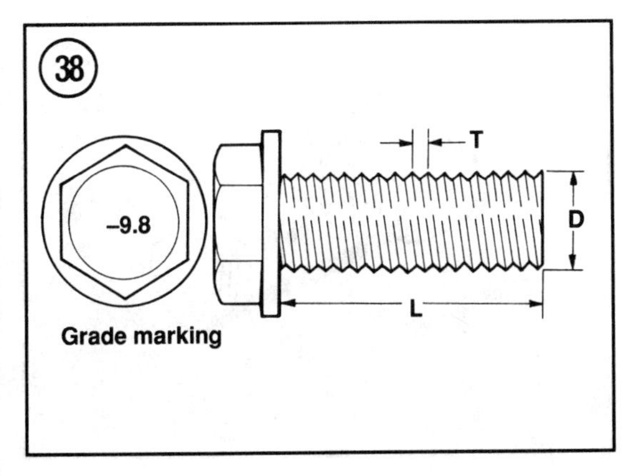

engine or equipment manufacturer specifies the tightening force (torque) to be used when tightening a specific screw or bolt. If not, recommended tightening torque specifications for general usage may be found in **Table 1**.

Screws and bolts are manufactured with a variety of head shapes to fit specific design requirements. Most machinery is equipped with the common hex and slotted head types, but other types, like those shown in **Figure 39** will also be encountered. As noted in the preceding *Basic Hand Tools* section, the proper tool must be used when turning a screw or bolt.

The most common nut used is the hex nut (**Figure 40**), often used with a lockwasher. Self-locking nuts have a nylon insert that prevents loosening; no lockwasher is required. Wing nuts, designed for fast removal by hand, are used for convenience in noncritical locations. Nuts are sized using the same system as screws and bolts. On hex-type nuts, the distance between two opposing flats indicates the proper wrench size to be used.

Self-locking screws, bolts and nuts may use a locking mechanism that utilizes an interference fit between mating threads. Interference is achieved in various ways: by distorting threads, coating threads with dry adhesive or nylon, distorting the top of an all-metal nut, using a nylon insert in the center or at the top of a nut, etc. Self-locking fasteners offer greater holding strength and better vibration resistance than standard fasteners. Some of these self-locking screws, bolts and nuts can be reused if in good condition; others, like the trilobial screw shown in **Figure 41**, are thread-rolling screws that form their own threads when installed and cannot be removed without displacement of the thread pattern. For greatest safety, self-locking fasteners should be discarded and new ones installed whenever components are disassembled.

Washers

There are two basic types of washers: flat washers and lockwashers. Flat washers are simple discs with

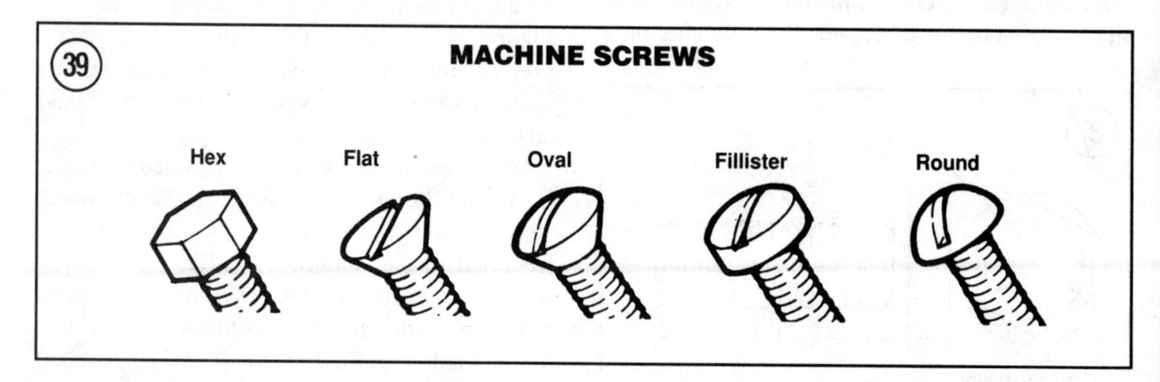

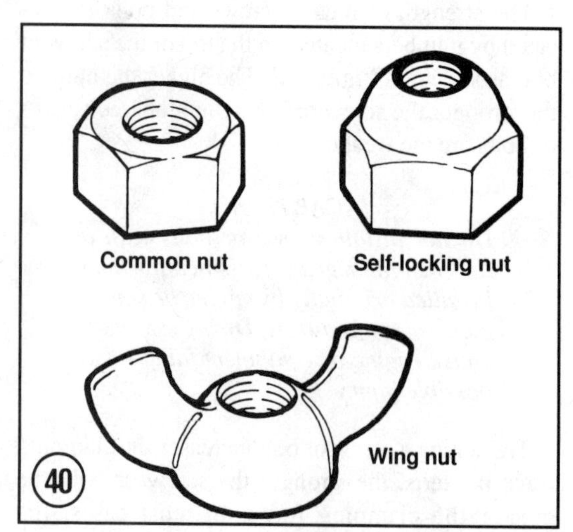

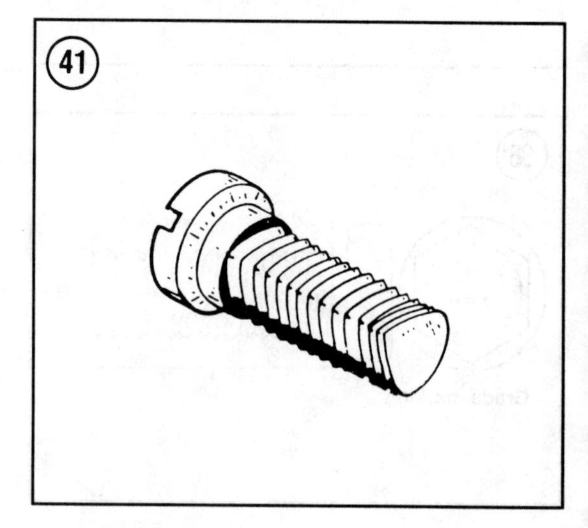

a hole to fit a screw or bolt. Lockwashers are designed to prevent a fastener from working loose due to vibration, expansion and contraction. Several types of washers are shown in **Figure 42**. Note that flat washers are often used between a lockwasher and a fastener to provide a smooth bearing surface. This allows the fastener to be turned easily with a tool.

Cotter Pins

Cotter pins (**Figure 43**) are used to secure special kinds of fasteners. The threaded stud must have a hole in it; the nut or nut lock piece has castellations around which the cotter pin ends wrap. Cotter pins should not be reused after removal.

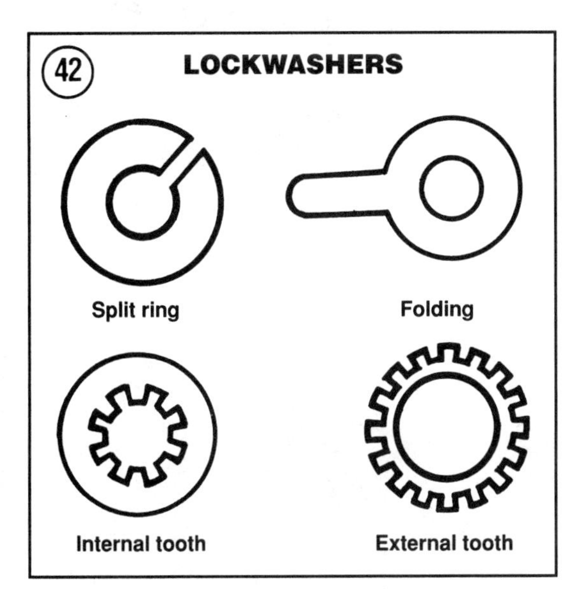

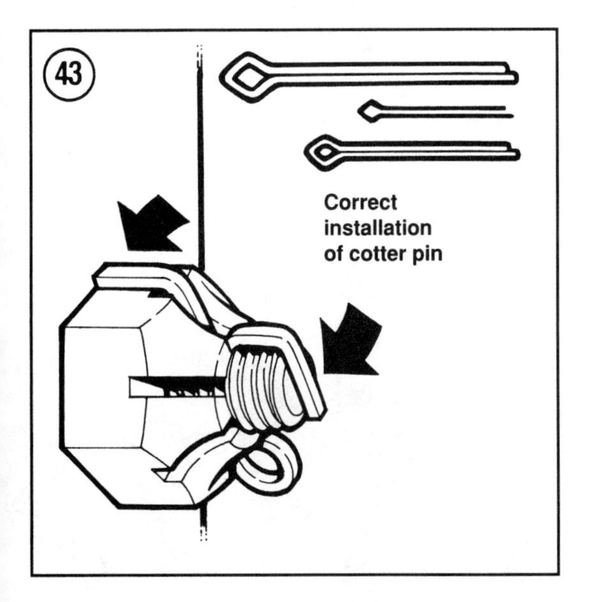

Retaining Rings

Retaining rings are designed to prevent or limit axial movement of shafts and bearings. Some examples of retaining rings are shown in **Figure 44**. A common type of retaining ring is known as a snap ring. Snap rings may have a rectangular or circular cross-section with plain ends or holes in the ends (Truarc) to accommodate special snap ring pliers. Another common retaining ring is the E-ring, which is usually found on linkage and other locations where appreciable force is not placed against the ring. Retaining rings fit in a groove that is machined to be used with a specific type retaining ring. Use of a retaining ring other than that specified by the manufacturer may cause engine damage and possible failure.

SEALANTS, CEMENTS AND CLEANERS

Sealants and Adhesives

Many mating surfaces of an engine require a gasket or seal between them to prevent fluids and gases from passing through the joint. At times, the

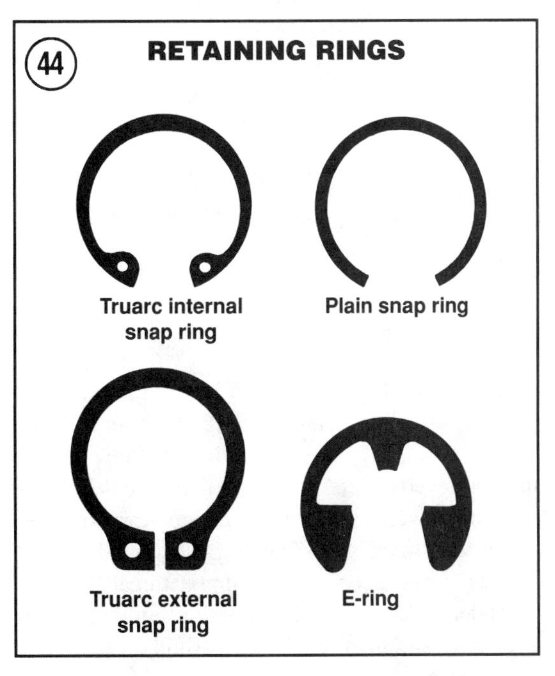

gasket or seal may be installed as is, however, most times some type of substance is applied to enhance the sealing capability of the gasket or seal. Note, however, that a sealing compound may be added to the gasket or seal during manufacture and adding sealant may cause premature failure of the gasket or seal.

RTV Sealants

One of the most common sealants is RTV (room temperature vulcanizing) sealant (**Figure 45**). This sealant hardens (cures) at room temperature over a period of several hours, which allows sufficient time to reposition parts if necessary without damaging the gaskets. RTV sealant is designed for different uses, including high temperatures. If in doubt as to correct type to use, ask vendor or read manufacturer's literature.

Cements and Adhesives

A wide variety of cements and adhesives is available (**Figure 46**), their use dependent on the type of materials to be sealed, and to some extent, the personal preference of the mechanic. Automotive parts stores offer the widest selection of cements and adhesives. Some points to consider when selecting cements or adhesives: the type of material being sealed (metal, rubber, plastic, etc.), the type of fluid contacting the seal (gasoline, oil, water, etc.) and whether the seal is permanent or must be broken periodically, in which case a pliable sealant might be desirable. Unless experienced in the selection of cements and adhesives, you should follow the engine or equipment manufacturer recommendation if a particular sealant is specified.

Thread Locking Compound

A thread locking compound is a fluid that is applied to fastener threads. After the fastener is tightened, the fluid dries to a solid filler between the mating threads, thereby locking the threads in position and preventing loosening due to vibration. The major manufacturer of locking compounds is the Loctite Corporation. The common thread locking compounds are Loctite 242 (a blue fluid) shown in **Figure 47** and Loctite 271 (a red fluid).

Loctite 242 (blue) has medium strength which allows unscrewing with normal hand tools. Loctite 271 (red) has high strength which may require special tools, such as a press or puller, as well as heat for disassembly.

Before applying Loctite, the contacting threads should be as clean as possible (aerosol electrical contact cleaner works well). Use only as much Loctite as necessary, usually one or two drops depending

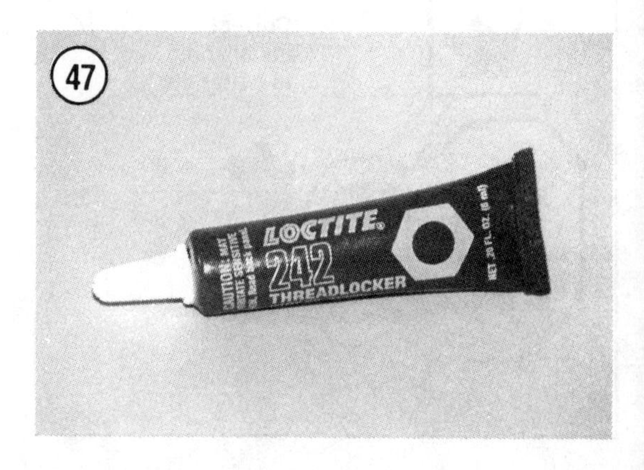

on the size of fastener. Excess fluid can work its way into adjoining parts.

Cleaners and Solvents

Cleaners and solvents are helpful in removing oil, grease and other residue found on engines and other equipment. Before purchasing cleaners and solvents, consider how they will be used and disposed of, particularly if they are not water soluble. Local ordinances may require special procedures for the disposal of certain cleaners and solvents.

WARNING
Some cleaners and solvents are harmful and may be flammable. Be sure to adhere to any safety precautions noted on the container or in the manufacturer's literature. Petroleum-resistant gloves are recommended to protect hands from the harmful effect of cleaners and solvents.

NOTE
An alternative for a big job is to take the parts to an automotive machine shop that will clean them for a few dollars, but be sure to ask them what they can clean. Some shops can't clean aluminum and the cleaning solution will damage plastic parts. When returned, always clean the parts again with an aerosol cleaner to remove any residue.

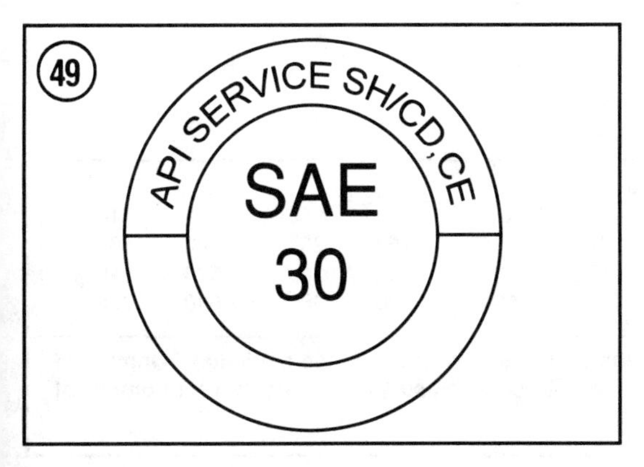

A variety of cleaners and solvents is shown in **Figure 48**. Cleaners designed for ignition contact cleaning are excellent for removing light oil from a part without leaving a residue. Cleaners designed to remove heavy oil and grease residue, called degreasers, contain a solvent that usually must "work" awhile. Some degreasers can be washed off with water. Removal of stubborn gaskets may be eased by using a solvent designed to remove gaskets.

One of the more powerful cleaning solutions is carburetor cleaner. It is designed to dissolve the varnish that may build up in carburetor jets and orifices. Good carburetor cleaner is usually expensive and requires special disposal. Carefully read directions before purchase; nonmetallic parts should not be immersed in carburetor cleaner.

LUBRICANTS

Lubricants are generally classified as oils if fluid, or greases if semi-solid. Grease is an oil which has been thickened by mixing with an additive.

Oil

Oil for four-stroke engines is graded by the American Petroleum Institute (API) and the Society of Automotive Engineers (SAE) in several categories. Oil containers display these ratings on the top of the oil container or on the label (**Figure 49**).

API oil grade is indicated by two letters, i.e., SE, SF, SG or SH. The first letter, "S," specifies that the oil is designed for use in a gasoline engine. The second letter indicates the API test grade that the oil meets. Based on engine performance and expected use, the engine manufacturer will specify the API grade that is approved for use in an engine.

Viscosity is an indication of the oil's thickness. The SAE uses numbers to indicate viscosity; thin oils have low numbers while thick oils have high numbers. A "W" after the number indicates that the viscosity testing was done at low temperature to simulate cold-weather operation. Engine oils fall into the 5W-30 and 20W-50 range.

Multigrade oils (for example 10W-40) are less viscous (thinner) at low temperatures and more viscous (thicker) at high temperatures than straight-grade oils (for example SAE 30). This allows the oil to perform efficiently across a wide range of engine operating conditions. The lower the number, the better the engine will start in cold climates. Higher numbers are usually recommended for engines running in hot weather conditions.

Grease

Greases are graded by the National Lubricating Grease Institute (NLGI). Greases are graded by number according to the consistency of the grease; these range from number 000 to number 6 with number 6 being the most solid. A typical multipurpose grease is NLGI number 2. For specific applications, equipment manufacturers may require grease with an additive such as molybdenum disulfide (MOS2).

In some instances, an antiseize lubricant may be specified (**Figure 50**). The antiseize lubricant prevents the formation of corrosion that may lock parts together.

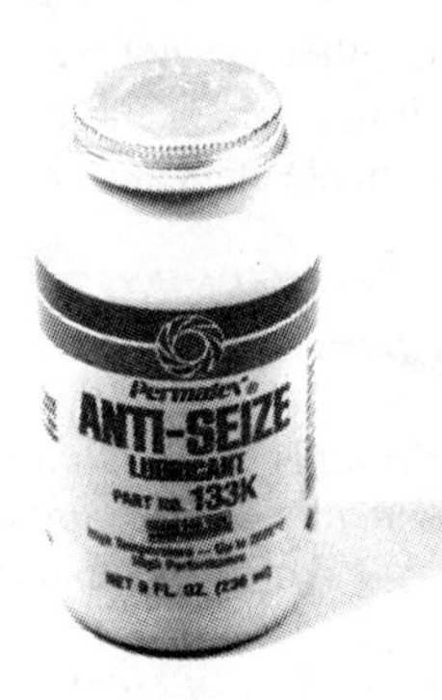

Table 1 GENERAL TORQUE SPECIFICATIONS

SAE 2 SAE 5 SAE 7 SAE 8

Type*	Body size or outside diameter (in.) 1/4	5/16	3/8	7/16	1/2	9/16	5/8	3/4	7/8	1
	Torque (ft.-lb.)									
SAE 2	6	12	20	32	47	69	96	155	206	310
SAE 5	10	19	33	54	78	114	154	257	382	587
SAE 7	13	25	44	71	110	154	215	360	570	840
SAE 8	14	29	47	78	119	169	230	380	600	700

* Fastener strength of SAE bolts can be determined by the bolt or screw head "grade markings." Unmarked bolt-heads and cap-screws are usually considered to be mild steel. Basically, the greater the number of "grade markings," the higher the fastener quality.

CHAPTER TWO

ENGINE FUNDAMENTALS

OPERATING PRINCIPLES

The Tecumseh engines covered in this manual are classified as internal combustion reciprocating engines.

The source of power is heat generated when a mixture of air and fuel is directed into the engine and ignited. Heat causes expansion of air trapped in the closed cylinder of the engine (**Figure 1** and **Figure 2**). A piston in the cylinder is forced through the cylinder in linear motion, which is translated into rotary motion through the connecting rod attached to the crankshaft crankpin. The process is repeated resulting in reciprocating piston movement.

There are five events that must occur for the engine to produce power. This series of events is called the "work cycle" and must be repeated for the engine to run. The events are named intake, compression, ignition, power and exhaust. The description and sequence are explained as follows:

1. *Intake*—As the piston moves downward, the exhaust valve is closed and the intake valve opens, allowing the new air:fuel mixture from the carburetor to be drawn into the cylinder (**Figure 3**). When the piston reaches the bottom of its travel, the intake valve closes, sealing the cylinder.
2. *Compression*—While the crankshaft continues to rotate, the piston moves toward the top of the cylinder, compressing the air:fuel mixture (**Figure 3**).

3. *Ignition*—As the piston almost reaches the top of its travel, the spark plug fires, igniting the compressed air:fuel mixture (**Figure 3**).

4. *Power*—The piston continues to top dead center, then is pushed downward by the rapidly expanding gases created as the air:fuel mixture burns in the cylinder (**Figure 3**).

5. *Exhaust*—When the piston reaches the bottom of its stroke, the exhaust valve opens. As the piston moves upward in the cylinder, combustion byproducts are forced out of the cylinder through the exhaust passage (**Figure 3**). After the piston has reached the top of its stroke, the exhaust valve closes and a new cycle begins with the intake event.

The succession of events in the work cycle may be accomplished in two or four strokes of the piston (the stroke is full piston travel in either direction). Note that the five events of the work cycle shown in **Figure 3** took place while the piston traveled through four strokes. Some engines are classified as two-stroke engines, performing the work cycle in two strokes of the piston. The Tecumseh engines covered in this manual are four-stroke cycle engines.

MAJOR ENGINE COMPONENTS

The engine is comprised of components and systems that must operate properly for efficient engine operation. The fuel and electrical systems are discussed in later chapters of this manual. Refer to **Figure 4** to identify the major components of a typical four-stroke engine described below.

1. *Air filter*—Prevents the entrance of dirt and other debris that will damage the engine if unfiltered.
2. *Carburetor*—Mixes fuel with air in the proper ratio to produce a combustible mixture. See Chapter Six.
3. *Intake manifold*—Directs air:fuel mixture from carburetor to engine.
4. *Flywheel*—Inertia produced by the rotating flywheel maintains crankshaft rotation when not on power stroke; may house magnets for ignition; fins move air for engine cooling; may be part of starter mechanism.
5. *Ignition system*—Generates electricity so a spark occurs at the spark plug gap just as the piston reaches a specified position in the cylinder. See Chapter Seven.
6. *Oil seal*—Prevents oil in the crankcase from escaping.
7. *Crankcase breather*—Contains a breather valve that maintains a vacuum in the crankcase. The breather assembly is located on the crankcase (8) on some engines or as a part of the valve tappet cover (7A) on other engines. See Chapter Five.
8. *Crankcase*—Houses internal engine components. Includes the cylinder which is cast as a one-piece unit with the crankcase.
9. *Spark plug*—Ignites the air:fuel mixture.
10. *Cylinder head*—May be removed for access to the cylinder, piston and valves.
11. *Head gasket*—Seals the surfaces of the cylinder head and the cylinder.
12. *Intake valve*—Used to open and close the intake passage.
13. *Exhaust valve*—Used to open and close the exhaust passage.

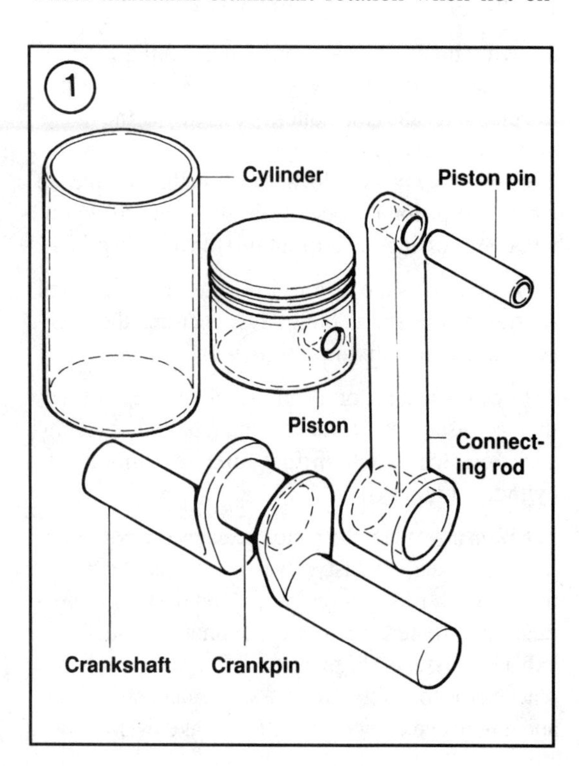

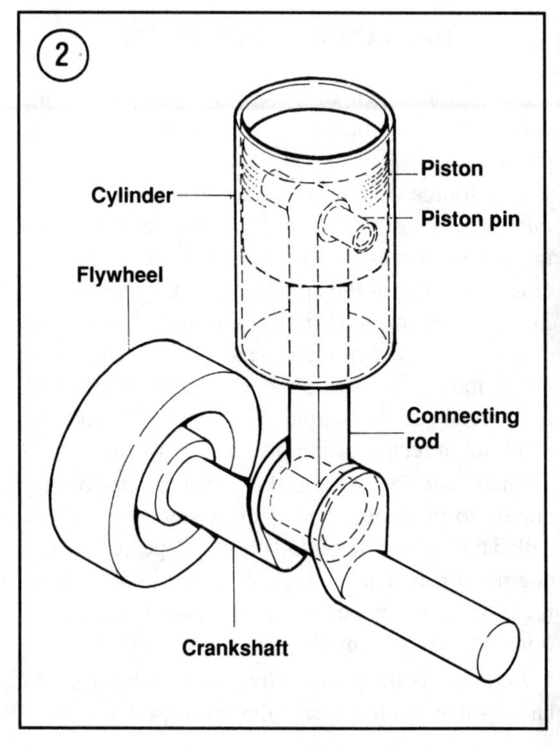

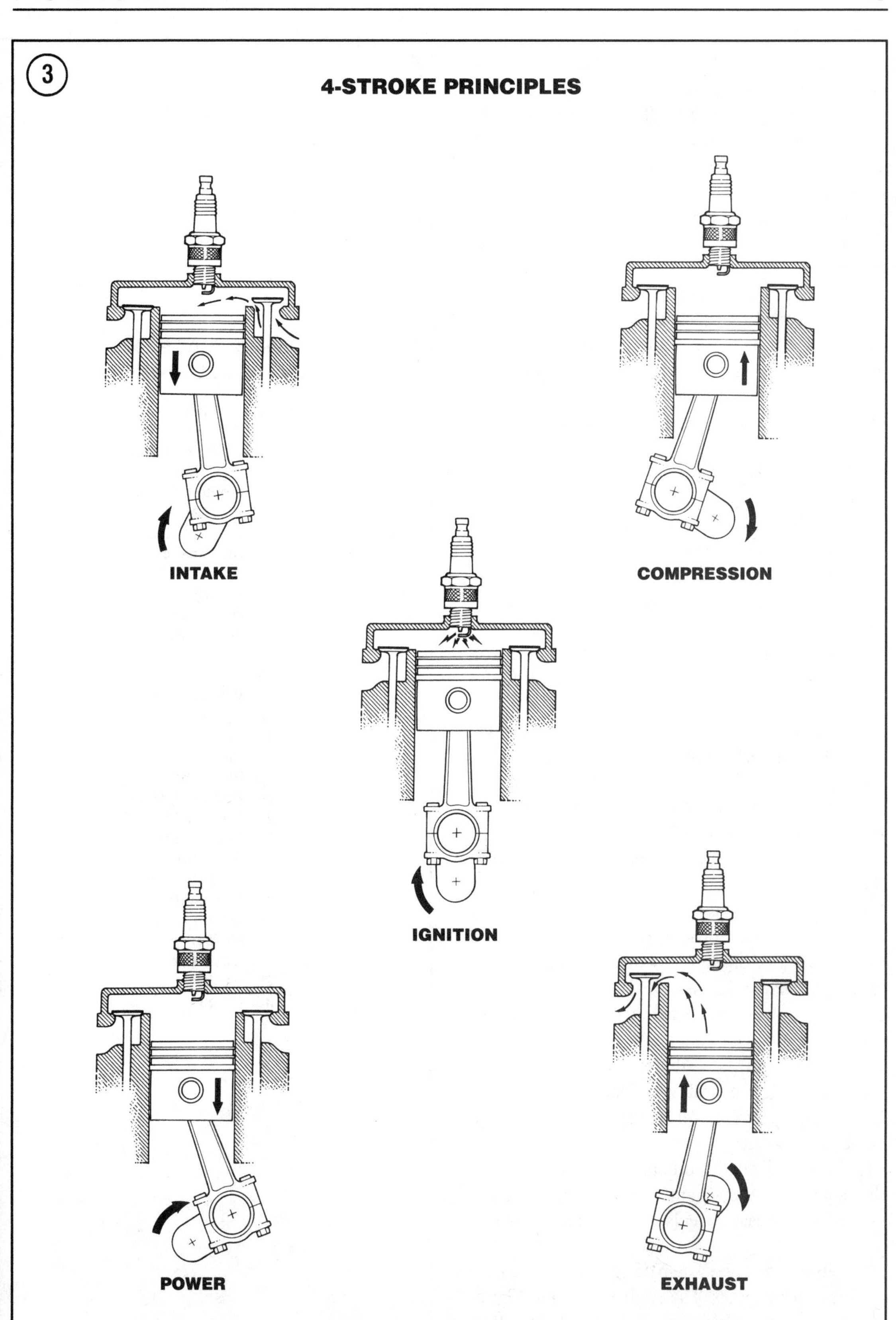
3
4-STROKE PRINCIPLES
INTAKE
COMPRESSION
IGNITION
POWER
EXHAUST

14. *Valve seal*—Prevents oil on the valve stem from entering the intake passage.
15. *Valve spring*—Used on the valves to hold the valves closed.
16. *Valve retainer*—Secures the valve and spring in position.
17. *Piston rings*—Upper piston rings maintain a seal between the piston and cylinder wall, while lower piston ring prevents oil in the crankcase from flowing past the piston into the combustion area. The piston is normally fitted with three piston rings, all of which are usually different in design.
18. *Piston*—Slides in the cylinder in reciprocating motion.
19. *Piston pin*—Connects the piston and connecting rod.
20. *Piston pin retaining clips*—A clip at each end of the piston pin prevents the pin from extending past the piston and contacting the cylinder wall.
21. *Connecting rod*—Transfers motion from the piston to the crankshaft.
22. *Crankshaft*—Converts reciprocating motion of the piston to rotary motion. One end of the crankshaft is designated the power take-off end (pto) so that power can be transmitted to the powered equipment.
23. *Governor linkage*—Transfers movement of the internal governor assembly (28) to the carburetor control linkage.
24. *Tappet*—Opens the valve according to the lobe profile on camshaft. May be called a lifter.
25. *Camshaft*—Lobes on the camshaft push against the tappets to open the valves.
26. *Oil pump*—Pumps oil in the oil pan to the engine components in the upper end of the engine. An oil pump is not used on engines with a horizontal crankshaft. See Chapter Eleven.
27. *Muffler*—Lessens noise in the exhaust system.
28. *Governor*—Monitors crankshaft speed and actuates linkage to the carburetor to maintain a set crankshaft speed regardless of load.
29. *Gasket*—Prevents oil leakage between the oil pan and crankcase.
30. *Oil pan*—Contains oil and supports one end of the crankshaft. On an engine with a horizontal crankshaft, the crankcase cover is used instead of a removable oil pan.

Two types of valve system designs have been used on Tecumseh engines. The engine shown in **Figure 4** is an L-head engine. The "L-head" design illus-

ENGINE ASSEMBLY (VERTICAL CRANKSHAFT)

1. Air filter
2. Carburetor
3. Intake manifold
4. Flywheel
5. Ignition system
6. Oil seal
7. Crankcase breather
8. Crankcase
9. Spark plug
10. Cylinder head
11. Head gasket
12. Intake valve
13. Exhaust valve
14. Valve seal
15. Valve spring
16. Valve retainer
17. Piston rings
18. Piston
19. Piston pin
20. Retaining clips
21. Connecting rod
22. Crankshaft
23. Governor linkage
24. Tappet
25. Camshaft
26. Oil pump
27. Muffler
28. Governor
29. Gasket
30. Oil pan

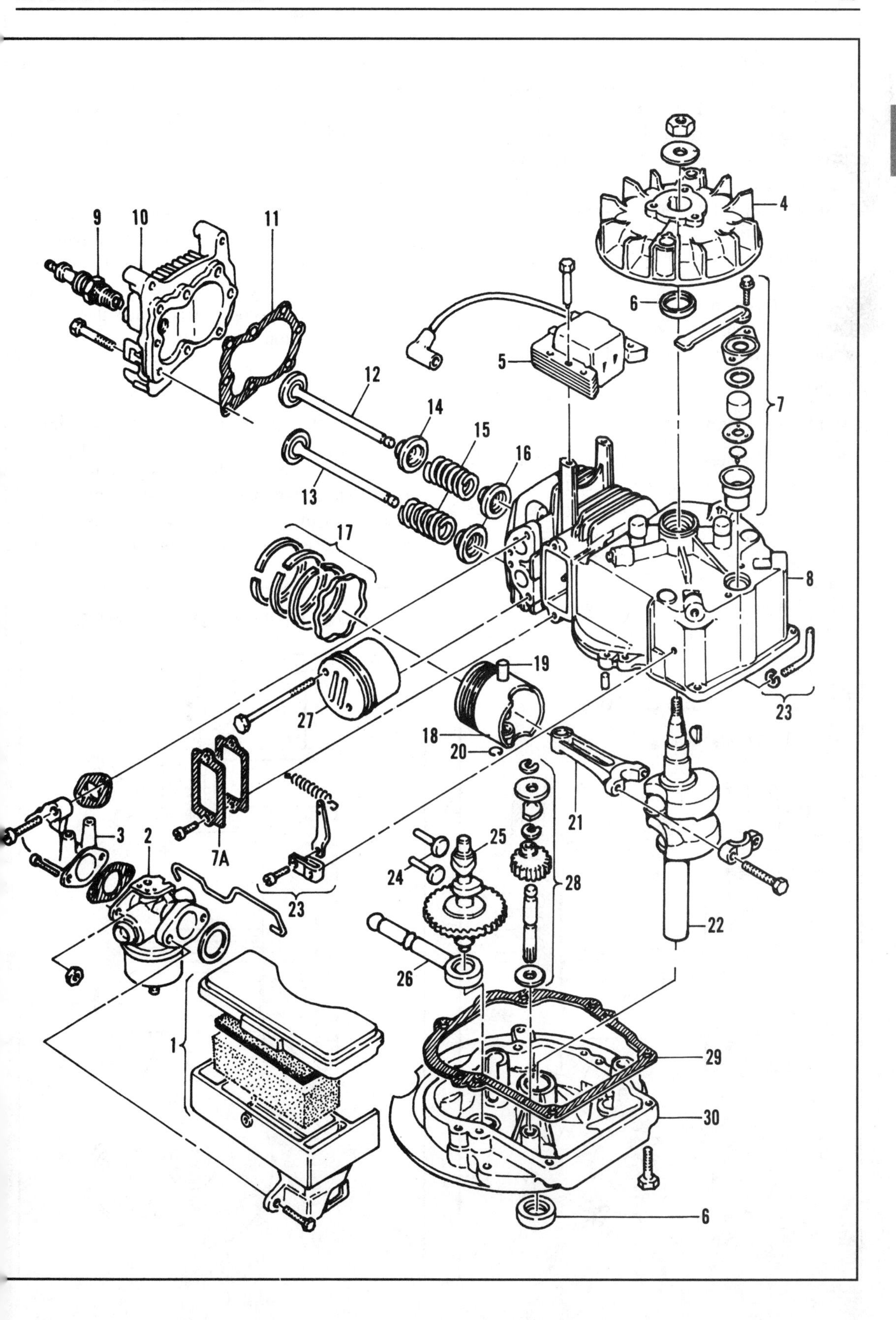
1
2
3
4
5
6
7
7A
8
9
10
11
12
13
14
15
16
17
18
19
20
21
22
23
24
25
26
27
28
29
30

trated in **Figure 5** incorporates the valve system in the side of the cylinder block. The overhead-valve design illustrated in **Figure 6** places the valve system in the cylinder head. The overhead-valve arrangement offers better engine cooling and more efficient combustion chamber design, but at the penalty of adding push rods and rocker arms to actuate the valves. Early Tecumseh engines were all built using the L-head design, while later engines are built using both designs. The L-head design engines are covered in this manual.

Additional information on major components and systems is outlined in the following.

Cylinder Head and Gasket

The cylinder head is removable and forms the closed end of the cylinder. The spark plug is mounted in the cylinder head. A cavity on the inside surface of the cylinder head, called the combustion chamber, is shaped to enhance combustion of the fuel:air mixture.

A gasket, located between the cylinder head and cylinder, seals the mating surfaces of the cylinder head and cylinder thereby preventing gases from escaping. The gasket material is designed to withstand the heat and pressure generated by combustion in the cylinder.

Piston

All pistons in Tecumseh engines are made of aluminum. When the piston is positioned with the closed end up as shown in **Figure 7**, the top of the piston is called the "crown" while the area of the piston below the piston pin is called the "skirt" portion of the piston.

Piston Rings

Three piston rings are used on the piston (**Figure 8**). The top piston ring (nearest the piston crown) is a compression type ring that is designed to contain the expanding gases during combustion and prevent leakage past the piston into the crankcase of the engine (**Figure 9**). The second piston ring (**Figure**

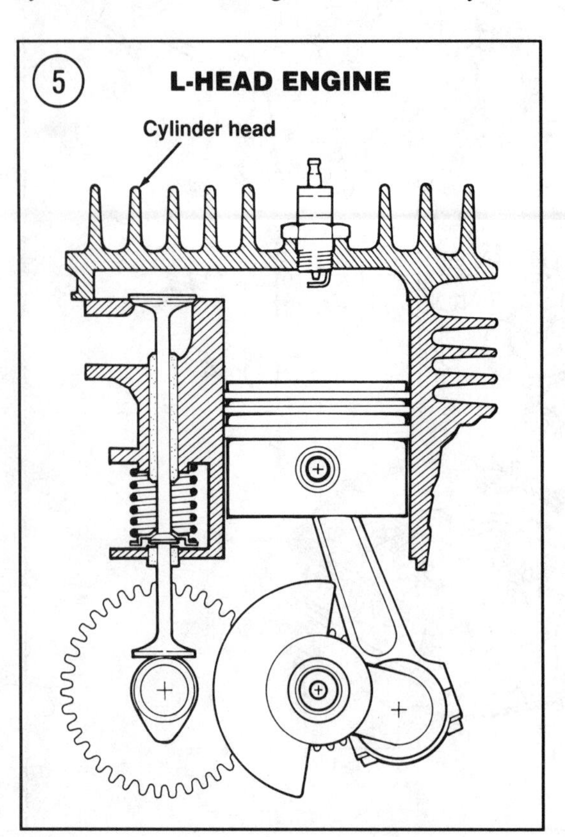

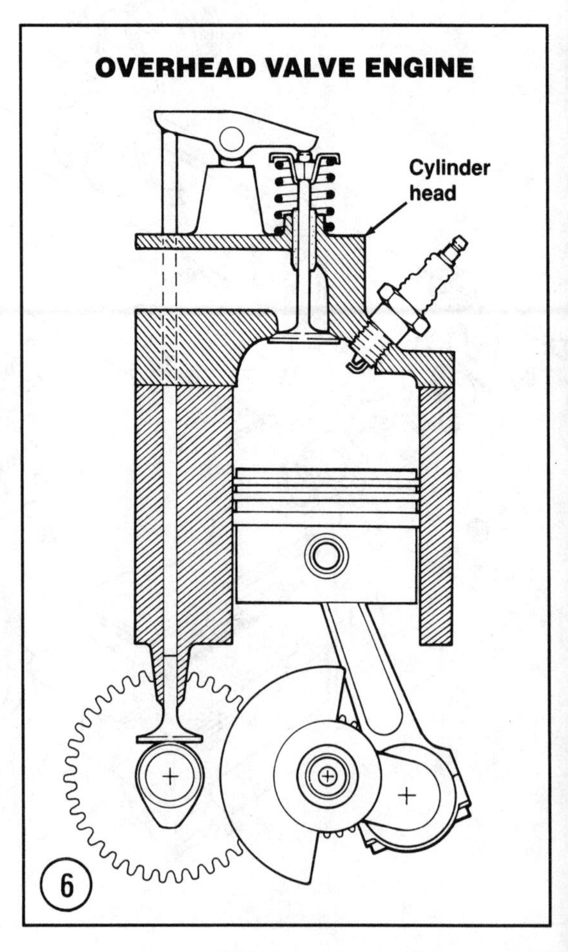

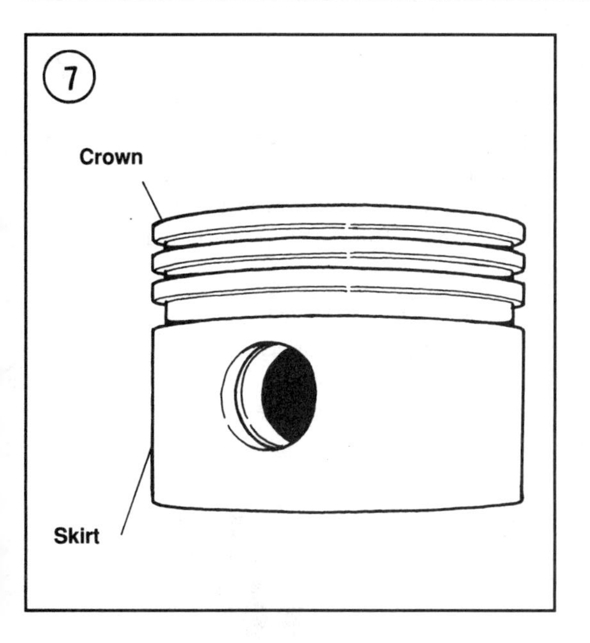

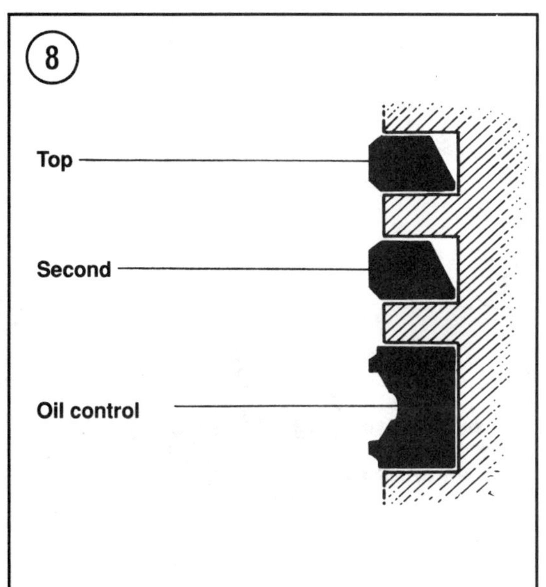

9) is also a compression type ring, but it serves the dual purpose of sealing gases that may have passed the top compression ring and also scraping oil off the cylinder wall that has passed the oil control ring (the second compression ring is sometimes called a scraper ring). The bottom piston ring (nearest the piston skirt) is called the oil control ring (**Figure 10**) and is designed to prevent oil in the crankcase from passing between the piston and cylinder wall into the combustion area. The oil control ring may be a single ring (**Figure 10**) or an assembled unit. The assembled oil control ring may consist of two pieces, the control ring plus an expander, or three pieces, two rails and an expander.

Due to their design, top and second compression rings must be installed on a piston in the correct groove with the specified side of the ring towards the piston crown. An identifying mark (**Figure 11**) is sometimes used to indicate which side must be towards the piston crown.

Piston Pin

A steel piston pin attaches the piston to the connecting rod. The piston pin is generally hollow and

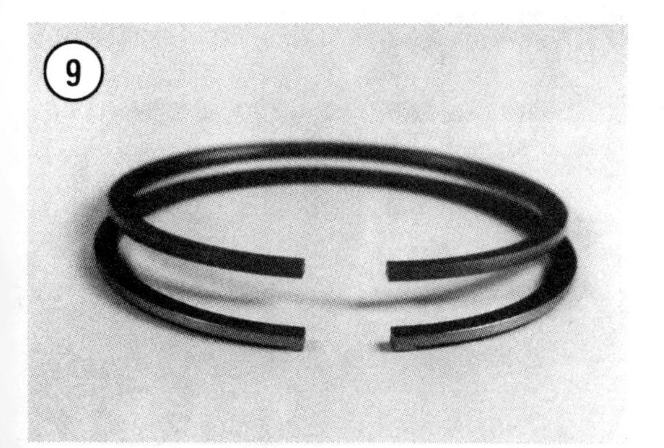

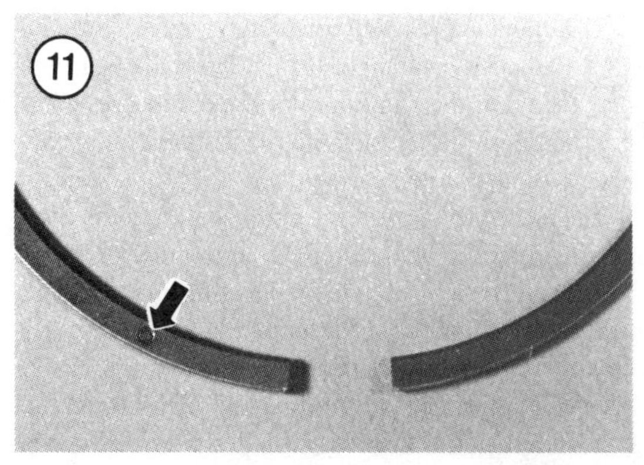

open at both ends, but some engines may be equipped with a piston pin that is closed at one end. Retaining clips are used to secure the pin in the piston, but some pistons are equipped with an internal stop on one side and a retaining clip on the opposite side. The retaining clip (**Figure 12**) fits in a groove in the piston.

Connecting Rod

The connecting rod serves as a link between the piston and crankshaft. For purposes of identification, the ends of the connecting rod are identified by their size, small end or big end. The steel piston pin connects the small end of the connecting rod to the piston, while the big end of the connecting rod attaches to the crankpin of the crankshaft. The big end is split to form a cap that is secured to the connecting rod either by screws or by nuts on studs.

The connecting rod on Tecumseh engines covered in this manual is made of aluminum and rides directly on the crankshaft, no bearing insert is present between the connecting rod and crankshaft. Aluminum is used for its lightness, because its relative softness will absorb tiny particles of foreign material rather than scratch the crankpin surface, and the metal is dissimilar to the iron of the crankshaft thereby preventing welding should the rod and crankpin touch without lubrication while the engine is running.

Crankshaft

The iron crankshaft converts reciprocating motion to rotary motion. A flywheel is attached to one end of the crankshaft while the opposite end provides an attachment point to transfer power from the engine to driven equipment.

The crankshaft is supported at machined journals (A, **Figure 13**) by main bearings. The main bearings may be either the aluminum of the crankcase, plain type bearings (called bushings) or anti-friction bearings (ball or needle bearings).

The machined area of the crankshaft that the connecting rod rides on is called the crankpin (B, **Figure 13**). The crankpin is hardened during manufacture to increase durability. The crankpin can be ground down to a smaller diameter if the damage is not severe. This is called regrinding and requires the use of a connecting rod with a smaller diameter at the big end to obtain the desired clearance between the crankpin and connecting rod.

The flywheel end of the crankshaft is tapered to fit a corresponding taper in the flywheel. A key is used to match the keyways in the crankshaft and flywheel so the flywheel is properly indexed to the crankshaft.

The power take-off (pto) end of the crankshaft may be shaped in a variety of configurations to match the driven equipment coupled to the engine. Common variations include external or internal

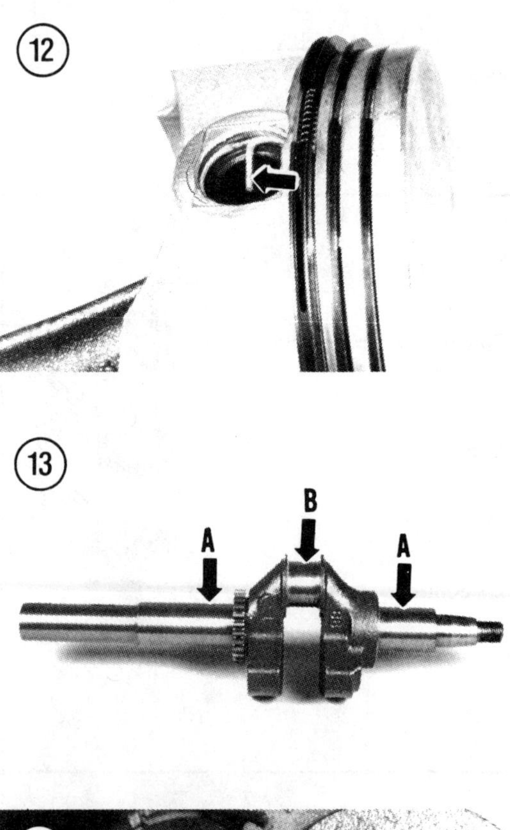

threads, with or without a keyway, and a plain shaft with a keyway.

Located on the crankshaft is a gear that drives the camshaft gear. The crankshaft gear and the gear on the camshaft must be "timed" so that the valves operate at the proper time according to piston position. The timing mark on the camshaft gear must align with a beveled tooth or indentation on the crankshaft gear or the drive key for the crankshaft gear (**Figure 14**). The marks must align on every other revolution of the crankshaft.

Main Bearings

The crankshaft is supported by main bearings that may be either the aluminum of the crankcase, crankcase cover or oil pan (if made of aluminum), plain type bearings (called bushings) or anti-friction bearings (ball or needle bearings). See **Figure 15**.

Some engines that are manufactured with an aluminum crankcase support the crankshaft directly in the aluminum, which is a good bearing material. In most cases, the bearing bore can be machined if damaged and a plain type bearing (bushing) can be installed.

Plain type bearings (bushings) and anti-friction bearings (ball or needle bearings) can be removed and replaced if necessary.

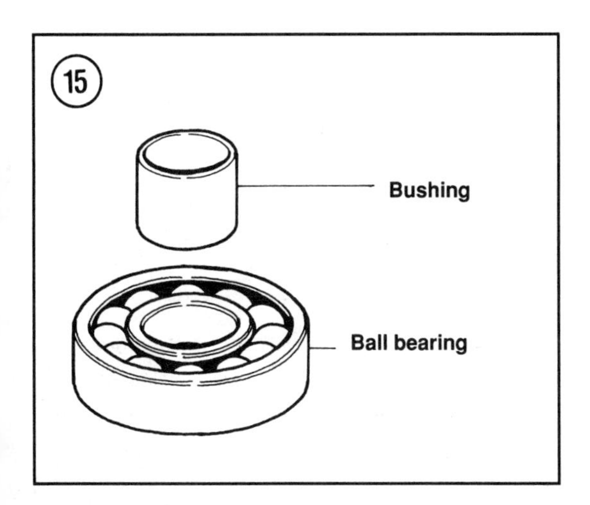

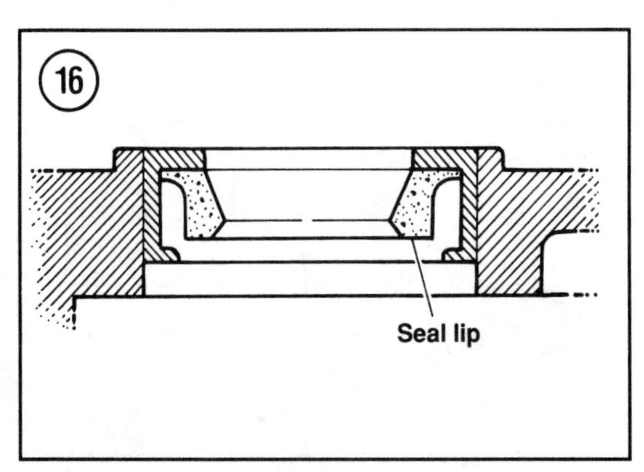

Oil Seals

Oil seals are designed to keep foreign material from entering the engine and prevent oil leakage along the crankshaft. The seal has a rubber or neoprene lip (**Figure 16**) that rests against the shaft to form a seal. Depending on the application, the seal may have one or more lips, as well as a garter spring behind the lip to increase pressure on the seal lip. Correct installation of an oil seal is important if the seal is to function properly.

Camshaft and Tappets

The camshaft converts rotary motion into reciprocating motion to operate the valves. A gear on the end of the camshaft is driven by the gear on the crankshaft (**Figure 17**). Machined on the camshaft are lobes for each of the valves. Riding against the cam lobe is a valve tappet, often called a valve lifter, that transfers motion from the cam lobe to the valve. When the camshaft turns, the raised portion of the cam lobe pushes against the tappet which in turn forces the valve to open. Each lobe is designed and machined so the respective valve opens and closes so that optimum engine performance is achieved. Excessive wear of either the cam lobe or tappet will cause a degradation in engine performance due to decreased valve action.

The valves must operate "in time" with the position of the piston for proper four-stroke engine operation. To achieve correct timing between the crankshaft and camshaft, timing marks are located on the crankshaft and camshaft gears. On most Tecumseh engines, the hole in the camshaft gear must align with a beveled tooth or indentation on the crankshaft gear or the drive key for the crankshaft gear (**Figure 14**). The marks must align on every other revolution of the crankshaft so that the valves operate at the proper time according to piston position.

Valve System

All Tecumseh engines are equipped with "poppet" type valves. On the engines covered by this manual, the valves are located in the cylinder block portion of the engine (**Figure 17**). This type of valve arrangement is called a side valve or "L-head" (also known as "flathead" due to the configuration of the cylinder head).

Two valves are used in the single-cylinder engine, an intake valve and an exhaust valve. Refer to **Figure 18** for terminology related to a valve. The intake and exhaust valves are usually constructed of different metals. The exhaust valve is manufactured from metal that is capable of withstanding the high temperature of the exhaust gas.

When closed, the valve rests against a seat (**Figure 17**). Contact with the seat closes the air passage, as well as cooling the valve by providing a path for the heat to transfer from the valve head to the cylinder block. The valve seat may be machined directly into the metal of the engine, or a separate valve seat insert may be installed (**Figure 19**). Tecumseh aluminum engines have nonrenewable seat inserts for both valves.

17

The valve is held in the closed position by a valve spring. The spring is held by a valve spring retainer (**Figure 17**), that has a slot that fits a groove in the end of the valve stem (**Figure 20**).

Each valve rides in a valve guide (**Figure 17**) that is machined directly in the aluminum crankcase or is a nonrenewable, insert type valve guide. If the valve guide is excessively worn, the guide diameter is increased using a reamer so a valve with an oversize stem can be installed. Since the guide centers the valve on its valve seat, worn guides cause gradual engine power loss due to poor contact between the valve and seat, eventually resulting in valve system overhaul.

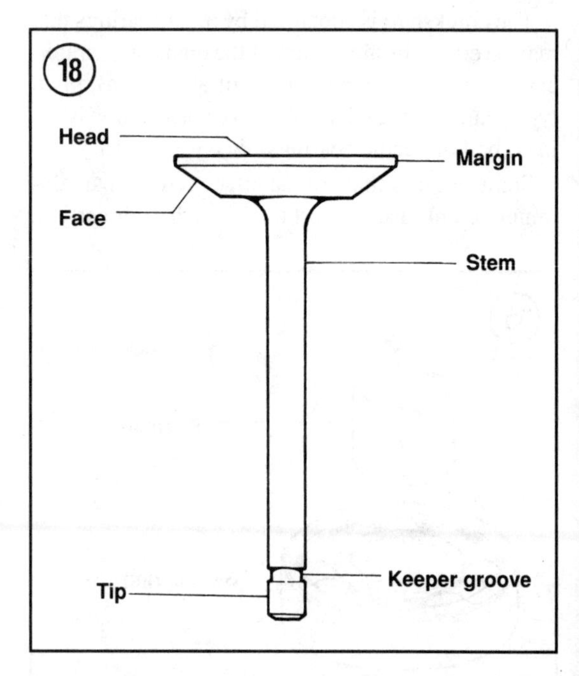

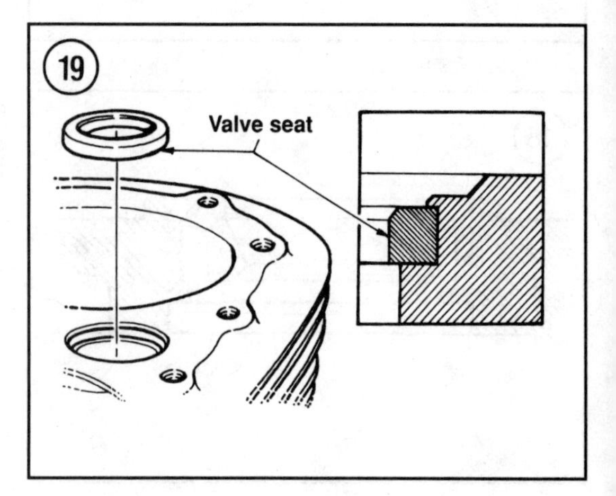

Compression Release

Single-cylinder engines can be difficult to start using a manual starter. One problem is developing sufficient engine speed so the engine will continue to run after firing and reach the next ignition event. While starting, the starter must work against compression pressure, which tends to slow down the engine. A compression release device is used to bleed off compression pressure so the engine will rotate faster during starting.

Tecumseh engines may be equipped with a mechanism on the camshaft that holds the exhaust valve open slightly during starting to reduce compression pressure. The compression release mechanism is mounted on the back of the camshaft gear. When the engine is stopped, the spring-loaded actuator (A, **Figure 21**) extends the pin (B) so it will contact the exhaust valve tappet during starting. When the engine is running, centrifugal force moves the actuator outward thereby withdrawing the pin and allowing the exhaust valve tappet to fully contact the cam lobe.

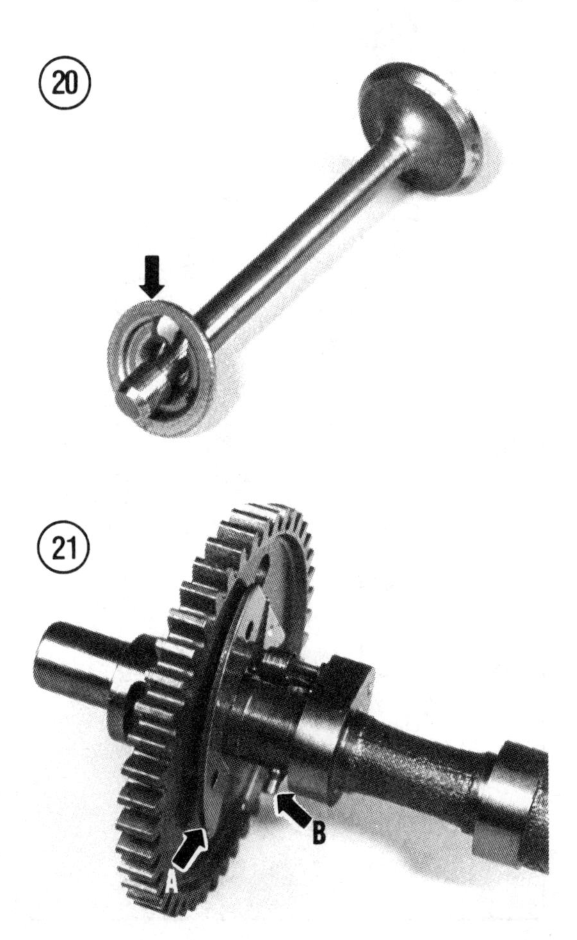

Governor System

The purpose of the governor system is to maintain a desired engine speed regardless of the load imposed (within the limits of the engine power range). The governor will open or close the throttle to adjust the engine power output to match the load, thereby maintaining the desired engine speed. On some engines, the engine speed may be varied (lawn mowers), while on other engines the engine speed is fixed (generators and pumps). Tecumseh engines are equipped with a mechanical type governor system.

The governor system utilizes a governor spring that is connected to the throttle linkage. The governor spring tension opposes the action of the governor and tends to open the throttle, increasing engine speed and power. For variable speed governors, the governor spring is connected to an adjustable control of some type so that the tension of the spring can be changed by the operator. Increasing the spring tension will raise the engine speed. Decreasing the spring tension will lower engine speed. For fixed speed governors, the spring is connected to a fixed point to provide a constant engine speed. The desired governed speed is maintained when the force created by the governor counterbalances the tension of the governor spring.

The mechanical governor system utilizes the centrifugal force of rotating flyweights to oppose the governor spring. A set of flyweights is mounted on a gear that is driven by the camshaft gear (**Figure 22**). When the engine speed increases, the flyweights are thrown outward by centrifugal force. When the engine speed decreases, the flyweights recede. Trapped between the flyweights is a flanged sleeve that moves in and out with the flyweights, pushing against a governor arm. Externally, the governor arm transfers motion to the governor lever, which is connected to the carburetor throttle linkage. Outward movement of the flyweights tends to close the carburetor throttle plate.

As the load on the engine is increased, the engine will start to slow down. When this happens the centrifugal force of the flyweights decreases, reducing the opposing force against the governor spring. This allows the governor spring to open the throttle, increasing engine power to compensate for the increased load and thus maintain the desired engine speed.

The opposite effect occurs when the load on the engine decreases. The engine speed starts to increase which increases the centrifugal force of the flyweights, increasing the opposing force against the governor spring. The movement of the governor linkage will stretch the governor spring and close the throttle to reduce engine power to match the load and maintain the desired engine speed.

Based on a variety of factors, including the intended application of the engine, an engine is designed to operate at a specific fixed governed speed or in an operating range. The most critical governed speed is maximum governed speed as it sets the upper limit of engine operation. Exceeding the maximum governed speed can cause overspeeding which may result in engine failure.

Fuel System

The fuel system consists of the carburetor, fuel pump, intake manifold and air inlet system. Refer to Chapter Six for a discussion of fuel system service procedures.

Ignition System

Service procedures covering the components that deliver spark to the engine are outlined in Chapter Seven.

FUEL SYSTEM

Carburetor Operating Principles

The function of the carburetor on a spark-ignition engine is to atomize the fuel and mix the atomized fuel in proper proportions with air flowing to the engine intake port or intake manifold. Carburetors used on engines that are to be operated at constant speeds and under even loads are of simple design since they only have to mix fuel and air in a relatively constant ratio. On engines operating at varying speeds and loads, the carburetor must be more complex because different air:fuel mixtures are required to meet the varying demands of the engine.

Air:fuel mixture ratio requirements

To meet the demands of an engine being operated at varying speeds and loads, the carburetor must mix fuel (gasoline) and air at different mixture ratios. Air:fuel mixture ratios required for different operating conditions are shown in **Table 1**.

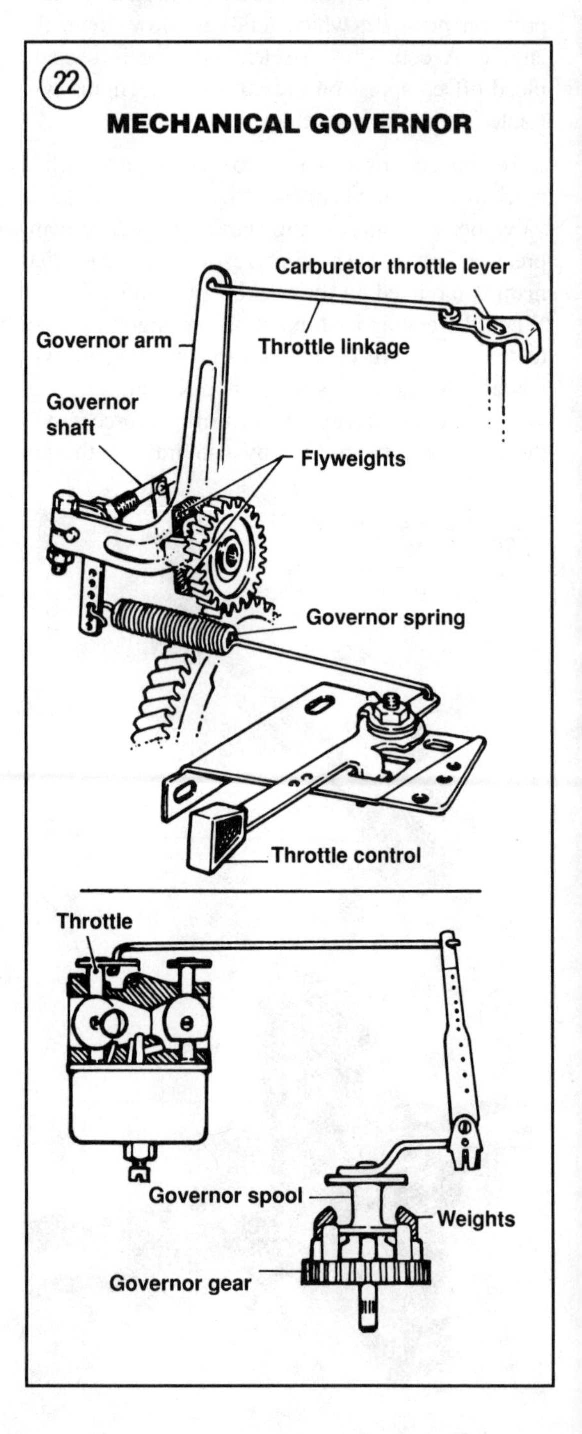

Basic design

Carburetor design is based on the venturi principle which simply means that a gas or liquid flowing through a necked-down section (venturi) in a passage undergoes an increase in velocity (speed) and a decrease in pressure as compared to the velocity and pressure in the full size sections of the passage.

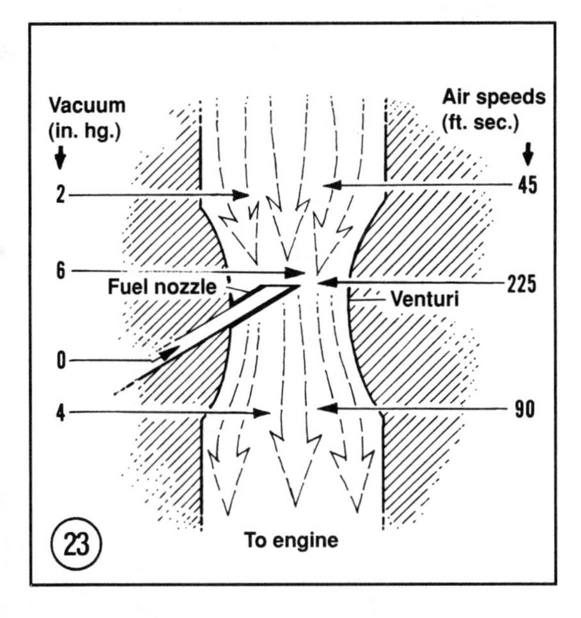

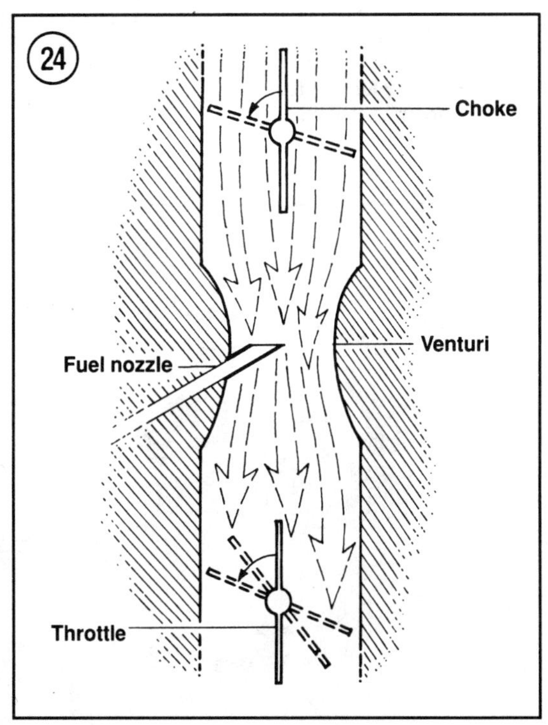

The principle is illustrated in **Figure 23**, which shows air passing through a carburetor venturi. The figures given for air speeds and vacuum are approximate for a typical wide-open throttle operating condition. Due to low pressure (high vacuum) in the venturi, fuel is forced out through the fuel nozzle by the atmospheric pressure (0 vacuum) on the fuel; as fuel is emitted from the nozzle, it is atomized by the high velocity air flow and mixes with the air.

In **Figure 24**, the carburetor choke plate and throttle plate are shown in relation to the venturi. Downward pointing arrows indicate air flow through the carburetor.

At cranking speeds, air flows through the carburetor venturi at a slow speed; thus, the pressure in the venturi does not usually decrease to the extent that atmospheric pressure on the fuel will force fuel from the nozzle. If the choke plate is closed as shown by the dotted line in **Figure 24**, air cannot enter into the carburetor and pressure in the carburetor decreases greatly as the engine is turned at cranking speed. Fuel can then flow from the fuel nozzle. In manufacturing the carburetor choke plate, a small hole or notch (**Figure 25**) is cut in the plate so that some air can flow through the plate when it is in closed position to provide air for the starting air:fuel mixture. In some instances after starting a cold engine, it is advantageous to leave the choke plate in a partly closed position as the restriction of air flow will decrease the air pressure in the carburetor venturi, thus causing more fuel to flow from the nozzle, resulting in a richer air:fuel mixture. The choke plate should be in fully open position for normal engine operation.

If, after the engine has been started, the throttle plate is in the wide-open position as shown by the solid line in **Figure 24**, the engine can obtain enough fuel and air to run at dangerously high speeds. Thus,

the throttle plate must be partly closed as shown by the dotted lines to control engine speed. At no load, the engine requires very little air and fuel to run at its rated speed and the throttle must be moved toward the closed position as shown by the dash lines. As more load is placed on the engine, more fuel and air are required for the engine to operate at its rated speed and the throttle must be moved closer to the wide open position as shown by the solid line. When the engine is required to develop maximum power or speed, the throttle must be in the wide open position.

A simple carburetor which relies only on the venturi principle will supply a progressively richer mixture as engine speed is increased, but the engine will not run at all at idle speeds. The carburetor must therefore be built with additional elements so the engine will run efficiently at varying speeds.

An idle or slow speed circuit uses a separate fuel mixing and metering system like that shown in **Figure 26**. An idle passage leads from the fuel chamber to the air horn at the approximate location of the throttle plate (1). When the throttle plate is closed, air flow is shut off. This reduces the pressure in the intake manifold, which results in insufficient air passing through the venturi to draw fuel from the main fuel nozzle. However, fuel is drawn up the idle passage through idle jet (5), then through primary idle orifice (2) into the intake manifold. At the same time, air enters through the secondary idle orifice (3) and air metering orifice (4) to mix with fuel in the idle passage. The sizes of the two orifices (2 and 3) and the idle jet (5) are carefully designed to obtain desired engine idle speed performance. An idle mixture screw (6) is used on some carburetors so the mixture can be adjusted to fine tune engine idle speed performance.

When the throttle plate (1, **Figure 26**) is opened slightly to a fast idle position (as indicated by the broken lines) both the primary and secondary idle orifices (2 and 3) are subjected to high manifold vacuum. The incoming flow of air through the secondary orifice (3) is cut off, which increases the speed of fuel flow through the idle jet (5). This supplies the additional fuel needed to properly mix with the greater volume of air passing around the throttle plate. As the throttle plate is further opened, the idle fuel system ceases to operate and the fuel mixture is again controlled by the venturi of the main fuel system.

The air:fuel mixture of the main fuel system (generally called the high speed mixture system) is determined by either a fixed jet (orifice) or an adjustment screw (**Figure 27**). The jet is often available in different sizes to accommodate different engine applications and operating conditions, while the adjustment screw may be turned to alter the mixture setting.

Although some simple carburetors use just the systems previously described, most carburetors re-

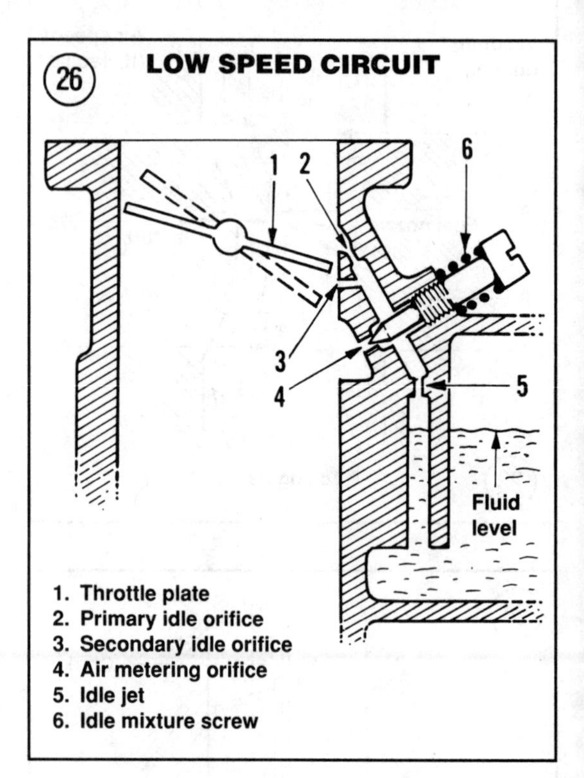

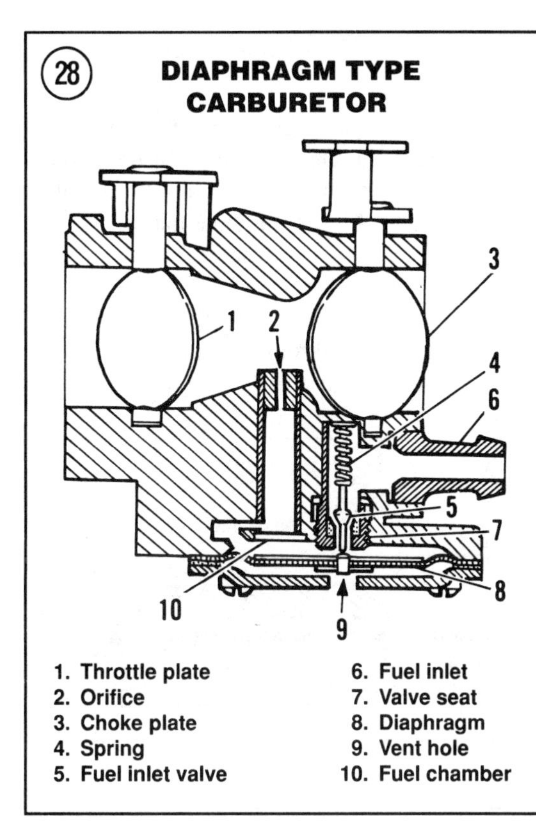

1. Throttle plate
2. Orifice
3. Choke plate
4. Spring
5. Fuel inlet valve
6. Fuel inlet
7. Valve seat
8. Diaphragm
9. Vent hole
10. Fuel chamber

FLOAT TYPE CARBURETOR

1
2
3
4
6
7
5

1. Throttle plate
2. Choke plate
3. Fuel inlet
4. Fuel inlet valve
5. Float
6. Nozzle
7. Fuel level

29

quire more complex features to supply a fuel:air mixture that will produce efficient engine operation over a wide range of engine speeds and in a variety of operating conditions. These design features are described in the following paragraphs which outline the different carburetor types.

Carburetor Types

Diaphragm carburetor

A drawing of a typical diaphragm type carburetor is shown in **Figure 28**. Fuel is routed from the fuel tank to the carburetor inlet (6). Atmospheric pressure is maintained on the lower side of diaphragm (8) through vent hole (9). When the choke plate (3) is closed and the engine is started, or when the engine is running, pressure at the orifice (2) in the carburetor bore is less than atmospheric pressure. This low pressure, or vacuum, is transmitted to the fuel chamber (10) above the diaphragm. The higher (atmospheric) pressure at the lower side of the diaphragm will push the diaphragm upward, compressing the spring (4) and allowing the fuel inlet valve (5) to open and admit fuel into the fuel chamber (10). Most carburetors are equipped with adjustable screws that control the air:fuel mixture. Refer to Chapter Six for operating information on specific diaphragm carburetors.

Float carburetor

The principle of float type carburetor operation is illustrated in **Figure 29**. Fuel is delivered to the carburetor inlet (3) by gravity when the fuel tank is located above the carburetor, or by a fuel lift pump when the tank is located below the carburetor inlet. Fuel flows through the open fuel inlet valve (4) until the fuel level (7) in the fuel bowl lifts the float (5) sufficiently to close the inlet valve (4). When the engine is running, fuel will be emitted from the nozzle (6), lowering the fuel level. When the fuel level drops, the float will also drop, thereby opening the inlet valve (4) and allowing more fuel into the carburetor fuel bowl. Most carburetors are equipped with adjustable screws that control the air:fuel mixture as well as vents, known as air bleed holes, that introduce air into the carburetor metering system.

Refer to Chapter Six for operating information on specific float type carburetors.

Fuel Pump

A fuel pump is used to transfer fuel from a fuel tank to the carburetor. The most common type of fuel pump is equipped with a diaphragm and valves. Although design of the pump may vary as to type of check valves, etc., all operate on the principle shown in **Figures 30-32**. A pulse passage connects one side of the diaphragm to the engine crankcase (**Figure 30**). When the engine's piston is on an upward stroke (**Figure 31**), vacuum is created in the pulse passage which deflects the diaphragm. Atmospheric pressure against fuel in the fuel tank forces the fuel into the inlet, past the inlet check valve and into the chamber below the diaphragm as shown in **Figure 31**. When the piston is on a downward stroke, crankcase pressure passes through the pulse passage against the diaphragm as shown in **Figure 32**. The diaphragm forces the fuel out of the chamber below the diaphragm, past the outlet check valve and out the pump outlet.

IGNITION SYSTEM

The Tecumseh engines covered in this manual are equipped with magneto ignition systems. The basic design is similar for all systems, with the greatest difference being the use of a breakerless solid-state design on later engines. All systems utilize flywheel magnets, a generating coil, a switching device (breaker points or solid-state module), a transformer (coil) and a spark plug to ignite the air:fuel mixture at the appropriate time in the cylinder. A typical breaker point-type magneto ignition system is shown in **Figure 33**. Fundamental operating principles are discussed in the following paragraphs.

Ignition System Operating Principles

The heart of the ignition system is the ignition coil and armature assembly. The ignition coil consists of two sets of windings, the primary and the secondary, which are wound around an armature. Each set of windings is connected to other components to form the primary and secondary circuits. The interaction of the flywheel magnets, primary and secondary circuits comprises the ignition event during engine operation.

The ignition event begins when magnets in the flywheel induce an electric current in the primary windings of the ignition coil through the armature legs as shown in **Figure 34**. Note that the breaker points are closed. As the flywheel continues to rotate (**Figure 35**), the magnetic field changes direction, although the primary current continues flowing because magnetic lines through the center remain the same due to the design of the armature. The primary

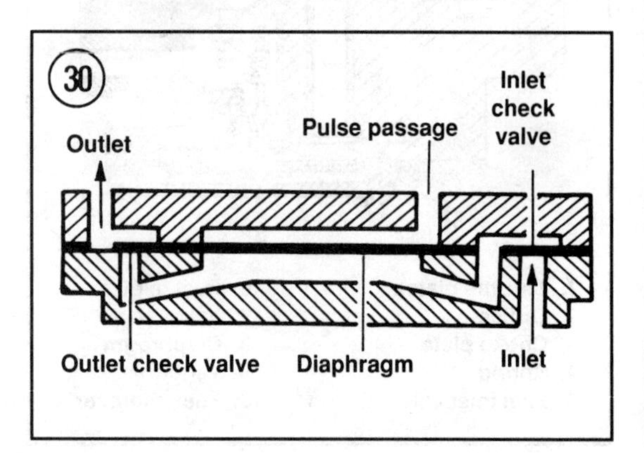

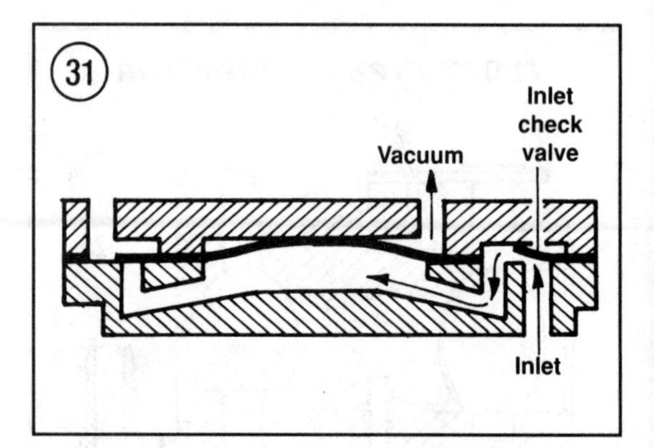

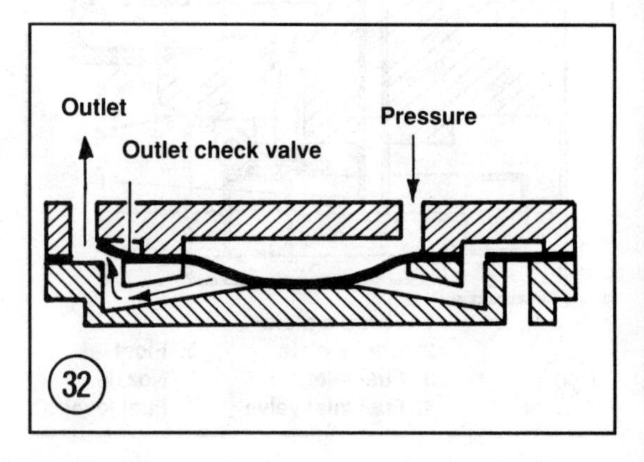

current also results in an electromagnetic field around the primary and secondary coil windings. The breaker points open, stopping primary circuit current flow (**Figure 36**). The condenser serves as a buffer to prevent arcing across the breaker points. When the primary current stops, the electromagnetic field collapses which induces high voltage in the secondary coil windings. The high voltage in the secondary windings passes through the secondary wiring (high tension spark plug lead) and fires the spark plug (**Figure 36**). The sequence of events is timed so the spark plug fires when the piston is located at a specified point in the cylinder usually noted as degrees of crankshaft rotation before top dead center (BTDC), i.e., 10° BTDC.

To prevent engine starting or to stop the engine, the primary circuit is grounded, usually through the ignition switch. Grounding the primary circuit prevents buildup of the electromagnetic field, thereby stopping operation of the secondary circuit and spark plug firing.

On early engines, the breaker points are operated by a cam on the crankshaft. On later engines, the breaker points and condenser are replaced by an ignition module. The electrical components in the module perform the same function electronically as the mechanical breaker points, switching the primary circuit on and off as required.

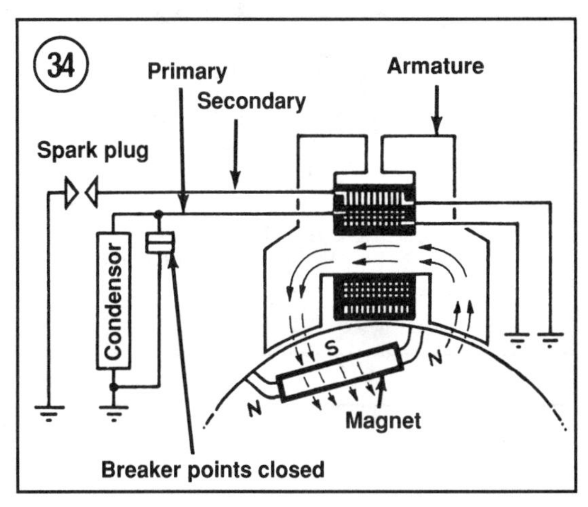

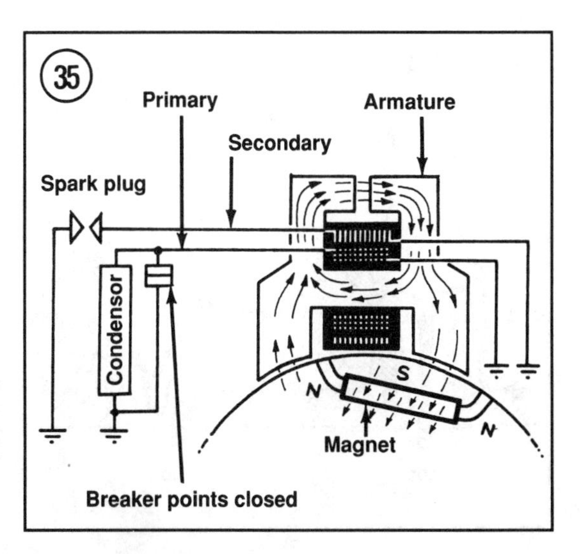

Spark Plug

In any spark-ignition engine, the spark plug (**Figure 37**) provides the means for igniting the compressed air:fuel mixture in the cylinder. Before an electric charge can pass across an air gap, the intervening air must be charged with electricity, or ionized. If the spark plug is properly gapped and the system is not shorted, not more than 7000 volts may

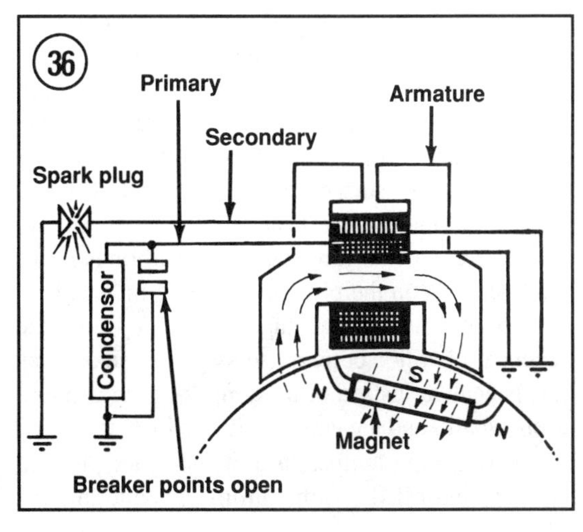

be required to initiate a spark. Higher voltage is required as the spark plug warms up, or if compression pressure or the distance of the air gap is increased. Compression pressures are highest at full throttle and relatively slow engine speeds, therefore, high voltage requirements or a lack of available secondary voltage most often shows up as a miss during maximum acceleration from a slow engine speed.

There are many different types and sizes of spark plugs which are designed for a number of specific requirements.

Thread size

The threaded, shell portion of the spark plug and the attaching hole in the cylinder are manufactured to meet certain industry standards. The diameter is called "thread size." Commonly used thread sizes are 10 mm, 14 mm, 18 mm, 7/8 in. and 1/2 in. pipe.

Reach

The length of thread and the thread depth in the cylinder head or wall are also standardized dimensions. The dimension is measured from the gasket seat on the spark plug to the end of the threads. Common sizes for spark plug reach are 3/8 in., 7/16 in., 1/2 in. and 3/4 in.

Heat range

During engine operation, part of the heat generated during combustion is transferred to the spark plug, and from the plug to the cylinder through the shell threads and gasket, if used. The operating temperature of the spark plug plays an important part in engine operation. If too much heat is retained by the plug, the fuel:air mixture may be ignited by contact with the heated surface before the ignition spark occurs. If insufficient heat is retained, partially burned combustion products (soot, carbon and oil) may build up on the plug tip resulting in "fouling" or shorting out of the plug. If this happens, the secondary ignition current is dissipated uselessly as it is generated instead of bridging the plug gap as a useful spark, and the engine will misfire.

The operating temperature of the spark plug tip can be controlled, within limits, by altering the length of the path the heat must follow to reach the threads and gasket of the plug. Thus a plug with a short, stubby insulator around the center electrode will run cooler than one with a long, slim insulator. Refer to **Figure 38**. Most spark plugs in the more popular sizes are available in a number of heat ranges. Engine manufacturers specify the correct

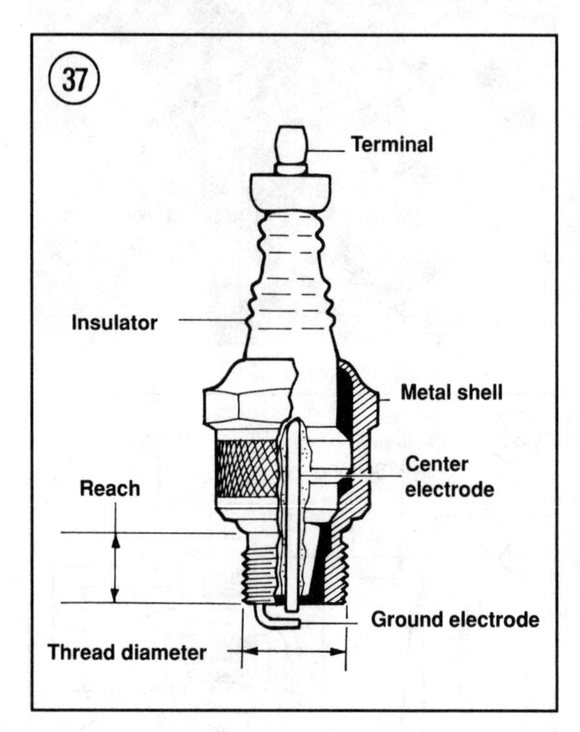

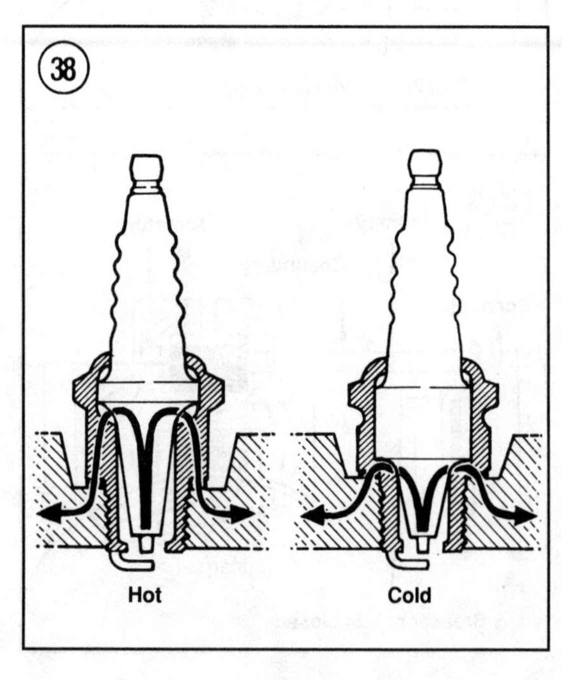

spark plug heat range for normal and severe engine usage.

ELECTRICAL SYSTEM

The electrical system consists of the alternator, regulator/rectifier, battery and electric starter motor. Additional components, such as lights and accessories, may be found on the engine-driven equipment and are connected to the engine's electrical system. The design of the engine's electrical system depends on the engine's application, some engines may be equipped with only an alternator, while a battery must be used on engines equipped with an electric starter, as well as the electrical devices required to charge the battery.

Alternator

The most common type of alternator used on small engines is the flywheel alternator. Magnets located in the flywheel induce a current in the alternator coils (stator) that are located under the flywheel. The electrical principle is the same as previously outlined in the *Ignition System* section. Refer to views in **Figure 39** and the following description:

View A—As the flywheel rotates, a magnetic field is induced in the stator.

View B—During flywheel rotation, the magnetic field reverses and an electric current is created in the wires surrounding the stator.

2

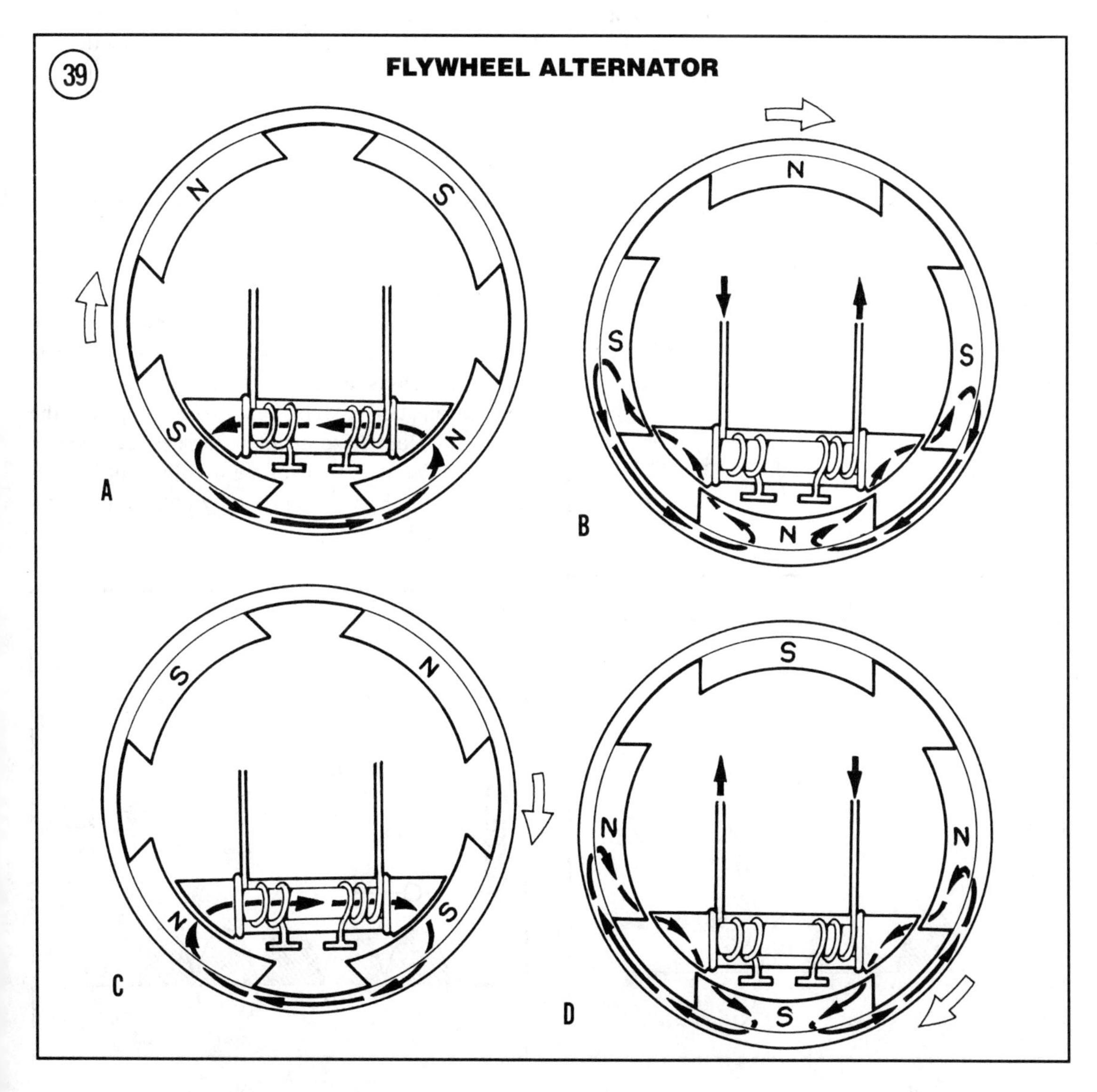

View C—At some point during flywheel rotation, the magnetic field reverses and electric current returns to zero.

View D—Continued flywheel rotation results again in the creation of electric current in the alternator, but the direction of the current is reversed.

The electric current from an alternator forms a sinusoidal wave as shown in **Figure 40** that is alternately oriented negative and positive, hence the term alternating current. The alternator is designed to produce the desired output to fit the requirements of the output device or circuit.

In some cases, the alternator circuit is connected directly to lights, which must be properly selected to match the alternator output. A bulb with insufficient capacity will burn out when alternator output exceeds bulb capacity. Conversely, a bulb with excessive capacity will burn dimly when connected to an alternator with lesser output.

Rectifier

Alternating currect must be converted to direct current for output to a battery or a device requiring direct current (batteries and DC equipment have negative and positive terminals). Alternating current may be converted to direct current by using a rectifier. The rectifier consists of one or more diodes that allow current to flow in one direction, but not in the reverse direction. A circuit using a simple diode is shown in **Figure 41**. The circuit consists of an alternator coil (A), rectifier diode (B), battery (C) and a load (D). After passing through the rectifier, a pulsating direct current is produced as shown by the graph wave. Note that half of the current is lost (shown by broken lines) because the rectifier diode allows current to flow in only one direction, which is half of the alternating current. This type of rectifier is often called a half-wave rectifier.

To use all of the current, a rectifier is designed with four diodes as shown in **Figure 42**. All current flows through the rectifier diodes (B) which results in the graph wave shown. This type of rectifier is often called a full-wave or bridge rectifier. All of the alternating current is converted to direct current.

Regulator

The regulator maintains the correct charging system voltage so the battery is not undercharged or overcharged regardless of variations in engine speed and load.

Virtually all later small engine models are equipped with a modular, solid-state regulator. The regulator senses the voltage in the charging system and automatically either opens or completes the circuit as required to maintain the desired voltage. The rectifier and regulator are usually contained in a single unit. Service is generally limited to testing and replacement.

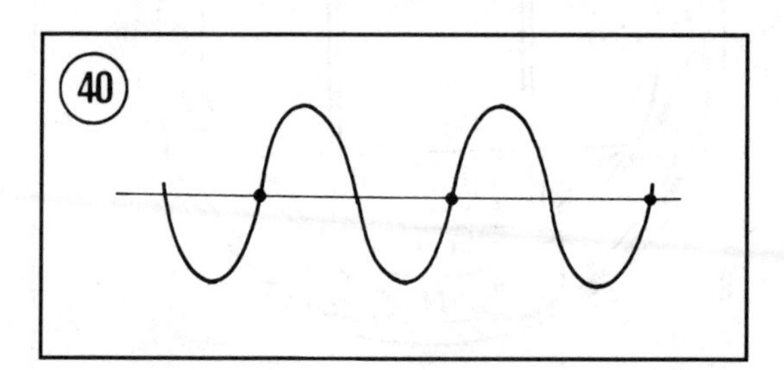

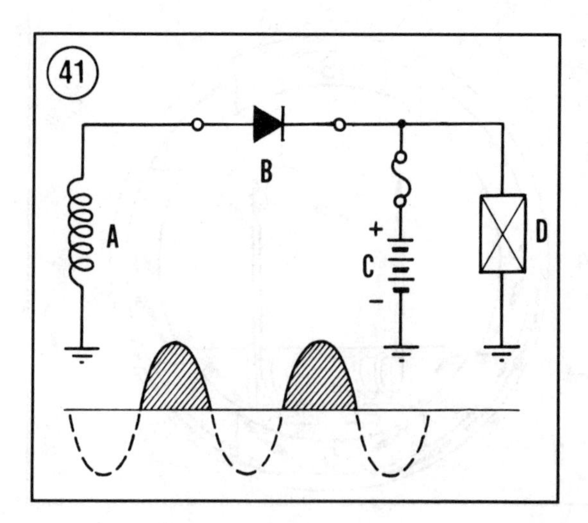

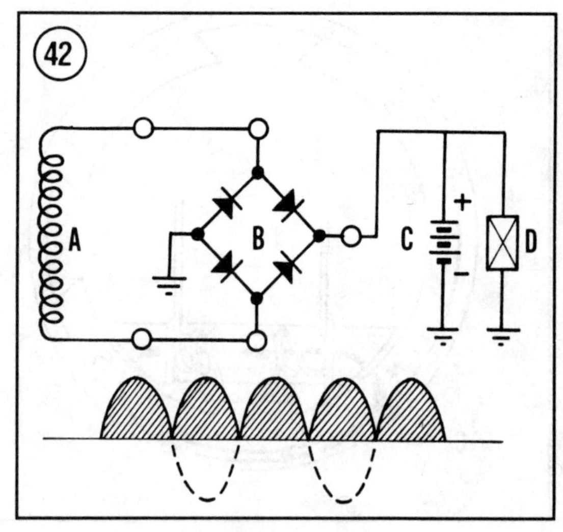

Table 1 AIR:FUEL MIXTURE RATIOS

	Air	Fuel
Starting, cold weather	7 lb.	1 lb.
Accelerating	9 lb.	1 lb.
Idling (no load)	11 lb.	1 lb.
Part open throttle	15 lb.	1 lb.
Full load, wide open throttle	13 lb.	1 lb.

CHAPTER THREE

GENERAL ENGINE INFORMATION

FUEL

The recommended fuel for Tecumseh engines is lead-free regular gasoline, although low-lead gasoline may be used. The fuel should be pure gasoline, although Tecumseh permits the use of gasoline that is blended with alcohol (ethanol) if the alcohol content is not greater than 10 percent. *Do not* use gasoline that contains methanol.

WARNING
Gasoline is extremely flammable. Do not smoke or allow sparks or open flame around fuel or in presence of fuel vapor. Be sure area is well-ventilated. Observe fire prevention rules.

Stored gasoline, either in the engine's fuel tank or in a gasoline can, degrades as time passes. The gasoline becomes less volatile and if present in an engine, harmful deposits form on engine parts causing erratic engine operation and possible damage. A stabilizing agent, such as Sta-Bil, that retards gasoline degradation should be added to gasoline that is to be stored more than one month. When in doubt concerning the quality of gasoline, drain the old gasoline and refill with fresh gasoline.

OIL

Check engine oil level after every five hours of operation or before initial start-up of engine. If the engine is equipped with an oil fill plug (**Figure 1**),

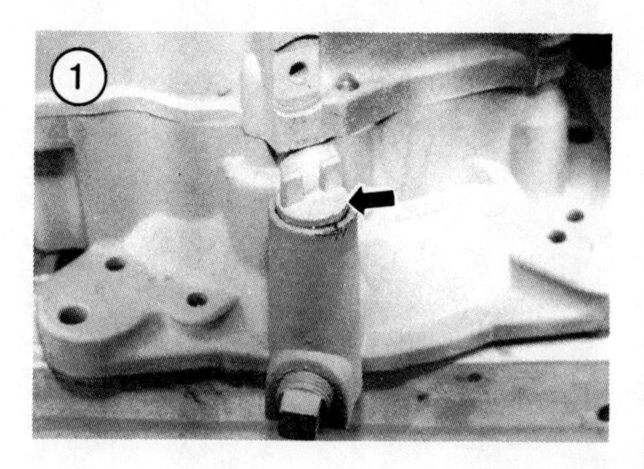

unscrew the plug and check the oil level. The oil should be even with the top threads in the plug hole (**Figure 2**).

CAUTION
Debris, dirt and other foreign material can easily enter the engine crankcase through the oil fill plug hole causing increased engine wear. Thoroughly clean around the oil fill plug before removing the plug.

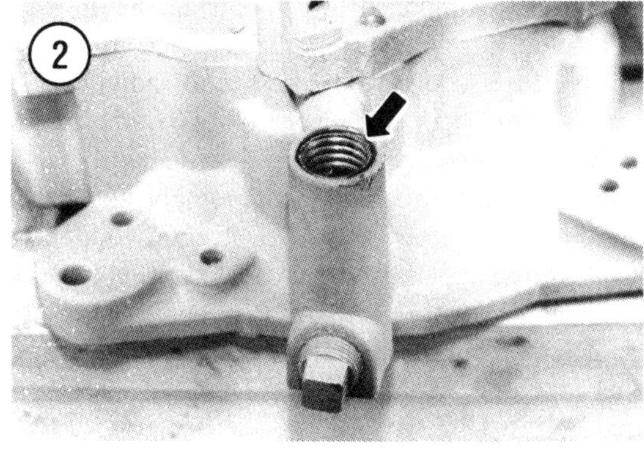

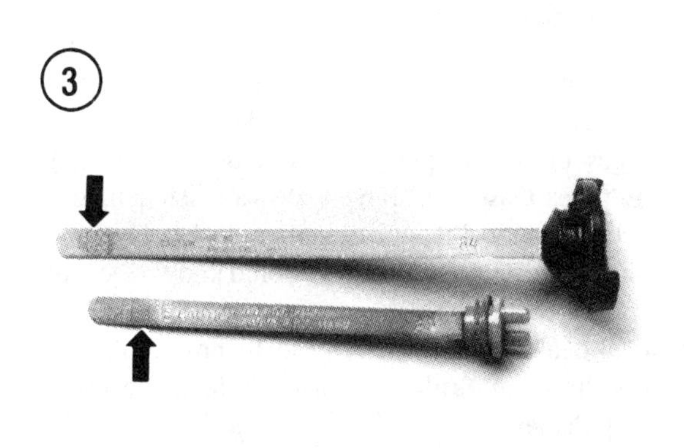

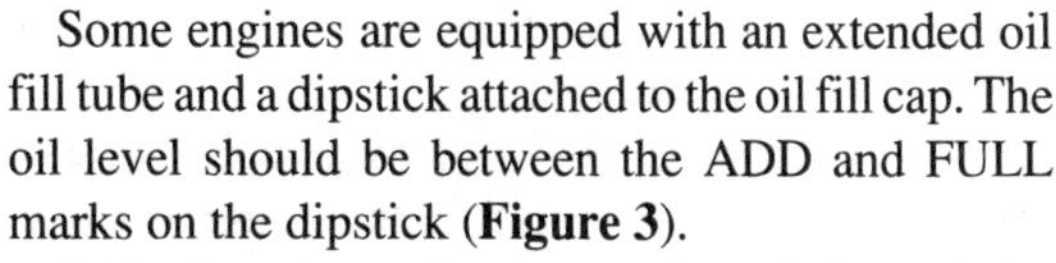

Some engines are equipped with an extended oil fill tube and a dipstick attached to the oil fill cap. The oil level should be between the ADD and FULL marks on the dipstick (**Figure 3**).

Add oil to the engine by pouring oil through the opening for the oil fill plug or the oil dipstick cap.

Tecumseh recommends using oil with an API service classification of SF or SG. Use SAE 30 or 10W-30 oil for temperatures above 32° F (0° C) and SAE 5W-30 or 10W oil for temperatures below 32° F (0° C). Tecumseh states that SAE 10W-40 oil *must not* be used in the engine.

If unfamiliar with oil grade and viscosity terms, refer to *Lubricants* section in Chapter One.

Oil should be changed after every 25 hours of normal operation, more frequently if operation is severe.

On any engine, do not overfill the engine with oil. Excess oil is forced into the cylinder where it burns, resulting in exhaust smoke and carbon deposits on the piston, cylinder head, valves and valve seats. Excess oil also prevents the oil dipper from working properly, which can cause engine damage due to inadequate lubrication in the top of the engine.

ENGINE IDENTIFICATION

Before servicing the engine and ordering parts, the Tecumseh engine model and specification numbers must be determined. Although engines may be similar in appearance, specific differences that affect service specifications and part configuration are only defined by the model and specification numbers. Although rarely required, provide the serial number when ordering parts.

Engine identification numbers, including the model number, specification number and serial number, are located on the blower housing around the flywheel or on a tag attached to the engine. The numbers are stamped in an identification plate or directly in the metal. See **Figure 4**.

The engine model number identifies the basic engine family. Refer to **Table 1** for a breakdown and example of a typical Tecumseh engine model number.

The specification number specifies the parts configuration of the engine, as well as cosmetic details such as paint color and decals. The specification number also determines governor speed settings depending on the engine's application, i.e., lawn mower, tractor, pump, etc.

The serial number provides information concerning the manufacturing of the engine. Refer to **Table 1** for a breakdown of a typical serial number.

Note in the table that some engines are classified using the term "frame." Tecumseh classifies the basic engine structure according to the type of metal, either aluminum or cast iron, used to manufacture the engine crankcase and the metal in the cylinder bore. A small frame engine is made of aluminum with an aluminum cylinder bore. A medium frame engine is made of aluminum and has a cast iron liner in the cylinder bore. A heavy frame engine is made of cast iron with a cast iron cylinder bore (heavy frame engines are not covered in this manual).

BASIC ENGINE SPECIFICATIONS

Table 2 lists basic engine specifications for Tecumseh engines covered in this manual.

PURCHASING PARTS

Virtually all parts necessary to maintain and repair a Tecumseh engine are available from Tecumseh dealers, including Tecumseh oil and filters. Parts are also available from outside vendors such as hardware stores, lawn equipment dealers and mass merchandise stores. The quality of aftermarket parts may not meet the design standards specified by Tecumseh, resulting in reduced engine life or performance. When in doubt as to the quality of an aftermarket part, ask the supplier if they will pay for costs incurred should the part not perform properly.

Before ordering parts, refer to the *Engine Identification* section and note all engine identifying information on engine. All of it may be necessary for the parts supplier to identify the part needed in the parts book.

If possible, the old part should be taken along when ordering parts. The old and new parts can then be matched to be certain the proper part is obtained. In some cases, a knowledgeable parts supplier can provide the correct part after seeing the old part, even if the engine identification numbers are unavailable.

There are components that are available only as an assembly, such as piston rings, which are available only as a set of piston rings and not individually. Some components, such as carburetors, may have individual components that can be ordered from the manufacturer, but the parts supplier only stocks the complete assembly to reduce inventoried items. When ordering parts, be aware that although only a single component is faulty, it may be necessary to replace the entire assembly. Contact the parts supplier for parts availability.

Where to Get Parts

Almost every town of medium size has a Tecumseh dealer. The first place to look is in the Yellow Pages of the telephone book under the category "Engines-Gasoline." If no dealers are listed, try contacting the local lawn and garden dealer. They may not service engines, but they should know where the nearest Tecumseh dealer is located. If there is no local dealer, contact one of the Tecumseh distributors shown in **Table 3** and ask for the nearest servicing dealer.

Table 1 ENGINE MODEL NUMBER BREAKDOWN

ECH	Exclusive Craftsman horizontal crankshaft
ECV	Exclusive Craftsman vertical crankshaft
H	Horizontal crankshaft
HH	Horizontal crankshaft heavy duty (cast iron)
HHM	Horizontal crankshaft heavy duty (cast iron) medium frame
HM	Horizontal crankshaft medium frame
HS	Horizontal crankshaft small frame
LAV	Lightweight aluminum vertical crankshaft
OHM	Overhead valve horizontal crankshaft medium frame
OVM	Overhead valve vertical crankshaft medium frame
OVRM	Overhead valve vertical crankshaft rotary mower
TNT	Toro N'Tecumseh
TVM	Tecumseh vertical crankshaft (medium frame)
TVS	Tecumseh styled vertical crankshaft
V	Vertical crankshaft
VH	Vertical crankshaft heavy duty (cast iron)
VLV	Vector lightweight vertical crankshaft
VM	Vertical crankshaft medium frame
EXAMPLE	
Engine model and specification numbers: TVS120-63624F	
TVS	Tecumseh styled vertical crankshaft
120	Indicates a 12 cubic inch displacement
63624F	Is the specification number used to identify engine parts
Engine serial number: 0109N	
0	First digit is the year of manufacture (1990)
109	Indicates calendar day of that year (109th day or April 19, 1990)
N	Represents the line and shift on which the engine was built at the factory

Table 2 ENGINE SPECIFICATIONS

Model	Bore	Stroke	Displacement	Power rating
ECH90	2.500 in. (63.50 mm)	1.844 in. (46.84 mm)	9.05 cu. in. (148 cc)	No rating
ECV100	2.625 in. (66.68 mm)	1.844 in. (46.84 mm)	9.98 cu. in. (164 cc)	No rating
ECV105	2.625 in. (66.68 mm)	1.938 in. (49.23 mm)	10.49 cu. in. (172 cc)	No rating
ECV110	2.750 in. (69.85 mm)	1.938 in. (49.23 mm)	11.50 cu. in. (189 cc)	No rating
ECV120	2.812 in. (71.43 mm)	1.938 in. (49.23 mm)	12.04 cu. in. (197 cc)	No rating
H25	2.313 in. (58.74 mm)	1.844 in. (46.84 mm)	7.75 cu. in. (127 cc)	2.5 hp (1.9 kW)
H30 after 1982	2.500 in. (63.50 mm)	1.844 in. (46.84 mm)	9.05 cu. in. (148 cc)	3 hp (2.2 kW)
H30 prior to 1983	2.313 in. (58.74 mm)	1.844 in. (46.84 mm)	7.75 cu. in. (127 cc)	3 hp (2.2 kW)

(continued)

Table 2 ENGINE SPECIFICATIONS (continued)

Model	Bore	Stroke	Displacement	Power rating
H35 after 1982	2.500 in. (63.50 mm)	1.938 in. (49.23 mm)	9.51 cu. in. (156 cc)	3.5 hp (2.6 kW)
H35 prior to 1983	2.500 in. (63.50 mm)	1.844 in. (46.84 mm)	9.05 cu. in. (148 cc)	3.5 hp (2.6 kW)
H40	2.500 in. (63.50 mm)	2.250 in. (57.15 mm)	11.04 cu. in. (181 cc)	4 hp (3 kW)
H50	2.625 in. (66.68 mm)	2.250 in. (57.15 mm)	12.18 cu. in. (229 cc)	5 hp (3.8 kW)
H60	2.625 in. (66.68 mm)	2.500 in. (63.50 mm)	13.53 cu. in. (222 cc)	6 hp (4.5 kW)
H70	2.750 in. (69.9 mm)	2.532 in. (64.3 mm)	15.04 cu. in. (246 cc)	7 hp (5.2 kW)
H80	3.062 in. (77.8 mm)	2.532 in. (64.3 mm)	18.65 cu. in. (305 cc)	8 hp (6 kW)
HM70[1]	2.938 in. (74.61 mm)	2.532 in. (64.3 mm)	17.16 cu. in. (281 cc)	7 hp (5.2 kW)
HM70[2]	3.125 in. (79.38 mm)	2.532 in. (64.3 mm)	19.4 cu. in. (318 cc)	8 hp (6 kW)
HM80[1]	3.062 in. (77.8 mm)	2.532 in. (64.3 mm)	18.65 cu. in. (305 cc)	8 hp (6 kW)
HM80[2]	3.125 in. (79.38 mm)	2.532 in. (64.3 mm)	19.4 cu. in. (318 cc)	8 hp (6 kW)
HM100[1]	3.187 in. (80.95 mm)	2.532 in. (64.3 mm)	20.2 cu. in. (330 cc)	10 hp (7.5 kW)
HM100[2]	3.313 in. (84.15 mm)	2.532 in. (64.3 mm)	21.82 cu. in. (358 cc)	10 hp (7.5 kW)
HMXL70	3.125 in. (79.38 mm)	2.532 in. (64.3 mm)	19.4 cu. in. (318 cc)	8 hp (6 kW)
HS40	2.625 in. (66.68 mm)	1.938 in. (49.23 mm)	10.49 cu. in. (172 cc)	4 hp (3 kW)
HS50	2.812 in. (71.43 mm)	1.938 in. (49.23 mm)	12.04 cu. in. (197 cc)	5 hp (3.8 kW)
LAV25	2.313 in. (58.74 mm)	1.844 in. (46.84 mm)	7.75 cu. in. (127 cc)	2.5 hp (1.9 kW)
LAV30	2.313 in. (58.74 mm)	1.844 in. (46.84 mm)	7.75 cu. in. (127 cc)	3 hp (2.2 kW)
LAV35, LV35	2.500 in. (63.50 mm)	1.844 in. (46.84 mm)	9.05 cu. in. (148 cc)	3.5 hp (2.6 kW)
LAV40	2.625 in. (66.68 mm)	1.938 in. (49.23 mm)	10.49 cu. in. (172 cc)	4 hp (3 kW)
LAV50	2.812 in. (71.43 mm)	1.938 in. (49.23 mm)	12.04 cu. in. (197 cc)	5 hp (3.8 kW)
TNT100	2.625 in. (66.68 mm)	1.844 in. (46.84 mm)	9.98 cu. in. (164 cc)	4 hp (3 kW)
TNT120	2.812 in. (71.43 mm)	1.938 in. (49.23 mm)	12.04 cu. in. (197 cc)	5 hp (3.8 kW)
TVM125	2.625 in. (66.68 mm)	2.250 in. (57.15 mm)	12.18 cu. in. (229 cc)	5 hp (3.8 kW)
TVM140	2.625 in. (66.68 mm)	2.500 in. (63.50 mm)	13.53 cu. in. (222 cc)	7 hp (5.2 kW)
TVM170	2.938 in. (74.61 mm)	2.532 in. (64.3 mm)	17.16 cu. in. (281 cc)	7 hp (5.2 kW)
TVM195, TVXL195	3.125 in. (79.38 mm)	2.532 in. (64.3 mm)	19.4 cu. in. (318 cc)	8 hp (6 kW)
TVM220, TVXL220	3.313 in. (84.15 mm)	2.532 in. (64.3 mm)	21.82 cu. in. (358 cc)	10 hp (7.5 kW)

(continued)

Table 2 ENGINE SPECIFICATIONS (continued)

Model	Bore	Stroke	Displacement	Power rating
TVS75	2.313 in. (58.74 mm)	1.844 in. (46.84 mm)	7.75 cu. in. (127 cc)	3 hp (2.2 kW)
TVS90	2.500 in. (63.50 mm)	1.844 in. (46.84 mm)	9.05 cu. in. (148 cc)	3.5 hp (2.6 kW)
TVS100	2.625 in. (66.68 mm)	1.844 in. (46.84 mm)	9.98 cu. in. (164 cc)	4 hp (3 kW)
TVS105	2.625 in. (66.68 mm)	1.938 in. (49.23 mm)	10.49 cu. in. (172 cc)	4 hp (3 kW)
TVS115	2.812 in. (71.43 mm)	1.844 in. (46.84 mm)	11.45 cu. in. (188 cc)	5 hp (3.8 kW)
TVS120	2.812 in. (71.43 mm)	1.938 in. (49.23 mm)	12.04 cu. in. (197 cc)	5 hp (3.8 kW)
V40	2.500 in. (63.50 mm)	2.250 in. (57.15 mm)	11.04 cu. in. (181 cc)	4 hp (3 kW)
V40 with external ignition	2.625 in. (66.68 mm)	1.938 in. (49.23 mm)	10.49 cu. in. (172 cc)	4 hp (3 kW)
V50	2.625 in. (66.68 mm)	2.250 in. (57.15 mm)	12.18 cu. in. (229 cc)	5 hp (3.8 kW)
V60	2.625 in. (66.68 mm)	2.500 in. (63.50 mm)	13.53 cu. in. (222 cc)	6 hp (4.5 kW)
V70	2.750 in. (69.9 mm)	2.532 in. (64.3 mm)	15.04 cu. in. (246 cc)	7 hp (5.2 kW)
V80	3.062 in. (77.8 mm)	2.532 in. (64.3 mm)	18.65 cu. in. (305 cc)	8 hp (6 kW)
VLV50, VLXL50	2.797 in. (71.04 mm)	2.047 in. (51.99 mm)	12.58 cu. in. (206 cc)	5 hp (3.8 kW)
VLV55, VLXL55	2.797 in. (71.04 mm)	2.047 in. (51.99 mm)	12.58 cu. in. (206 cc)	5.5 hp (4.1 kW)
VM70	2.750 in. (69.9 mm)	2.532 in. (64.3 mm)	15.04 cu. in. (246 cc)	7 hp (5.2 kW)
VM80[1]	3.062 in. (77.8 mm)	2.532 in. (64.3 mm)	18.65 cu. in. (305 cc)	8 hp (6 kW)
VM80[2]	3.125 in. (79.38 mm)	2.532 in. (64.3 mm)	19.4 cu. in. (318 cc)	8 hp (6 kW)
VM100	3.187 in. (80.95 mm)	2.532 in. (64.3 mm)	20.2 cu. in. (330 cc)	10 hp (7.5 kW)

1. HM70, HM80 or VM80 models prior to specification letter E, and early HM100 models.
2. HM70, HM80 or VM80 models specification letter E and after, and later HM100 models.

Table 3 TECUMSEH CENTRAL PARTS DISTRIBUTORS (Arranged Alphabetically by State)

Charlie C. Jones, Inc.
Phone (602) 272-5621
2440 W. McDowell Rd.
P.O. Box 6654
Phoenix, Arizona 85005

Billou's, Inc.
Phone (209) 784-4102
1343 S. Main
Porterville, California 93257

Pacific Power Equipment Co.
Phone (303) 744-7891
1441 Bayaud Ave., Unit 4
Denver, Colorado 80223

Radco Distributors, Inc.
Phone (904) 731-7957
6261 Powers Ave.
P.O. Box 5459
Jacksonville, Florida 32207

Small Engine Clinic, Inc.
Phone (808) 488-0711
98019 Kam Highway
P.O. Box 427
Pearl City, Hawaii 96782

Industrial Engine & Parts
Phone (708) 263-0500
50 Noll Street
Waukegan, Illinois 60085

(continued)

Table 3 TECUMSEH CENTRAL PARTS DISTRIBUTORS (continued)
(Arranged Alphabetically by State)

Medart Engines of Kansas
Phone: (913) 888-8828
15500 West 109th Street
Lenexa, Kansas 66219

Engines Southwest
Phone (318) 222-3871
215 Spring St., P.O. Box 67
Shreveport, Louisiana 71161

W.J. Connell Co.
Phone (508) 543-3600
65 Green St., Rt. 106
Foxboro, Massachusetts 02035

Central Power Distributors, Inc.
Phone (612) 633-5179
2976 N. Cleveland
St. Paul, Minnesota 55113

Medart Engines of St. Louis
Phone: (314) 343-0505
100 Larkin Williams Industrial Ct.
Fenton, Missouri 63026

Original Equipment, Inc.
Phone: (406) 245-3081
905 Second Avenue, North
Billings, Montana 59103

Gardner, Inc.
Phone: (609) 860-8060
12 Melrich Rd., Rd. 3
Cranbury, New Jersey 08512

Smith Engines & Irrigation, Inc.
Phone (704) 392-3100
4250 Golf Acres Dr.
P.O. Box 668985
Charlotte, North Carolina 28266

Gardner, Inc.
Phone (614) 488-7951
1150 Chesapeake Ave.
Columbus, Ohio 43212

Medart Engines of Tulsa
Phone: (918) 627-1448
7450 East 46th Place
Tulsa, Oklahoma 74115

Power Equipment Systems
Phone (503) 585-6120
2257 McGilchrist S.E.
P.O. Box 629
Salem, Oregon 97308

Pitt Auto Electric Co.
Phone: (412) 766-9112
2900 Stayton Street
Pittsburgh, Pennsylvania 15212

Medart Engines of Memphis
Phone: (901) 795-4365
4365 Old Lamar
Memphis, Tennessee 38118

Frank Edwards Co.
Phone (801) 972-0128
1284 S. 500 West
Salt Lake City, Utah 84101

Power Equipment Systems
Phone (206) 763-8902
88 South Hudson
P.O. Box 3901
Seattle, Washington 98124

CANADIAN DISTRIBUTORS

CPT Canada Power Technology
Phone (403) 453-5791
13315 146th Street
Edmonton, Alberta T5L 4S8

CPT Canada Power Technology
Phone (416) 890-6900
161 Watline Ave.
Mississauga, Ontario L4W 2T7

CHAPTER FOUR

TROUBLESHOOTING

4

For any engine to operate, several conditions must be met. Before troubleshooting an engine, be sure the following items are checked and found satisfactory.

FUEL

The engine requires clean, fresh gasoline to operate at maximum efficiency. See Chapter Three for fuel recommendation. Two fuel-related problems usually cause the most cases of erratic engine operation or an inability to start an engine: poor fuel and water in the fuel.

The engines covered in this manual use gasoline or alcohol/gasoline for fuel. Stored gasoline, either in the engine's fuel tank or in a gasoline can, degrades as time passes. The gasoline becomes less volatile and if present in an engine, harmful deposits form on engine parts causing erratic engine operation and possible damage. When in doubt concerning the quality of gasoline, drain the old gasoline and refill with fresh gasoline.

Fuel that is contaminated with water will cause erratic engine operation and may prevent the engine from starting. Water can enter the fuel system from a contaminated fuel can, open fuel tank cap, or condensation in the fuel tank. The water may be difficult to detect in the fuel tank, although a significant amount will appear as a bubble in the gasoline. If water-contaminated gasoline is a continuing problem, identify the cause. Testing kits are available from some small-engine shops that will detect water in gasoline.

CONTROLS

Before operating the engine, be sure all controls are in good, safe operating condition. Maladjusted, disconnected, bent, or otherwise damaged controls can prevent engine starting, cause erratic engine operation, damage the engine and possibly injure the operator.

WARNING

Do not operate engine or any equipment unless all safety-related devices are functional.

If the engine is mounted on a piece of equipment and will not start or operate properly, the problem may be caused by electrical or mechanical problems on the equipment that affect the engine. For instance, faulty ignition switches and safety interlocks may prevent engine starting. It may be necessary to isolate the engine from the equipment to determine whether the problem is with the engine or with the equipment.

STARTER

The starter system, whether manual or electrical, must be capable of turning the engine's crankshaft with sufficient speed to start the engine. A manual starter should operate freely without binding or slippage so the crankshaft is turned through several revolutions. The electric starter system must be in good operating condition. If so equipped, the battery should be charged periodically if the engine is not used frequently enough to maintain battery charge.

TROUBLESHOOTING

Every small engine requires an uninterrupted supply of fuel and air, proper ignition and adequate compression. If any of these are lacking, the engine will not run.

Diagnosing mechanical problems is relatively simple if you use orderly procedures and keep a few basic principles in mind.

The troubleshooting procedures in this chapter analyze typical symptoms and show logical methods of isolating causes. These are not the only methods. There may be several ways to solve a problem, but only a systematic approach can guarantee success.

Never assume anything. Do not overlook the obvious. If the engine suddenly quits, check the easiest, most accessible problem spots first. Is there gasoline in the tank? Has the spark plug wire fallen off?

If nothing obvious turns up in a quick check, look a little further. Learning to recognize and describe symptoms will make repairs easier for you or a mechanic at the shop. How fast was the engine running when it quit? How long had the engine been running before it quit? If the engine was smoking, what color was the smoke?

Gather as many symptoms as possible to aid in diagnosis. Note whether the engine lost power gradually or all at once. Remember that the more complicated a machine is, the easier it is to troubleshoot because symptoms point to specific problems.

After the symptoms are defined, areas which could cause problems are tested and analyzed. Guessing at the cause of a problem may provide the solution, but it can easily lead to frustration, wasted time and a series of expensive, unnecessary parts. When analyzing a problem, be aware that an underlying problem may be the root cause. For instance, replacing a soot-fouled spark plug may get the engine running again. However, a maladjusted carburetor is the real problem and the engine will continue stopping until the carburetor is fixed.

You do not need fancy equipment or complicated test gear to determine whether repairs can be attempted. A few simple checks could save a large repair bill and lost time should the engine be sent to a professional shop. On the other hand, be realistic and do not attempt repairs beyond your abilities. Professional shops tend to charge heavily for putting together a disassembled engine that may have been abused. Some shops won't even take on such a job—so use common sense, don't get in over your head.

Some common problems and remedies are outlined in the following paragraphs. Also refer to the troubleshooting information found in **Table 1**.

Engine Will Not Start

Failure to start is usually caused by fuel not reaching the cylinder or spark not occurring at the spark plug. Do not neglect to check obvious causes (fuel tank empty, stop switch actuated, safety switch malfunctioning, etc.). Be sure the problem is not due to a faulty starter or an internal problem is causing sufficient drag that the crankshaft will not rotate normally. If the engine is mounted on equipment, be sure the clutch is operating properly or devices connected directly to the engine are not slowing crankshaft rotation. On lawn mower engines, the blade must be secured tightly to the crankshaft. If the ambient temperature is very cold, the use of incorrect viscosity oil can cause excessive drag against internal engine parts.

Test for fuel and spark

Fuel and a timed spark are essential for the engine to run. Remove the spark plug and check the condition of the spark plug. The spark plug should have an odor of gasoline and may be slightly damp appearing. If the spark plug insulator is not tan or light gray, refer to **Figure 1** which depicts defective spark plugs and lists possible causes.

To check for ignition spark, hold the spark plug against the cylinder head fins with the spark plug lead connected as shown in **Figure 2**. A spark plug tester like that shown in **Figure 3** can also be used.

(1)

SPARK PLUG CONDITION

NORMAL

- Identified by light tan or gray deposits on the firing tip.
- Can be cleaned.

GAP BRIDGED

- Identified by deposit buildup closing gap between electrodes.
- Caused by oil or carbon fouling. If deposits are not excessive, the plug can be cleaned.

OIL FOULED

- Identified by wet black deposits on the insulator shell bore and electrodes.
- Caused by excessive oil entering combustion chamber through worn rings and pistons, excessive clearance between valve guides and stems or worn or loose bearings. Can be cleaned. If engine is not repaired, use a hotter plug.

CARBON FOULED

- Identified by black, dry fluffy carbon deposits on insulator tips, exposed shell surfaces and electrodes.
- Caused by too cold a plug, weak ignition, dirty air cleaner, too rich a fuel mixture or excessive idling. Can be cleaned.

LEAD FOULED

- Identified by dark gray, black, yellow or tan deposits or a fused glazed coating on the insulator tip.
- Caused by highly leaded gasoline. Can be cleaned.

WORN

- Identified by severely eroded or worn electrodes.
- Caused by normal wear. Should be replaced.

FUSED SPOT DEPOSIT

- Identified by melted or spotty deposits resembling bubbles or blisters.
- Caused by sudden acceleration. Can be cleaned.

OVERHEATING

- Identified by a white or light gray insulator with small black or gray brown spots and with bluish-burnt appearance of electrodes.
- Caused by engine overheating, wrong type of fuel, loose spark plugs, too hot a plug or incorrect ignition timing. Replace the plug.

PREIGNITION

- Identified by melted electrodes and possibly blistered insulator. Metallic deposits on insulator indicate engine damage.
- Caused by wrong type of fuel, incorrect ignition timing or advance, too hot a plug, burned valves or engine overheating. Replace the plug.

Set engine control to run position and operate starter. A bright electrical spark should jump across the gap between the spark plug lead and spark plug. A spark indicates the ignition is operating.

WARNING
The engine can start during test. Be sure all safety precautions concerning starting the equipment are observed.

If the ignition will not provide a spark, check the ignition electrical system further to locate cause for lack of spark. Especially check the condition of the engine stop switch wiring for shorts, correct operation of safety switch and condition of breaker points and condenser on models so equipped.

If the test indicates a good spark, but the removed spark plug is dry with no evidence of fuel, be sure the engine is getting fuel. The carburetor choke valve should be closed when attempting to start engine. Some engines are equipped with a primer (**Figure 4**) that injects additional fuel for starting when the primer is pushed. Check the carburetor for proper operation.

NOTE
A prime cause of failure to start or erratic operation is water in the fuel. Be sure there is no water in the fuel. Drain the fuel tank if necessary.

If testing indicates a good spark and the removed spark plug is damp, the spark plug may be faulty or the plug wire connector is corroded. Clean the connector and recheck. If the problem remains, install a new spark plug and attempt to start the engine (be sure the spark plug electrode gap is correct).

If testing indicates a good spark and the removed spark plug is damp, but the engine will not start or runs erratically with a new spark plug, remove the flywheel and check for a sheared flywheel key (**Figure 5**) and excessive wear in the flywheel and crankshaft keyways. A sheared flywheel key or excessive keyway wear will affect ignition timing.

If testing indicates a good spark and the removed spark plug was exceedingly wet with fuel, the engine may be flooded due to a carburetor malfunction. Check for proper adjustment of carburetor controls, and if correct, check carburetor adjustment. If all adjustments are correct, inspect, and if required, overhaul the carburetor.

NOTE
To clear a flooded engine, set the engine control so the throttle is in wide open position and operate the starter. DO

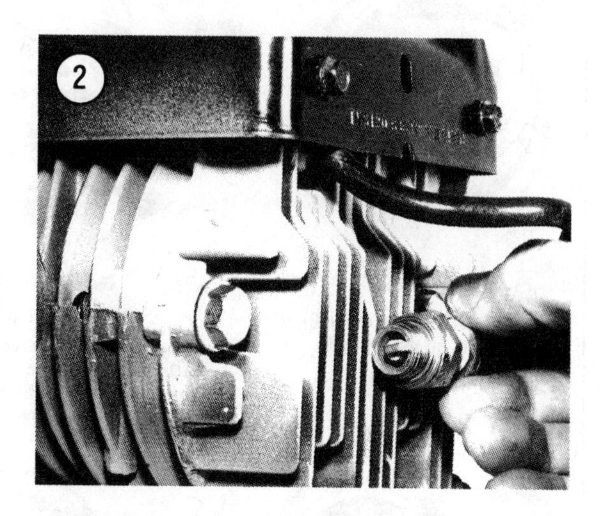

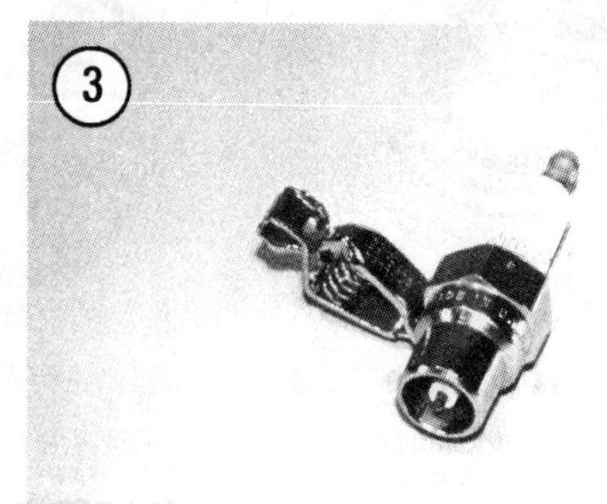

4

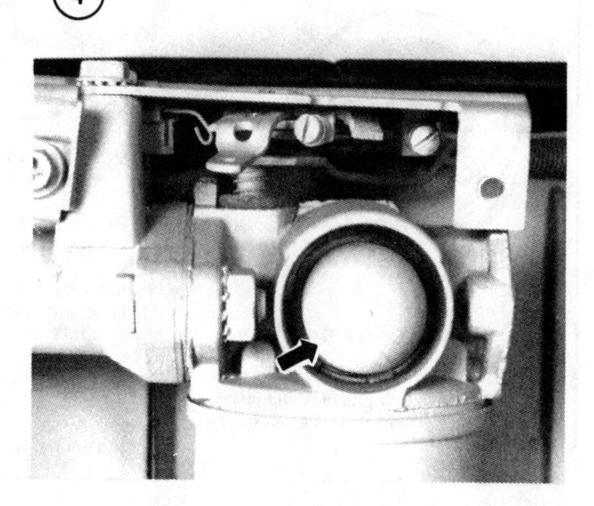

NOT attempt to expel fuel in the cylinder by removing the spark plug as fuel may be ignited by ignition spark or other sources.

Check compression

If the ignition spark is good and the fuel system is operating properly, compression in the cylinder may be insufficient for engine operation. To check compression, disconnect the spark plug wire from the spark plug and remove the spark plug.

WARNING
When grounding the spark plug wire, be sure that the wire end is securely held against the cylinder head, otherwise, a loose wire end may emit a spark that can ignite fuel expelled from the cylinder.

Disengage the flywheel brake if so equipped. Hold a finger against the spark plug hole in the cylinder head and operate the starter. If the finger is blown off the spark plug hole each time the engine completes its compression stroke, compression is adequate for the engine to start. If pressure is minimal or nonexistent, one or more of the following problems may exist:

a. Defective cylinder head gasket.
b. Warped cylinder head.
c. Worn or broken piston rings.
d. Sticking valves.
e. Worn valve guides.
f. Worn or damaged valves or seats.
g. Insufficient valve spring tension.
h. Mistimed camshaft gear.

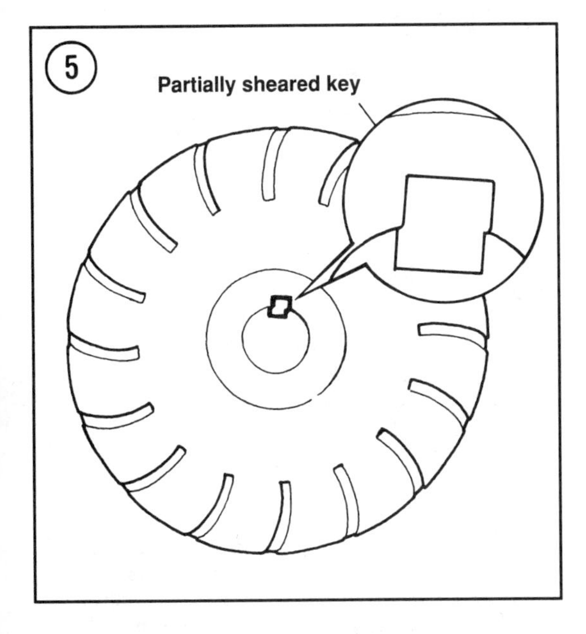

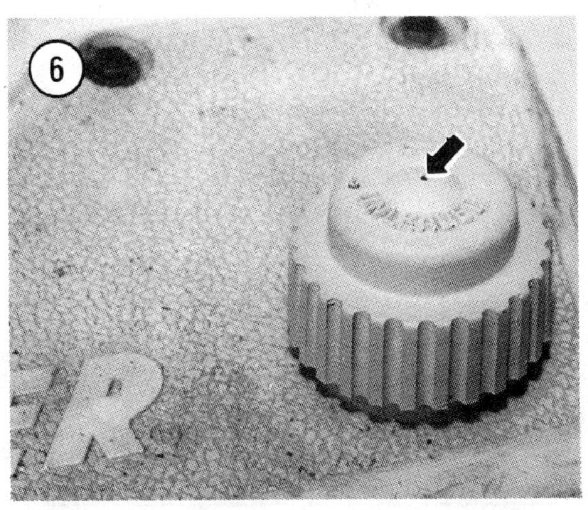

Engine Hard to Start

The same conditions that prevent the engine from starting can also cause hard starting. See preceding section.

If the engine sounds like it is firing, but just won't continue to run, some of the more common causes are a fouled spark plug, contaminated fuel, improperly adjusted carburetor, and improperly adjusted engine controls.

Engine Starts Then Stops

If the engine starts, runs a minute or two, then dies, check the fuel tank cap. The cap must have a vent so fuel will properly feed into the carburetor or the fuel system. If the cap has a vent hole like that shown in **Figure 6**, the hole must be open so air can enter the fuel tank. If the vent is absent (an incorrect fuel cap may have been installed) or plugged, fuel will cease to flow when sufficient vacuum builds in the tank.

Engine Runs Rough or Lacks Power

An intermittent miss may be caused by a malfunctioning ignition system or fuel system. Conduct a spark test as outlined in *Engine Will Not Start* section.

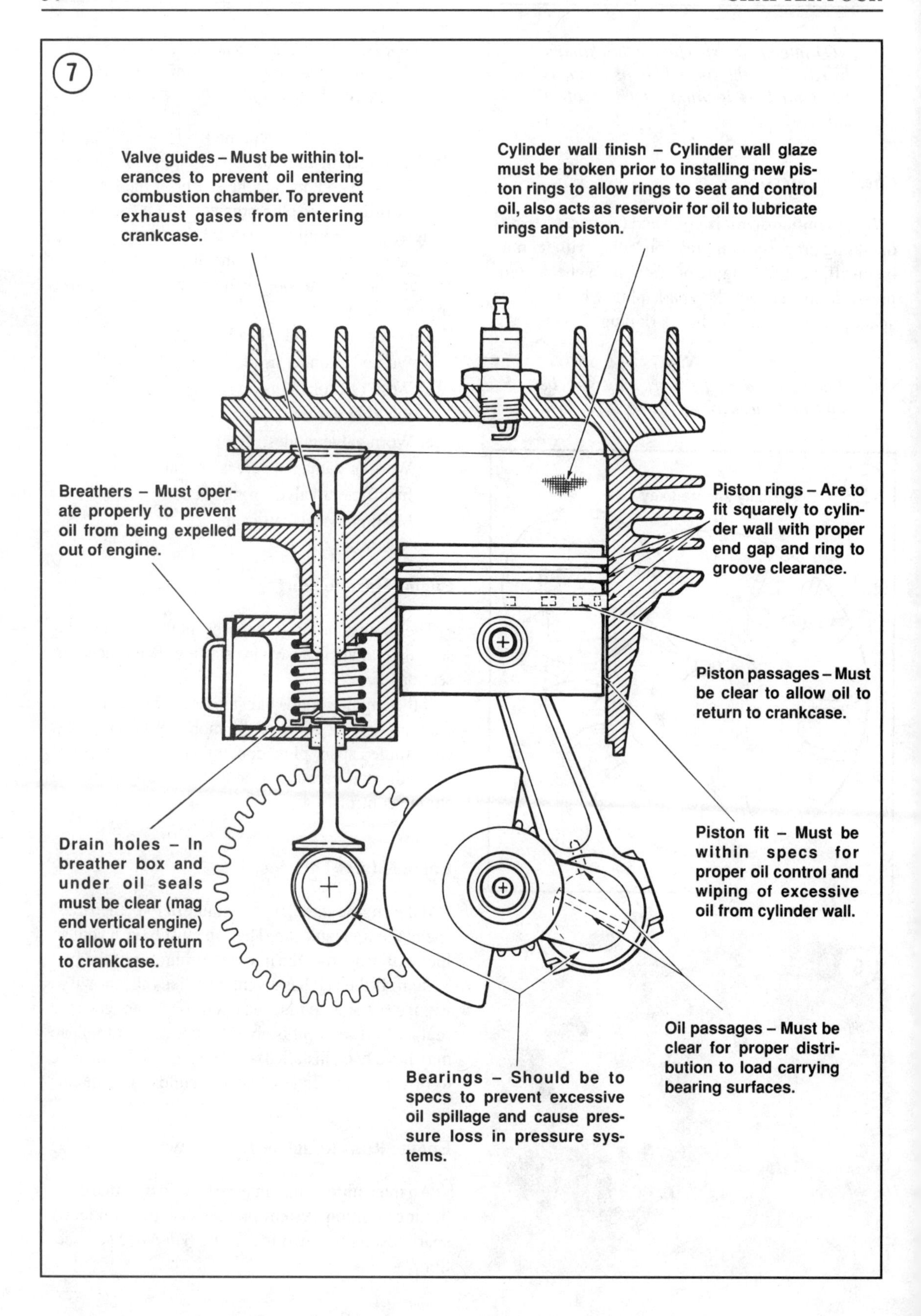
7
Valve guides – Must be within tolerances to prevent oil entering combustion chamber. To prevent exhaust gases from entering crankcase.
Cylinder wall finish – Cylinder wall glaze must be broken prior to installing new piston rings to allow rings to seat and control oil, also acts as reservoir for oil to lubricate rings and piston.
Breathers – Must operate properly to prevent oil from being expelled out of engine.
Piston rings – Are to fit squarely to cylinder wall with proper end gap and ring to groove clearance.
Piston passages – Must be clear to allow oil to return to crankcase.
Drain holes – In breather box and under oil seals must be clear (mag end vertical engine) to allow oil to return to crankcase.
Piston fit – Must be within specs for proper oil control and wiping of excessive oil from cylinder wall.
Oil passages – Must be clear for proper distribution to load carrying bearing surfaces.
Bearings – Should be to specs to prevent excessive oil spillage and cause pressure loss in pressure systems.

A spark plug may produce a good spark when tested outside the engine, but produce a weak or sporadic spark when subjected to compression pressure inside the engine. The weak spark may be due to a defective spark plug or ignition coil. Install a new spark plug and run the engine. If an intermittent miss continues, the ignition coil may be faulty. Refer to Chapter Seven.

An intermittent miss caused by the fuel system may be due to an improperly adjusted carburetor, sticking choke causing an excessively rich mixture (exhaust smoke will be black), a dirty carburetor, contaminated fuel, or a clogged fuel filter. Check the fuel and the fuel system lines and filters before servicing carburetor.

Lack of power without an apparent miss may be caused by improper carburetor adjustment, incorrect governor adjustment, incorrect ignition timing, a dirty air cleaner, a plugged muffler, a dirty carburetor or a malfunctioning governor. Also check equipment connected to the engine which may cause a seeming lack of power, such as a dull lawn mower blade.

The engine may be run *temporarily* without an air cleaner to check performance.

CAUTION
Running an engine for even a short time in dusty conditions causes rapid engine wear.

Engine power may be decreased due to overheating, which can be caused by a lean air-fuel mixture or inadequate cooling. A lean air-fuel mixture is usually caused by an improperly adjusted carburetor, but may also be a result of an air intake leak (bad gaskets) or limited fuel supply. Inadequate cooling is usually due to debris covering the cooling fins on the cylinder and cylinder head. Broken flywheel fan blades, as well as running without proper air shrouds will also cause overheating.

Reduced engine compression also results in a lack of power, usually producing hard engine starting as well. Perform a compression test as outlined previously in this chapter.

Engine Surges

If the engine surges (rhythmically speeds up and slows down) while running, the trouble is probably caused by incorrect carburetor adjustment, incorrect governor adjustment or malfunctioning carburetor or governor. Refer to the appropriate section in this manual for adjustment or service.

Excessive Engine Vibration

The usual cause of excessive engine vibration is a bent crankshaft.

WARNING
Do not attempt to straighten a bent crankshaft. The crankshaft may have incurred stress fractures when bent and attempting to straighten the crankshaft will increase the possibility of the crankshaft breaking, which can cause personal injury if the engine is running.

Engine vibration may also be caused by loose mounting bolts and out-of-balance equipment attached to the crankshaft, such as a loose mower blade.

Excessive Oil Consumption

An engine that uses too much oil may be damaged due to a low oil level. Oil serves as both lubricant and coolant in an engine and when the oil supply is reduced to a critical level, engine damage will result. Leaking oil seals are usually readily apparent, as well as leaking gaskets. Overfilling the engine with oil will cause high oil consumption, as will excessive engine speeds. Malfunctioning internal components can result in high oil consumption, usually evidenced by blue exhaust smoke. Refer to **Figure 7** for a drawing showing internal engine areas that can cause high oil consumption.

Table 1 is on the following page.

Table 1 COMMON TROUBLESHOOTING PROBLEMS

Symptom	Cause	Correction
Fails to Start		
No fuel to carburetor	No fuel in tank	Fill tank
	Shut-off valve closed	Open valve
	Intake screen plugged	Clean screen
	Fuel line plugged	Clean line
No fuel to cylinder	No fuel to carburetor	See above
	Float stuck	Clean carburetor
	Inlet valve stuck	Clean carburetor
	Incorrect float level	Adjust float
	Choke inoperative	Overhaul carburetor
	Fuel passages clogged	Overhaul carburetor
Engine flooded	Float stuck	Clean carburetor
	Incorrect float level	Adjust float
	Overchoked	Purge engine
	No spark at plug	See below
No spark at plug	Plug fouled	Clean plug
	Incorrect plug gap	Adjust gap
	No spark at plug wire	See below
No spark at plug wire	Breaker points coated	Clean points
	Flywheel key sheared	Install new key
	Breaker points not opening	Adjust points
	Breaker points not closing	Adjust points
	Breaker points burned	Install new points
	Kill switch shorted	Repair switch
	Condenser shorted	Install new condenser
	Ignition coil shorted	Install new coil
	Flywheel magnets coated	Clean magnets
	Incorrect air gap	Adjust gap
	Ignition module bad	Install new module
	Safety switch malfunction	Repair/renew switch
	Cranking speed slow	Repair as needed
No compression	Valves stuck	Free valves
	Valves burned	Install new valves
	Piston damaged	Overhaul engine
	Cylinder damaged	Overhaul engine
	Connecting rod damaged	Overhaul engine
Lacks Power		
Engine smokes	Rich fuel mixture	Adjust carburetor
	Air filter plugged	Clean air filter
	Worn piston rings	Overhaul engine
Lean fuel mixture	Improper carburetor adjustment	Adjust carburetor
	Manifold gasket leaking	Install new gasket
	Fuel tank vent plugged	Clean vent
	Fuel tank screen plugged	Clean screen
Partial miss	Weak spark	Service ignition system
	Lean fuel mixture	See above
Engine overheats	Engine dirty	Clean engine
	Low oil level	Add oil
	Engine overloaded	Reduce load
	Cooling fins missing/broken	Install new parts
	Shrouds missing	Install shrouds
	Lean fuel mixture	See above

CHAPTER FIVE

LUBRICATION AND MAINTENANCE

LUBRICATION

Engines with a horizontal crankshaft are splash-lubricated by oil thrown off an oil dipper on the connecting rod. Vertical crankshaft engines are lubricated by oil pumped from a reciprocating plunger type oil pump.

All engines are equipped with a breather system that prevents a pressure build-up in the engine crankcase (just as the piston creates pressure in the cylinder on the up stroke, it also creates pressure in the crankcase on the down stroke). Excessive pressure in the crankcase can cause oil leakage past gaskets and seals. A check valve in the breather permits gases to flow out of the engine only, which creates a vacuum in the engine crankcase. Proper operation of the breather system is necessary to prevent excessive oil consumption and leakage.

Oil Requirements

Tecumseh recommends using oil with an API service classification of SF or SG. Use SAE 30 or 10W-30 oil for temperatures above 32° F (0° C) and SAE 5W-30 or 10W oil for temperatures below 32° F (0° C). Tecumseh states that SAE 10W-40 oil *must not* be used in the engine.

If unfamiliar with oil grade and viscosity terms, refer to *Lubricants* section in Chapter Three.

Refer to the following section for oil change intervals and procedure.

MAINTENANCE

Most machinery requires periodic maintenance so it will operate efficiently for as long as possible. Engine manufacturers specify the maintenance work that must be performed on the engine and when the work should be accomplished. If the engine is not maintained as specified, engine life and performance will be reduced. Neglecting maintenance work may save time and money initially, but the cost of early engine replacement or overhaul will cost more than any early cost savings.

The maintenance program prescribed by the manufacturer may only apply to normal operating conditions. The engine operator will have to determine if operating conditions are more severe and adjust the maintenance schedule accordingly. If in doubt as to recommended maintenance procedures and frequency for abnormally severe usage, contact a dealer or the manufacturer's service department.

Maintenance should include cleaning and inspection of the engine and any related equipment, such as controls and drive components. Any debris on or around the engine, particularly in or on the cooling fins, must be removed or the engine will overheat. Inspection will reveal problems that can be corrected before a more costly, major problem results. Any malfunctioning safety components must be corrected before further operation.

NOTE
Keep a log and write down the time of service as well as any abnormalities found, then refer to the log the next time maintenance is performed so minor problems can be spotted and cured before major, costly repairs are necessary.

SAFETY DEVICES

The engine and the equipment the engine is mounted on may be equipped with safety devices to protect the operator. The equipment should be periodically checked to be sure all safety devices operate as designed. Any faulty devices should be replaced or repaired before the engine and/or equipment are operated.

Some later Tecumseh engines are equipped with a flywheel brake that simultaneously stops the flywheel and grounds the ignition. The brake should stop the engine within three seconds when the operator releases the mower safety control and the speed control is in high speed position. Refer to Chapter Seven for service information.

OIL CHANGE

Tecumseh specifies that the engine oil should be changed after the first two hours of operation for a new or rebuilt engine. Thereafter, the oil should be changed after 25 hours of normal operation. If engine operating conditions are severe, i.e., extremely dusty or under heavy load, the oil should be changed more frequently.

The engine oil should be changed with the engine warm. Unscrew the drain plug and remove the oil. The drain plug is usually located on the underside of the crankcase on vertical crankshaft engines (**Figure 1**) and on the side of horizontal crankshaft engines (**Figure 2**). The drain plug (A, **Figure 3**) may also be located adjacent to the oil fill plug (B) on vertical crankshaft engines for greater accessibility. Some engines may have more than one drain plug.

NOTE
Properly discard used oil. Oil should not be placed in the trash for collection or poured on the ground. Used oil should be taken to a recycling center or to a retailer with an oil disposal container.

Fill the engine with oil specified in the preceding *Oil Requirements* section. Pour oil through the opening for the oil fill plug or the oil dipstick cap. If the engine is equipped with an oil fill plug (B, **Figure 3**), unscrew the plug and add oil until the oil level is even with the top threads in the plug hole (**Figure 4**).

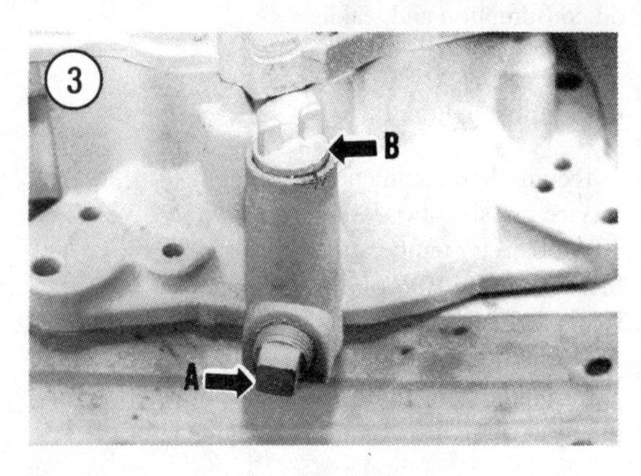

CAUTION
Debris, dirt and other foreign material can easily enter the engine crankcase through the oil fill plug hole, causing increased engine wear. Thoroughly clean around the oil fill plug before removing the plug.

Some engines are equipped with an extended oil fill tube and a dipstick attached to the oil fill cap. The oil level should be between the ADD and FULL marks on the dipstick (**Figure 5**).

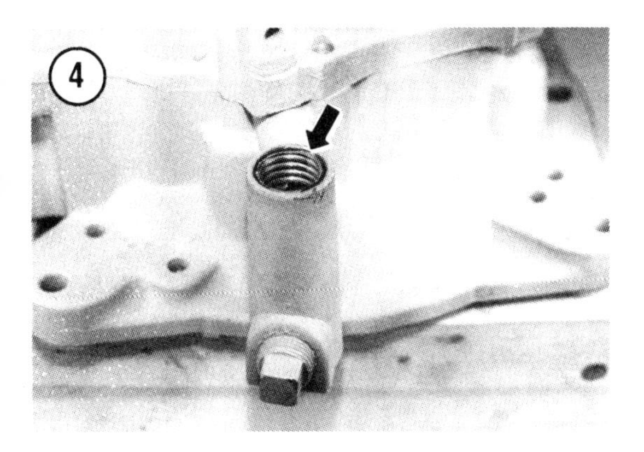

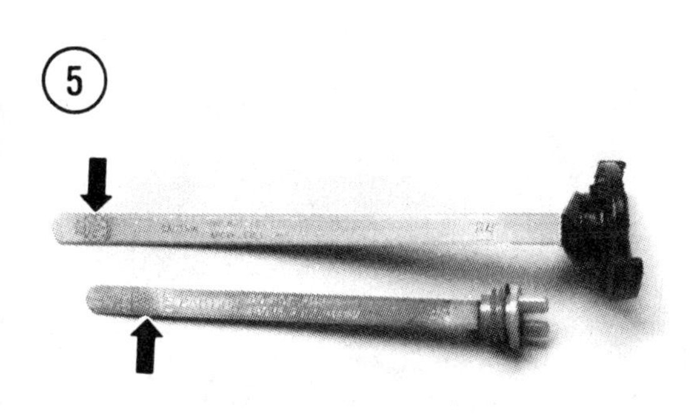

On any engine, do not overfill the engine with oil. Excess oil is forced into the cylinder where it burns, resulting in exhaust smoke and carbon deposits on the piston, cylinder head, valves and valve seats.

AIR CLEANER

Grit is one of the major causes of premature engine wear, and most grit enters the engine through the intake tract. The air cleaner is designed to prevent the entrance of grit and other debris into the engine, but the air cleaner can only function effectively if it is maintained properly. A dirty air cleaner will affect engine performance by restricting air flow while an improperly serviced or assembled air cleaner will allow grit to enter the engine. A filter is relatively inexpensive when compared to engine repairs and it should be discarded if it cannot be cleaned, if it has holes or tears, or if it does not seal properly against the canister.

NOTE
When obtaining a replacement air filter, be sure the filter meets Tecumseh specifications for fit and filtering capability. Installing a low quality filter can shorten engine life.

CAUTION
Never run an engine without an air filter. Ingesting only a small amount of grit will damage an engine.

The air cleaner consists of a canister and the filter element it contains. The canister is secured by one or two wing nuts, screws or knobs. The filter element may be made of foam or paper, or a combination of both foam and paper. The recommended maintenance interval depends on the type of filter element.

Foam Type Filter

Foam type filter elements (**Figure 6**) should be cleaned, inspected and re-oiled after every 25 hours of engine operation, or after three months, whichever occurs first. Clean the filter in soapy water then squeeze the filter until dry (don't twist the filter). Inspect the filter for tears and holes or any other opening. Discard the filter if it cannot be cleaned satisfactorily or if the filter is torn or otherwise damaged. Pour clean engine oil into the filter, then squeeze the filter to remove the excess oil and distribute oil throughout the filter. Clean the filter canister. Inspect and, if necessary, replace any defective gaskets.

When installing the filter element in the canister, be sure the filter seals properly. If the filter does not seal properly, grit can bypass the filter and enter the engine.

Paper Type Filter

Paper type filter elements (**Figure 7**) should be replaced annually or more frequently if the engine operates in a severe environment, such as extremely dusty conditions. A dirty filter element cannot be cleaned and must be discarded. Attempting to clean the filter element may dislodge embedded grit that may enter the engine if the element is reinstalled, or the filter element may be damaged so it is not effective. Clean the filter canister. Inspect and, if necessary, replace any defective gaskets.

Combination Type Filter

The combination type air cleaner consists of a foam or flock type filter, known as the precleaner, wrapped around or in front of a paper or foam type filter. See **Figure 8**. Flock or paper type elements cannot be cleaned and must be replaced. Clean a foam type element using the cleaning methods previously outlined for foam filters.

COOLING SYSTEM

The Tecumseh engines covered in this manual use air to cool the engine. Any stoppage or disruption of the flow of air around the engine can cause overheating of the engine and subsequent damage. The fins on the flywheel force air around the cooling fins on the cylinder head and cylinder, while shrouds are used to direct cooling air as needed.

Depending on the operating environment of the engine, the cooling system should be inspected periodically. Inspect the engine and remove any shrouds or other components as necessary to remove any dirt or debris on the engine. All foreign matter in the cooling fins must be removed. Remove the blower shroud (may require removal of rewind starter) surrounding the flywheel and look for debris (**Figure 9**). Check the flywheel for broken or missing fins.

WARNING
Do not operate the engine if flywheel fins are broken or missing. The flywheel can crack when the engine is running and hurl harmful pieces. Replace a damaged flywheel.

HOSES AND WIRES

Whenever maintenance is performed, check all hoses and wires for looseness at connections and any damage. Old, hard hoses should be replaced with new hoses. Be sure any clamps and locating clips are in place. If hoses or wires are worn or damaged by moving parts, determine and correct the problem. Wire connections should be clean and tight.

OPERATING CABLE

Some engines are controlled by a remote control. Periodically apply a dry lubricant such as liquid

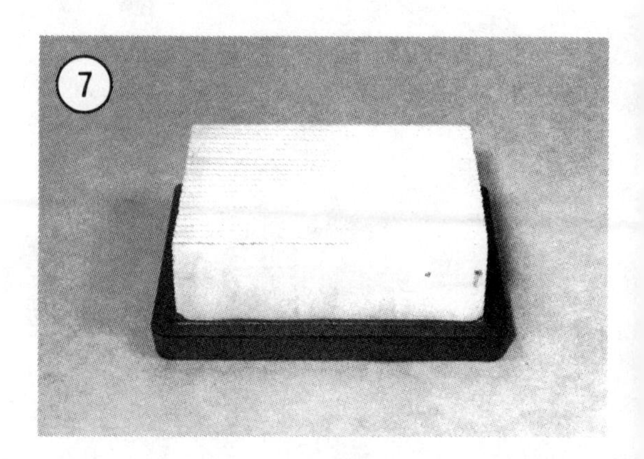

graphite to the cable to lubricate between the housing and steel wire inside. Oil can be used, but dirt and debris will adhere to the oil film.

Check all operating controls for proper operation. The choke should close fully during starting and the throttle should open fully when full throttle operation is required. The engine should stop when directed by the operator. If any malfunctions are found, refer to the appropriate chapter and correct the problem before further engine operation.

MUFFLER

Periodically remove and inspect the muffler. A blocked muffler or exhaust pipe can significantly decrease engine performance.

COMPRESSION TEST

Tecumseh does not specify the compression pressure for its engines. To check compression, disconnect the spark plug wire from the spark plug and ground the lead to the engine. Remove the spark plug. Disengage the flywheel brake if so equipped. Hold a finger against the spark plug hole in the cylinder head and operate the starter. If the finger is blown off the spark plug hole each time the engine completes its compression stroke, compression is adequate for the engine to start.

Low or nonexistent compression pressure may be due to one or more of the following engine problems.

a. Defective cylinder head gasket.
b. Warped cylinder head.
c. Worn or broken piston rings.
d. Sticking valves.
e. Worn valve guides.
f. Worn or damaged valves or seats.

g. Insufficient valve spring tension.
h. Mistimed camshaft gear.

SPARK PLUG

The spark plug is an important item in a good tune-up because it is the point of ignition as well as an indicator of how well combustion is occurring in the engine. Refer to Chapter Two for a discussion of spark plug operation and terminology.

The spark plug should be removed periodically, then inspected, cleaned and gapped. If spark plug fouling is a problem, the inspection interval must be adjusted accordingly. Refer to **Figure 1** in Chapter Four for possible engine problems when inspecting the spark plug.

The recommended spark plug may be found listed in a spark plug manufacturer's catalog, which can be found at most stores selling spark plugs.

NOTE
Special engine applications may require the use of a spark plug other than that listed for normal usage. Contact the equipment manufacturer or consult a Tecumseh parts manual for the recommended spark plug.

Before removing a spark plug, clean the area around the spark plug so dirt or debris will not enter through the spark plug hole. Use a suitable spark plug socket and wrench to remove the spark plug.

NOTE
The spark plug should "break free" when it is turned initially with moderate force. If the spark plug will not turn or is hard to turn, it may be frozen or tearing the threads in the cylinder head. Do not turn further. Apply penetrating oil such as WD-40 or Liquid Wrench around the base of the spark plug and let stand for 10-20 minutes. If the threads in the cylinder head are damaged, a thread repair insert can be installed (see Chapter Ten).

After removing the spark plug, note the condition of the plug. The material on the end of the plug will offer clues on the combustion event. Compare the spark plug with the examples shown in Chapter Four and note the possible causes.

The electrodes of a spark plug must be clean and sharp. A contaminated spark plug will "leak" voltage rather than producing an intense spark, causing misfiring. Worn electrodes require greater voltage to produce a spark, which may be greater than the ignition system is capable of generating, and misfiring results. A spark plug with light, soft deposits (such as a plug with soot from an excessively rich mixture) can be cleaned with a soft wire brush, but a plug with heavy deposits should be discarded and a new plug installed.

NOTE

Tecumseh does not recommend using abrasive blasting to clean spark plugs as this may introduce abrasive material into the engine which could cause extensive damage.

On both new and old spark plugs, the gap between the spark plug electrodes must be adjusted to the dimension specified by the engine manufacturer. Tecumseh specifies a gap of 0.030 in. (0.76 mm) for all engines. Measure the electrode gap (**Figure 10**) with a round feeler gauge (a flat feeler gauge will not accurately measure the gap, particularly on a worn plug) as shown in **Figure 11**. If the gap is correct, a slight drag will be felt when the gauge is pulled through the electrode gap. If the gap is incorrect, use a suitable gap setting tool (**Figure 12**) to adjust the electrode gap.

Before installing the spark plug, be sure that the area where the spark plug seats on the cylinder head is clean. Any dirt or debris in this area can prevent solid seating of the plug, which can cause damage from hot escaping combustion gases.

Although often overlooked, it is a good practice to apply a small amount of antiseize compound to the spark plug threads to lessen the possibility of thread-tearing during removal.

When installing the spark plug, screw the plug in by hand. If the plug binds, unscrew it and determine the cause. Excessive force may cause cross-threading. Hand-tighten the spark plug, then use a torque wrench and tighten the plug to 15 ft.-lb. (20 N•m).

If a torque wrench isn't available, screw the spark plug in by hand as far as possible. If installing a new spark plug, use a spark plug wrench to turn the spark plug 1/2 turn further. If installing a used spark plug, tighten the spark plug 1/8 to 1/4 turn further.

NOTE

Do not overtighten the spark plug. Overtightening may damage the spark plug hole threads.

Reconnect the spark plug lead, being sure it fits securely.

IGNITION BREAKER POINTS AND CONDENSER

Some Tecumseh engines are equipped with ignition breaker points and a condenser. Properly operating breaker points are critical to engine operation on Tecumseh engines so equipped.

NOTE

On some engines originally equipped with breaker points, an electronic

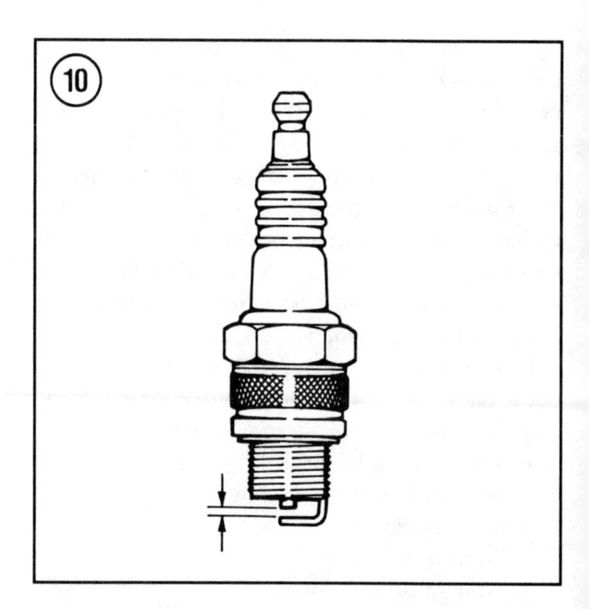

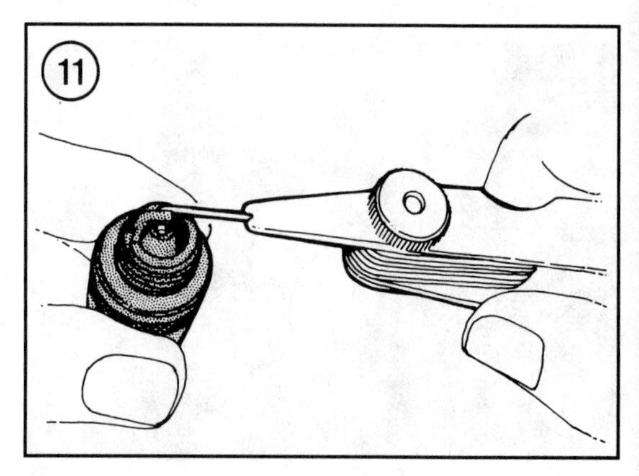

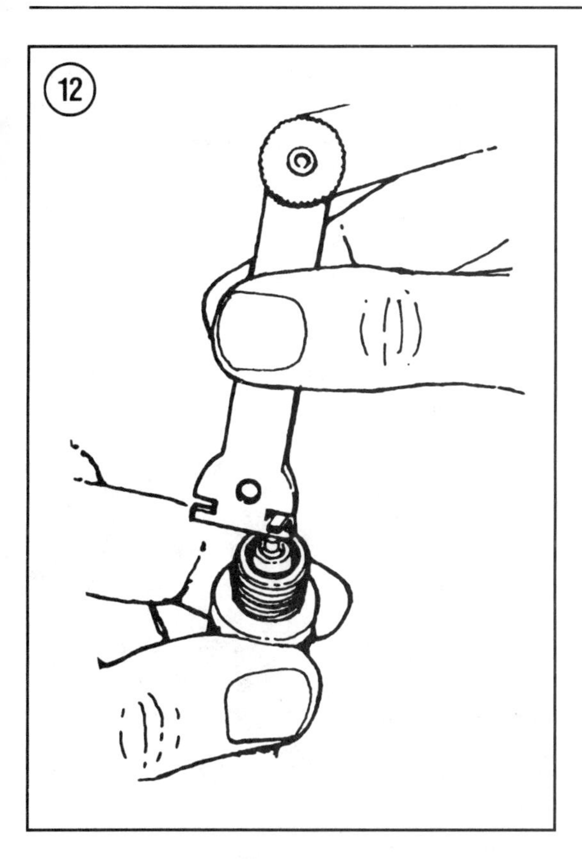

breakerless ignition system can be retrofitted. Contact a Tecumseh dealer for applicable engine models.

Removal

The breaker points are contained under a metal cover (A, **Figure 13**) behind the flywheel. The condenser (B) is mounted on the stator plate. A cam on the crankshaft actuates the movable breaker point arm.

1. Refer to Chapter Eleven for flywheel removal procedure.
2. Detach the cover retaining clip (C, **Figure 13**) and remove the metal cover (A) and gasket.
3. Unscrew the nut securing the wires to the breaker point terminal (A, **Figure 14**) and disconnect the wires.
4. Lift out the movable breaker point (B, **Figure 14**) and the terminal block. The terminal block fits in a groove in the housing and may be snug.
5. Unscrew retaining screw (A, **Figure 15**) and remove the fixed breaker point (B).
6. Spray the breaker point compartment with electrical contact cleaner and clean out the compartment.
7. Unscrew the condenser retaining screw to remove the condenser.

NOTE
Although the condenser is rarely faulty, it is a good practice to replace the condenser whenever the breaker points are replaced.

Installation

1. Apply a small drop of oil to the lubricating felt (C, **Figure 14**).

5

2. When installing the condenser, note that the wire should pass inside the mounting post for the stator plate.
3. Install the fixed breaker point (B, **Figure 15**) and tighten the retaining screw finger tight.
4. Install the movable breaker point (B, **Figure 14**) and the terminal block.
5. Connect the wires to the breaker point terminal (A, **Figure 14**) and secure with the retaining nut.
6. Adjust the breaker point gap.
7. Install the metal cover (A, **Figure 13**).
8. Install the flywheel.

Service Breaker Points

Virtually all problems concerning breaker points are related to the contacting surfaces of the breaker points. If the surfaces are not clean and sufficiently aligned so good contact is made, the primary circuit current will not flow and the engine will run erratically or not at all. Poor electrical contact between the surfaces is usually due to oxidation, erosion or oil contamination.

Breaker points that are excessively worn or pitted should be replaced. If there is excessive protrusion on one breaker point, then the condenser is faulty and must be replaced. Replace both breaker points, not just one.

If the breaker points are coated with oil, spray electrical contact cleaner on the points to clean them. Determine the source of the oil. Oil may be leaking from the crankshaft oil seal which should be serviced if the breaker points must be cleaned frequently. Refer to Chapter Eleven for the procedure to replace the crankshaft oil seal.

Unless excessively worn or pitted, the breaker points can be dressed using a breaker point file. The surface should be nearly flat and slightly rounded. Some small pits are acceptable, but there must be no protrusions. Do not use emery cloth or sandpaper to dress points as material may be embedded in point surface. Burnish the points by rubbing with stiff paper, such as a business card, then spray with electrical contact cleaner.

Breaker Point Gap Adjustment

The breaker point gap should be set as accurately as possible. If the breaker points are eroded, oxidized or dirty, service or replace them. The gap between worn points cannot be measured accurately.

1. Remove the flywheel and breaker point cover (A, **Figure 13**).
2. Note the highest point on the breaker cam on the crankshaft (A, **Figure 16**), then rotate the crankshaft (temporarily install the flywheel to rotate the crankshaft) so the arm of the movable breaker point is at the midpoint of the breaker cam raised portion as shown in **Figure 16**. The breaker point gap will now be maximum.

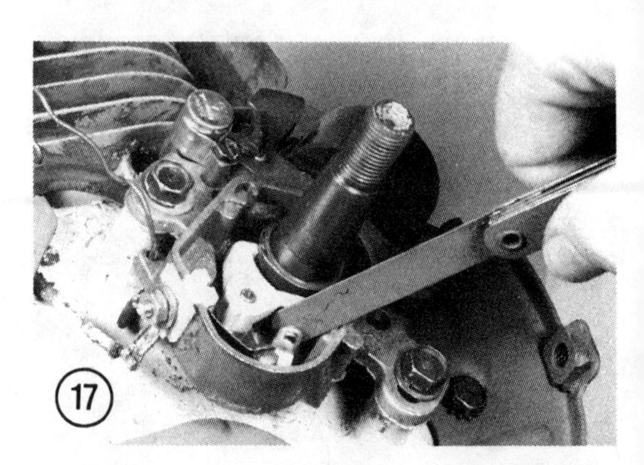

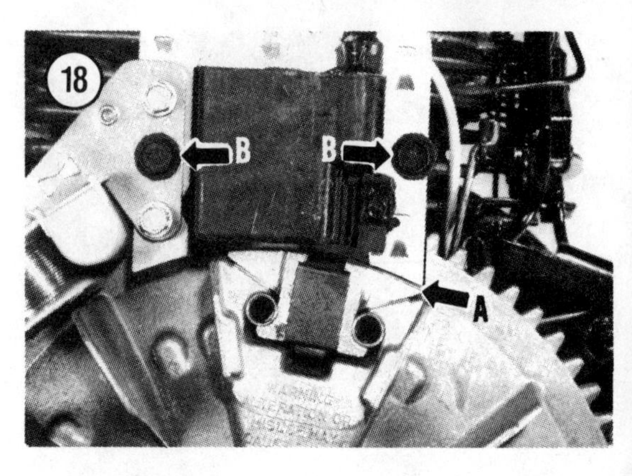

3. Loosen the fixed breaker point retaining screw (B, **Figure 16**) so that the fixed breaker point will move with some resistance.

4. Place a 0.020 in. (0.5 mm) feeler gauge between the breaker points as shown in **Figure 17**.

NOTE
Be sure the feeler gauge is clean. Oily residue on the feeler gauge will contaminate the breaker points.

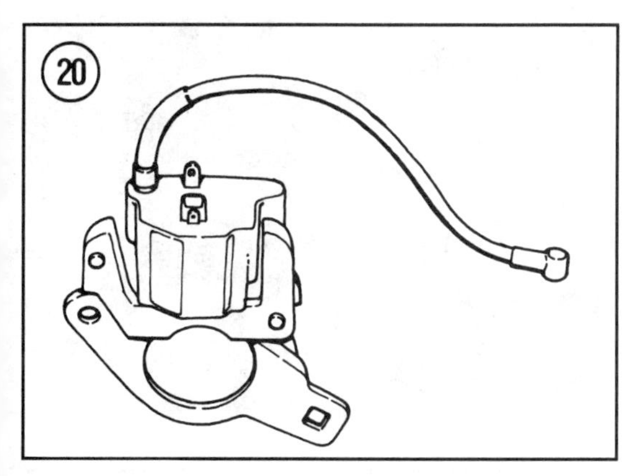

5. Move the fixed breaker point so there is a slight drag when the feeler gauge is withdrawn from between the breaker points, but be sure the movable breaker point does not move.
6. Tighten the fixed breaker point retaining screw and recheck the breaker point gap.

IGNITION COIL ARMATURE AIR GAP (MODELS WITH EXTERNAL IGNITION COIL)

On engines that have an ignition coil mounted outside the flywheel, Tecumseh specifies the desired air gap (A, **Figure 18**) between the ignition coil armature legs and the magnets in the flywheel. Specified armature air gap is 0.0125 in. (0.32 mm). The armature air gap can be adjusted, using a plastic or metal feeler gauge, using the following procedure.

1. Disconnect and properly ground the spark plug lead.
2. Remove the blower shroud so the flywheel and ignition coil are exposed.
3. Loosen the armature retaining screws (B, **Figure 18**), move the armature away from the flywheel and hand-tighten the screws.
4. Rotate the flywheel so the magnets are directly opposite the armature legs.
5. Insert a 0.0125 in. (0.32 mm) feeler gauge (such as Tecumseh feeler gauge 670297) between the flywheel and armature legs (**Figure 19**).
6. Loosen the armature retaining screws and push the armature legs against the feeler gauge, then tighten the armature retaining screws.
7. Extract the feeler gauge.

IGNITION TIMING

Ignition timing is not adjustable on engines with ignition systems that are mounted outside the flywheel. Ignition timing is not adjustable if the ignition system is a modular electronic ignition system mounted under the flywheel and appears as shown in **Figure 20**.

On engines with ignition components under the flywheel, ignition timing is adjusted by rotating the stator plate (A, **Figure 21**). Note that both electronic and breaker-point type ignition systems are adjustable, however, ignition timing is not specified on engines with electronic ignition.

Ignition Timing Adjustment on Engines Equipped With Breaker Points

Tools needed to check or adjust ignition timing include a tool to measure the position of the piston in the cylinder and, if equipped with breaker points, a tool to determine when the breaker points open. Completely read the following procedure before proceeding.

1. Remove the flywheel as outlined in Chapter Eleven.
2. Remove the breaker point cover (B, **Figure 21**) and disconnect the wires from the fixed breaker point terminal (**Figure 22**).

3A. Use the following method to determine piston position with a dial indicator.

NOTE
*Due to the location of the spark plug hole, an adapter must be attached to the dial indicator stem to measure piston movement. Tecumseh offers an adapter (**Figure 23**) that will fit most dial indicators. Note that the dial indicator must have a long stem so that the adapter will fit through the spark plug hole.*

a. Remove the spark plug.
b. Temporarily install the flywheel and turn the flywheel clockwise.
c. Place a finger over the spark plug hole while turning the flywheel. The piston is rising on the compression stroke when pressure is felt in the spark plug hole.
d. Install a dial indicator in the spark plug hole (**Figure 24**). Be sure the indicator "foot" properly contacts the piston.
e. Continue to turn the flywheel clockwise until the top of piston travel is determined. Rock the flywheel back and forth to determine the exact top of piston travel.

CAUTION
Be sure the piston does not "bottom" against the dial indicator. If the flywheel stops or becomes difficult to turn, unscrew the dial indicator and determine the cause.

f. Zero the dial indicator at top dead center by moving the indicator scale to "0."
g. Rotate the flywheel counterclockwise while reading the dial indicator and stop when the piston position is before top dead center at the dimension specified in **Table 1**. The piston is now in the position on the compression stroke when ignition should occur.

3B. Use the following method to determine piston position with a Vernier caliper (A, **Figure 25**) or depth micrometer.

a. Remove the cylinder head as outlined in Chapter Eleven.
b. Turn the flywheel clockwise so the piston is at top dead center on the compression stroke (note that both valves are closed). Measure the distance from the top of the cylinder to the piston (**Figure 25**).

NOTE
*If the piston extends above the top of the cylinder at top dead center, install the cylinder head gasket or another spacer (B, **Figure 25**) and measure from the gasket or spacer to the top of the piston.*

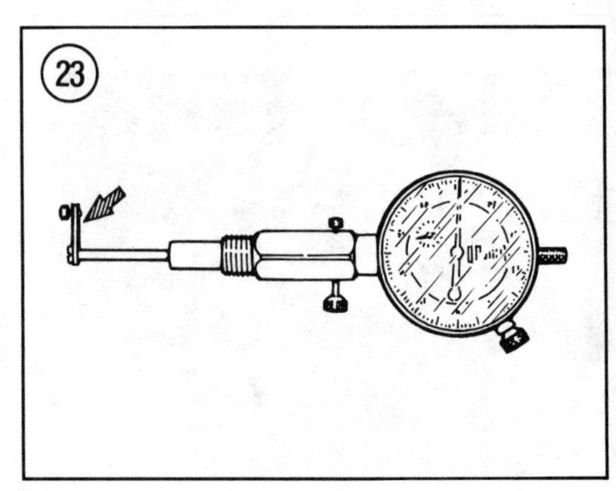

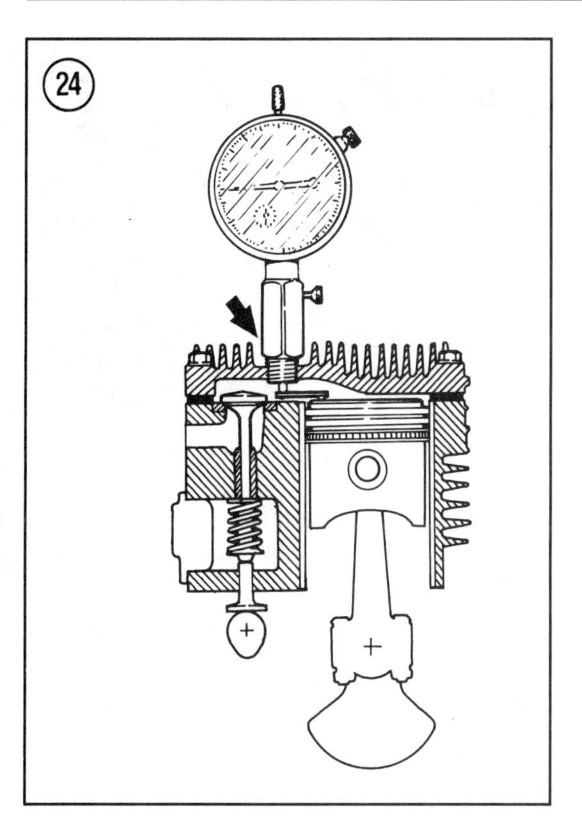

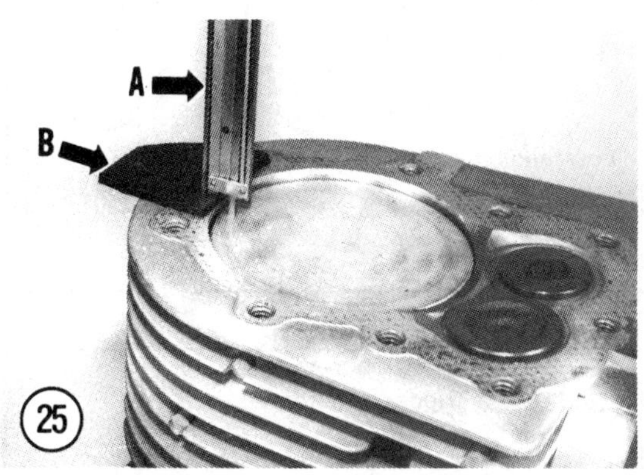

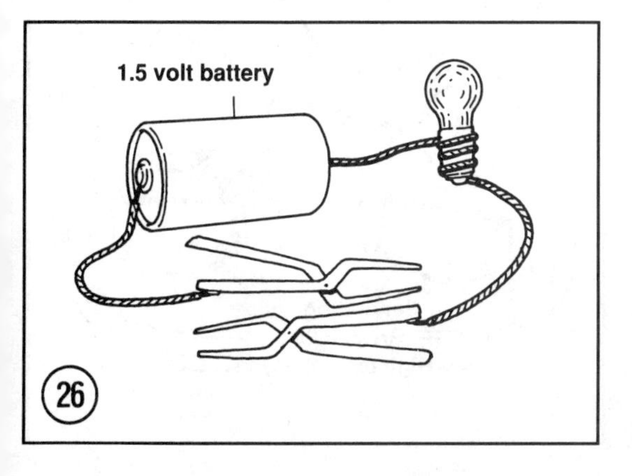

c. Turn the flywheel counterclockwise so the piston withdraws into the cylinder approximately 1/4 in. (6.4 mm).

d. Note the specified piston position dimension shown in **Table 1**, then add the measured distance (substep b) to the ignition timing specification. For example, add 0.010 in. (0.25 mm) (measured dimension with piston at top dead center) to 0.035 in. (0.89 mm) (timing dimension from **Table 1**). Set the measuring device (Vernier caliper, depth micrometer or other tool) to the correct timing dimension, which in this example is 0.045 in. (1.14 mm).

e. Rotate the flywheel clockwise until the piston just contacts the measuring device (**Figure 25**). The piston is now in the position on the compression stroke when ignition should occur.

4. Connect the leads of an ohmmeter to the fixed breaker point terminal (**Figure 22**) and ground the remaining ohmmeter lead against the engine crankcase.

NOTE

*If an ohmmeter is not available, a continuity tester can be used. A tester can be constructed using a penlight and battery (**Figure 26**). Attach the leads the same as for an ohmmeter. When the light goes out, the breaker points have opened.*

5. Loosen the stator plate retaining screws (A, **Figure 27**) and rotate the stator plate (B) counterclockwise so the breaker points are closed (the ohmmeter should indicate no resistance, or the continuity tester should show a complete circuit).

6. Rotate the stator plate clockwise slowly until the breaker points open as indicated by an infinite read-

ing on the ohmmeter or a broken circuit by the continuity tester.

7. Without disturbing the position of the stator plate, tighten the stator plate retaining screws.
8. For future reference, make alignment marks with a chisel on the stator plate and mounting post (**Figure 28**). If the stator plate is moved in the future, ignition timing should be correct if the stator plate is positioned so the marks are aligned.
9. Reconnect wires and reinstall removed parts.

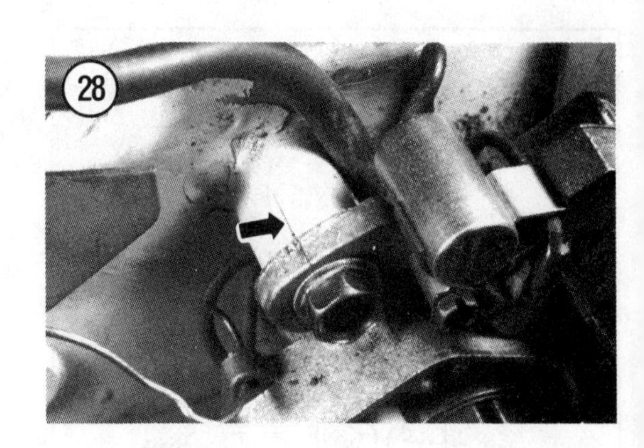

Ignition Timing Adjustment on Engines with Electronic Ignition Located behind the Flywheel

Some engines may be equipped with an electronic ignition like that shown in **Figure 29** or **Figure 30**. The ignition system shown in **Figure 29** has a trigger module that is located outside the flywheel, while the remaining components are mounted on the stator plate behind the flywheel. The ignition components for the system shown in **Figure 30** are all located on the stator plate. Use the following procedure to adjust the ignition timing.

1. Remove the flywheel as outlined in Chapter Eleven.
2. Loosen the stator retaining screws (similar to A, **Figure 27**).
3. If equipped with the ignition system shown in **Figure 29**, rotate the stator plate fully counterclockwise. If equipped with the ignition system shown in **Figure 30**, rotate the stator plate fully clockwise.
4. Tighten the stator retaining screws and reinstall removed parts.

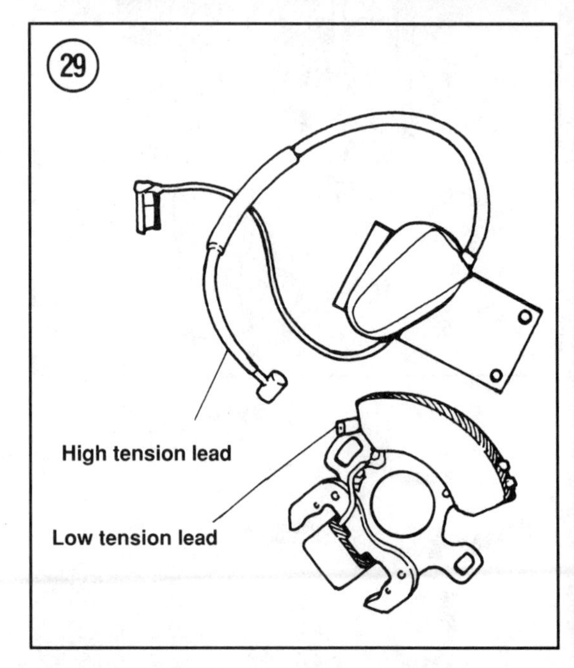

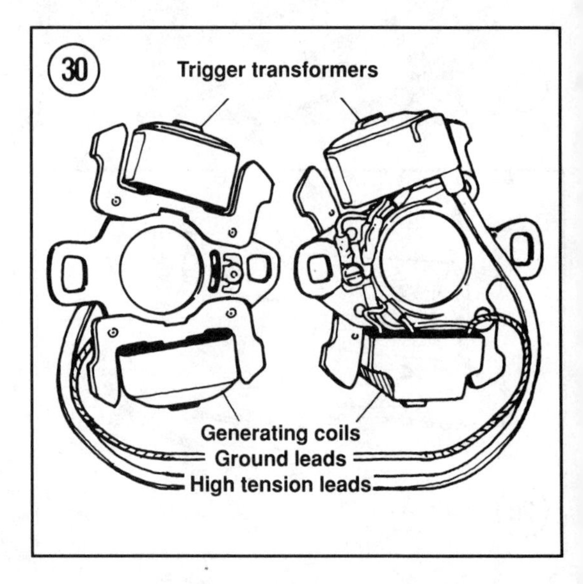

CARBURETOR ADJUSTMENT (EXCEPT VECTOR VLV50, VLXL50, VLV55 AND VLXL55)

Some carburetors are equipped with an idle speed screw and mixture adjustment screws that should be adjusted for optimum engine performance.

NOTE

*On some engines, the idle speed screw (**Figure 31**) is mounted on the speed control panel rather than on the carburetor. Engine idle speed should be adjusted to the idle speed specified by the equipment manufacturer, or if unavailable, adjust the idle speed so the engine* *idles smoothly (approximately 1500 rpm).*

Before attempting carburetor adjustment, check the following items.

1. The air cleaner is properly serviced (a dirty filter will affect carburetor mixture).
2. The fuel tank contains clean, fresh gasoline.
3. The ignition system is operating properly.
4. Carburetor and governor linkage operates properly.

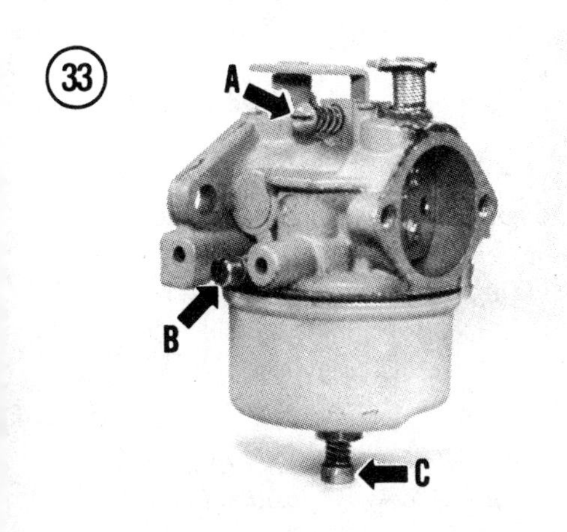

5. The engine is at normal operating temperature.
6. The engine is installed on equipment with all engine-driven components installed, otherwise, readjust carburetor after engine is installed.

NOTE
The carburetor must operate properly for successful adjustment. If abnormal mixture screw adjustments are required or the engine does not respond to mixture screw adjustments, the carburetor may be dirty, excessively worn or otherwise damaged. Overhaul the carburetor as outlined in Chapter Six.

5

Most carburetors are equipped with three adjustment screws: the idle mixture adjusting screw, the high speed mixture adjusting screw and the idle speed adjusting screw. Some carburetors are not equipped with any of the mixture adjusting screws, in which case, the mixture is controlled by a fixed jet or orifice.

Initial settings for mixture screws are given as turns out from a lightly seated position, for instance, 1 1/2 turns out. The initial setting will usually allow the engine to start and run so final adjustments can be performed with the engine at normal operating temperature.

CAUTION
Do not tighten mixture screws against their seats. Doing so may damage the tapered needle point of the screw tip, which will affect fuel flow, as well as possibly damaging the seat, which will require carburetor replacement.

The engines covered in this manual may be equipped with a diaphragm type carburetor (**Figure 32**) or a float type carburetor (**Figure 33**) manufactured by either Tecumseh or Walbro. Identify the carburetor manufacturer, then proceed to the appropriate following section. Tecumseh carburetors may be stamped with "TECUMSEH" or "LAUSON" (Lauson is the four-stroke engine division of Tecumseh). Walbro carburetors have "WALBRO" cast on the carburetor body.

The carburetor may be equipped with both the idle mixture screw and high speed mixture screw, with just the idle mixture screw, with just the high speed mixture screw, or without either mixture screw. A fixed nonadjustable jet is used if the mixture screw is absent.

Tecumseh Diaphragm Type Carburetor (Equipped With Idle and High Speed Mixture Screws)

Note in **Figure 32** the location of the idle speed screw (A), the idle mixture screw (B) and the high speed mixture screw (C).

1. Adjust the idle speed screw so it just contacts the throttle arm, then turn the screw in one turn. This will provide an engine idle speed adequate for mixture adjustment. The idle speed screw should be adjusted so the engine, when installed, runs at the idle speed recommended by the equipment manufacturer. If unavailable, adjust the idle speed so the engine idles smoothly (approximately 1500 rpm).
2. Turn the idle mixture screw and high speed mixture screw clockwise until they are lightly seated, then turn both screws counterclockwise one turn.
3. Run the engine until it is at normal operating temperature (approximately 5-7 minutes). The choke must be open.
4. Run the engine with the speed control in fast position.
5. Turn the high speed mixture screw clockwise until the engine begins to stumble and note the screw position.
6. Turn the high speed mixture screw counterclockwise until the engine begins to stumble and note the screw position.

NOTE

When turning the high speed mixture screw from clockwise position (lean) to counterclockwise position (rich), the engine will run smoother and engine speed will increase until the engine slows and stumbles again.

7. Turn the high speed mixture screw clockwise to a position that is midway between the clockwise (lean) and counterclockwise (rich) positions.
8. Run the engine with the speed control in slow position.
9. Turn the idle mixture screw clockwise until the engine begins to stumble and note the screw position.
10. Turn the idle mixture screw counterclockwise until the engine begins to stumble and note the screw position.

NOTE

When turning the idle mixture screw from clockwise position (lean) to counterclockwise position (rich), the engine will run smoother and engine speed will increase until the engine slows and stumbles again.

11. Turn the idle mixture screw clockwise to a position that is midway between the clockwise (lean) and counterclockwise (rich) positions.
12. With the engine running at idle, rapidly move the speed control to full throttle position. If the engine stumbles or hesitates, slightly turn the idle mixture screw counterclockwise and repeat the test.

NOTE

The operation of the idle and high speed mixture screws is related. After adjusting one mixture screw it may be necessary to adjust the other mixture screw.

Tecumseh Diaphragm Type Carburetor (Equipped With High Speed Mixture Screw Only)

1. Adjust the idle speed screw so it just contacts the throttle arm, then turn the screw in one turn. This will provide an engine idle speed adequate for mixture adjustment. The idle speed screw should be adjusted so the engine, when installed, runs at the idle speed recommended by the equipment manufacturer. If unavailable, adjust the idle speed so the engine idles smoothly (approximately 1500 rpm).
2. Turn the high speed mixture screw clockwise until it is lightly seated, then turn the screw counterclockwise one turn.
3. Run the engine until it is at normal operating temperature (approximately 5-7 minutes). The choke must be open.
4. Run the engine with the speed control in fast position.
5. Turn the high speed mixture screw clockwise until the engine begins to stumble and note the screw position.
6. Turn the high speed mixture screw counterclockwise until the engine begins to stumble and note the screw position.

NOTE

When turning the high speed mixture screw from clockwise position (lean) to counterclockwise position (rich), the engine will run smoother and engine speed will increase until the engine slows and stumbles again.

7. Turn the high speed mixture screw clockwise to a position that is midway between the clockwise (lean) and counterclockwise (rich) positions.

Tecumseh Diaphragm Type Carburetor (Equipped With Idle Mixture Screw Only)

1. Adjust the idle speed screw so it just contacts the throttle arm, then turn the screw in one turn. This will provide an engine idle speed adequate for mixture adjustment. The idle speed screw should be adjusted so the engine, when installed, runs at the idle speed recommended by the equipment manufacturer. If unavailable, adjust the idle speed so the engine idles smoothly (approximately 1500 rpm).
2. Turn the idle mixture screw clockwise until it is lightly seated, then turn the screw counterclockwise one turn.
3. Run the engine until it is at normal operating temperature (approximately 5-7 minutes). The choke must be open.
4. Run the engine with the speed control in slow position.
5. Turn the idle mixture screw clockwise until the engine begins to stumble and note the screw position.
6. Turn the idle mixture screw counterclockwise until the engine begins to stumble and note the screw position.

NOTE

When turning the idle mixture screw from clockwise position (lean) to counterclockwise position (rich), the engine will run smoother and engine speed will increase until the engine slows and stumbles again.

7. Turn the idle mixture screw clockwise to a position that is midway between the clockwise (lean) and counterclockwise (rich) positions.
8. With the engine running at idle, rapidly move the speed control to full throttle position. If the engine stumbles or hesitates, slightly turn the idle mixture screw counterclockwise and repeat the test.

Tecumseh Float Type Carburetor (Equipped With Idle and High Speed Mixture Screws)

Note in **Figure 33** the location of the idle speed screw (A), the idle mixture screw (B) and the high speed mixture screw (C).

1. Adjust the idle speed screw so it just contacts the throttle arm, then turn the screw in one turn. This will provide an engine idle speed adequate for mixture adjustment. The idle speed screw should be adjusted so the engine, when installed, runs at the idle speed recommended by the equipment manufacturer. If unavailable, adjust the idle speed so the engine idles smoothly (approximately 1500 rpm).
2. Turn the idle mixture screw and high speed mixture screw clockwise until they are lightly seated, then turn both screws counterclockwise the number of turns indicated in **Table 2**.
3. Run the engine until it is at normal operating temperature (approximately 5-7 minutes). The choke must be open.
4. Run the engine with the speed control in fast position.
5. Turn the high speed mixture screw clockwise until the engine begins to stumble and note the screw position.
6. Turn the high speed mixture screw counterclockwise until the engine begins to stumble and note the screw position.

NOTE

When turning the high speed mixture screw from clockwise position (lean) to counterclockwise position (rich), the engine will run smoother and engine speed will increase until the engine slows and stumbles again.

7. Turn the high speed mixture screw clockwise to a position that is midway between the clockwise (lean) and counterclockwise (rich) positions.
8. Run the engine with the speed control in slow position.
9. Turn the idle mixture screw clockwise until the engine begins to stumble and note the screw position.
10. Turn the idle mixture screw counterclockwise until the engine begins to stumble and note the screw position.

NOTE
When turning the idle mixture screw from clockwise position to counterclockwise position, the engine will run smoother and engine speed will increase until the engine slows and stumbles again.

NOTE
On most engines with less than 8 horsepower (6 kW), the idle mixture screw controls air entering the idle circuit. Turning the idle mixture screw clockwise enriches the idle mixture and vice versa. On engines with 8 horsepower (6 kW) or more, the idle mixture screw controls fuel entering the idle circuit. Turning the idle mixture screw clockwise leans the idle mixture and vice versa.

11. Turn the idle mixture screw clockwise to a position that is midway between the clockwise (lean) and counterclockwise (rich) positions.
12. With the engine running at idle, rapidly move the speed control to full throttle position. If the engine stumbles or hesitates, slightly turn the idle mixture screw counterclockwise and repeat the test.

NOTE
The operation of the idle and high speed mixture screws is related. After adjusting one mixture screw it may be necessary to adjust the other mixture screw.

Tecumseh Float Type Carburetor (Equipped With Idle Mixture Screw Only)

1. Adjust the idle speed screw so it just contacts the throttle arm, then turn the screw in one turn. This will provide an engine idle speed adequate for mixture adjustment. The idle speed screw should be adjusted so the engine, when installed, runs at the idle speed recommended by the equipment manufacturer. If unavailable, adjust the idle speed so the engine idles smoothly (approximately 1500 rpm).
2. Turn the idle mixture screw clockwise until it is lightly seated, then turn the screw counterclockwise the number of turns indicated in **Table 3**.
3. Run the engine until it is at normal operating temperature (approximately 5-7 minutes). The choke must be open.
4. Run the engine with the speed control in slow position.
5. Turn the idle mixture screw clockwise until the engine begins to stumble and note the screw position.
6. Turn the idle mixture screw counterclockwise until the engine begins to stumble and note the screw position.

NOTE
When turning the idle mixture screw from clockwise position (lean) to counterclockwise position (rich), the engine will run smoother and engine speed will increase until the engine slows and stumbles again.

NOTE
On most engines with less than 8 horsepower (6 kW), the idle mixture screw controls air entering the idle circuit. Turning the idle mixture screw clockwise enriches the idle mixture and vice versa. On engines with 8 horsepower (6 kW) or more, the idle mixture screw controls fuel entering the idle circuit. Turning the idle mixture screw clockwise leans the idle mixture and vice versa.

34 **WALBRO CARBURETOR**

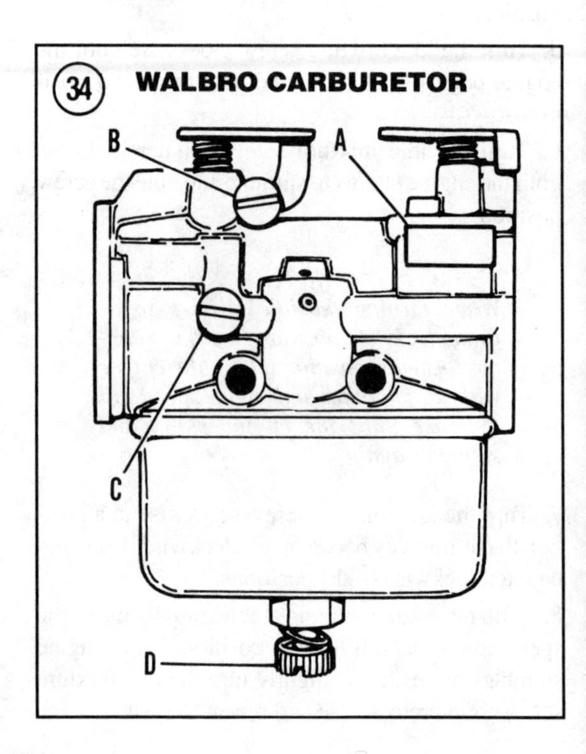

7. Turn the idle mixture screw clockwise to a position that is midway between the clockwise (lean) and counterclockwise (rich) positions.

8. With the engine running at idle, rapidly move the speed control to full throttle position. If the engine stumbles or hesitates, slightly turn the idle mixture screw counterclockwise and repeat the test.

Walbro Float Type Carburetor

Tecumseh engines may be equipped with a Walbro model LME carburetor. The carburetor model is stamped on a boss (A, **Figure 34**). Refer to **Figure 34** for the location of the idle speed screw (B), idle mixture screw (C) and high speed mixture screw (D). Some carburetors are not equipped with a high speed mixture screw; a fixed nonadjustable jet controls the high speed mixture.

1. Adjust the idle speed screw so it just contacts the throttle arm, then turn the screw in one turn. This will provide an engine idle speed adequate for mixture adjustment. The idle speed screw should be adjusted so the engine, when installed, runs at the idle speed recommended by the equipment manufacturer. If unavailable, adjust the idle speed so the engine idles smoothly (approximately 1500 rpm).

2. Turn the idle mixture screw and high speed mixture screw, if so equipped, clockwise until they are lightly seated, then turn both screws 1 1/4 turns counterclockwise.

3. Run the engine until it is at normal operating temperature (approximately 5-7 minutes). Be sure the choke is open.

4. Run the engine with the speed control in slow position.

5. Turn the idle mixture screw clockwise until the engine begins to stumble and note the screw position.

6. Turn the idle mixture screw counterclockwise 1/4 to 3/8 turn, which should provide the best idle mixture setting.

7. If the carburetor is not equipped with a high speed mixture screw, disregard adjustment Steps 8-10 and proceed to Step 11.

8. Run the engine with the speed control in the fast position with the engine under load (driving machinery normally connected to the engine).

9. Turn the high speed mixture screw clockwise until the engine begins to stumble and note the screw position.

10. Gradually turn the high speed mixture screw counterclockwise in small increments until the engine runs smooth. If excessive smoke is emitted, turn screw in slightly.

NOTE
Do not turn the high speed mixture screw rapidly. Allow the engine to run for a moment at one screw setting before proceeding so the engine has time to respond to a specific mixture setting.

11. With the engine running at idle, rapidly move the speed control to the full throttle position. If the engine stumbles or hesitates, slightly turn the idle mixture screw counterclockwise and repeat the test.

NOTE
If the carburetor is equipped with both idle and high speed mixture screws, note that the operation of the screws is related. After adjusting one mixture screw it may be necessary to adjust the other mixture screw.

CARBURETOR ADJUSTMENT (VECTOR VLV50, VLXL50, VLV55 AND VLXL55)

Fuel mixture is not adjustable on Vector VLV50, VLXL50, VLV55 and VLXL55 engines. Engine idle speed is adjusted by turning the idle speed screw (**Figure 35**—panel removed for clarity). The screw is accessible through the slot in the panel. Engine idle speed should be adjusted to the idle speed specified by the equipment manufacturer, or if unavailable, adjust the idle speed so the engine idles smoothly (approximately 1500 rpm).

GOVERNOR LINKAGE

The governor is designed to maintain a constant engine speed while the engine is subjected to varying loads, as well as limiting maximum engine speed. A malfunctioning governor can cause poor operation and possible engine damage.

> *CAUTION*
> *Modifying components of the governor mechanism or installing parts other than those specified by the manufacturer can cause engine failure.*

Inspection

Check the following when inspecting the governor mechanism.

1. Operate the governor linkage and check for binding. Any obstruction to free movement will cause erratic governor operation.
2. Check for missing or damaged pieces in the mechanism.
3. Check for excessively worn components that will affect governor operation.
4. Check the governor spring for stretching; the coils should be evenly spaced.

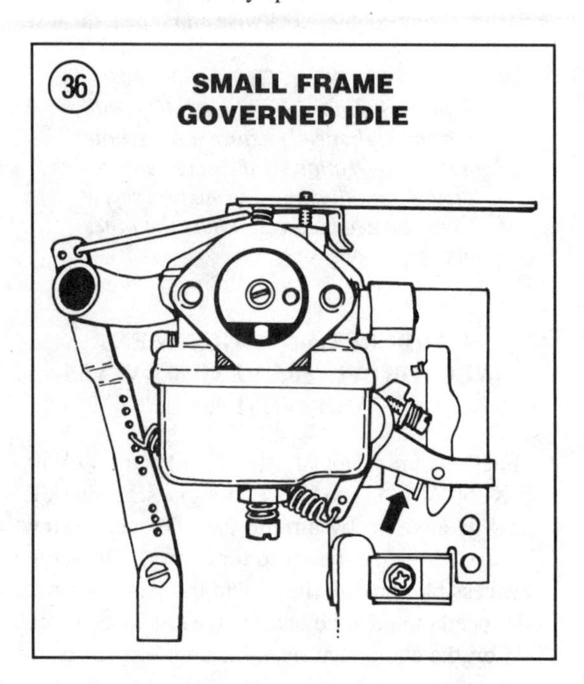

36 SMALL FRAME GOVERNED IDLE

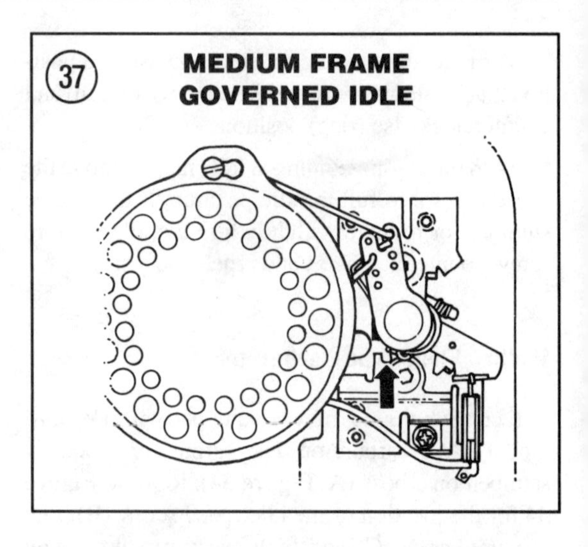

37 MEDIUM FRAME GOVERNED IDLE

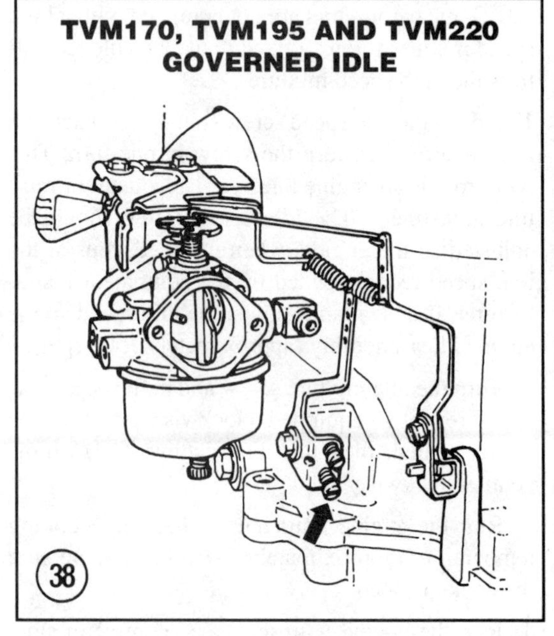

38 TVM170, TVM195 AND TVM220 GOVERNED IDLE

39

5. If the governor linkage is faulty or malfunctioning, refer to Chapter Six and correct the problem.

Governed Idle Speed Adjustment

On some engines, the engine idle speed is regulated by the governor. On these models, the idle speed screw on the carburetor is set to a speed lower than the desired governed idle speed, then the governed idle speed is adjusted. Some typical applications are described in the following paragraphs.

Some remote controlled engines with a horizontal crankshaft, may be equipped with a governed idle speed. Refer to **Figure 36** and **Figure 37** for two examples. The governed idle speed is adjusted by bending the linkage stop tab.

Some TVM170, TVM195 and TVM220 engines are equipped with a governed idle speed. The governed idle speed is adjusted by turning the lower screw (**Figure 38**) on the governor override lever.

COMBUSTION CHAMBER

Engines are susceptible to the formation of deposits on the cylinder head, piston crown and valves (see **Figure 39**). These deposits, which are primarily carbon, reduce engine performance. On all engines the deposits should be removed periodically or whenever the cylinder head is removed. The cylinder head must be removed as outlined in Chapter Eleven. Use a wooden or plastic scraper to remove the deposits. Spray a suitable solvent on hardened deposits to soften them. Be careful not to damage engine surfaces. Reinstall the cylinder head as outlined in Chapter Eleven.

Table 1 IGNITION TIMING

Engine model	Piston position before top dead center
ECV100, ECV105, ECV110, ECV120, HS40, LAV40, TNT100, TNT120, TVS105	0.035 in. (0.89 mm)
H40, HS50, LAV50, TVS120, V40	0.050 in. (1.27 mm)
ECH90, H25, H30, H35, LAV25, LAV30, LAV35, LV35, TVS75, TVS90	0.065 in. (1.65 mm)
H50, H60, H70, TVM125, TVM140, V50, V60, V70, VM70	0.080 in. (2.03 mm)
Engines with greater than 16.0 cu. in. (262 cc) displacement	0.090 in. (2.29 mm)

Table 2 CARBURETOR MIXTURE SCREW INITIAL SETTING (Float Carburetor With Two Mixture Screws)

Engine model prefix	Idle mixture screw	High speed mixture screw
H, HS, LAV	1	1 1/4
HM, TVM, VM	1 1/2	1 1/2

Table 3 CARBURETOR MIXTURE SCREW INITIAL SETTING (Float Carburetor With Single Mixture Screw)

Engine model prefix	Idle mixture screw
H, HS, LAV	1
HM, TVM, VM	1 1/2

CHAPTER SIX

FUEL AND GOVERNOR SYSTEMS

The fuel system consists of the carburetor, fuel pump, if so equipped, and operating linkage. If carburetor mixture and idle speed adjustments are required, refer to Chapter Five.

OPERATING LINKAGE

The engine control linkage may consist of engine-mounted operating levers or remote control linkage. In either case, the apparatus should move without binding (although some friction may be desirable to maintain a set position) and operate as designed, which includes controlling the choke, engine speed and stop switch on some models.

It may be necessary to loosen the retaining clamp on a cable and relocate the cable so full travel is achieved.

Engines with a vertical crankshaft may be equipped with a speed control panel that is a separate piece adjacent to the carburetor, except on Vector VLV50, VLXL50, VLV55 and VLXL55 engines, on which the panel is part of the blower housing baffle. The idle speed screw (A, **Figure 1**) and maximum governed speed screw (B) on Vector VLV50, VLXL50, VLV55 and VLXL55 engines are located on the panel. On all other engines so equipped, an idle speed screw (A, **Figure 2**) as well as the maximum governed speed screw (B) may be mounted on the panel.

Except on Vector VLV50, VLXL50, VLV55 and VLXL55 engines, the position of the panel must be adjusted so the linkage is synchronized. To adjust the panel position, loosen the mounting screws (**Figure 3**). Insert a rod through the holes in the panel, choke actuating lever and choke as shown, then retighten the mounting screws. Check operation.

DIAPHRAGM CARBURETORS

Two types of carburetors are used on Tecumseh engines, diaphragm type and float type. The float type carburetors are identified by the fuel bowl (**Figure 4**) that surrounds the internal float. Float type carburetors manufactured by Tecumseh and Walbro have been used. Diaphragm type carburetors

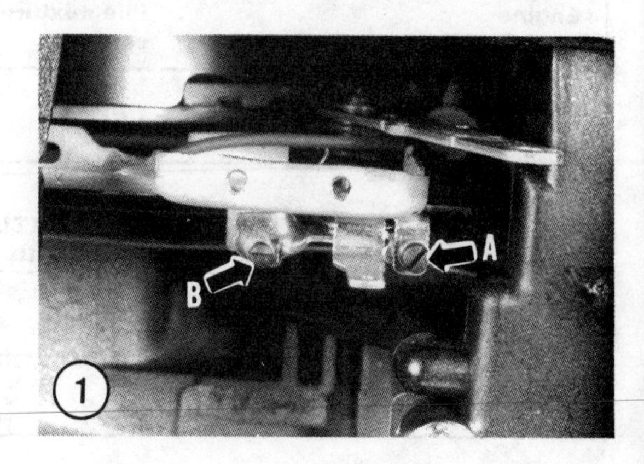

do not have a fuel bowl, instead, a flat cover (**Figure 5**) encloses the diaphragm on the bottom of the carburetor.

Overhaul kits are available from Tecumseh that contain the most frequently replaced parts when performing a carburetor overhaul. Gaskets, O-rings, fuel inlet valve and seat, mixture screws and Welch plugs are typically included in the kit. Diaphragms are included in kits for diaphragm carburetors. A float height gauge is included in kits for float carburetors. Contact a Tecumseh dealer for kit contents and availability.

Tecumseh offers "Standard Service" carburetors that are complete carburetors except that they do not include the throttle and choke assemblies, which must be transferred from the old carburetor. Contact a Tecumseh dealer for application, availability and price.

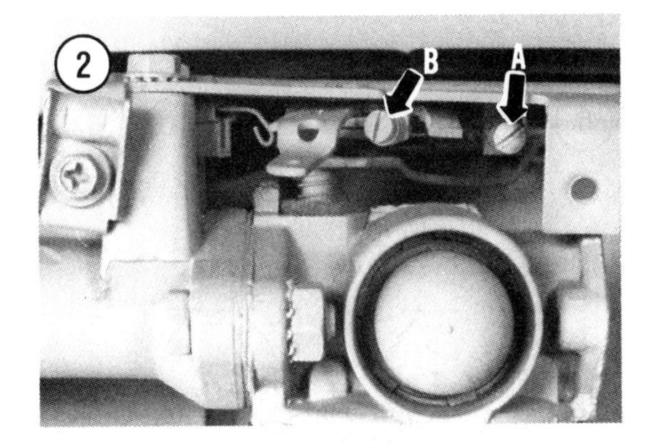

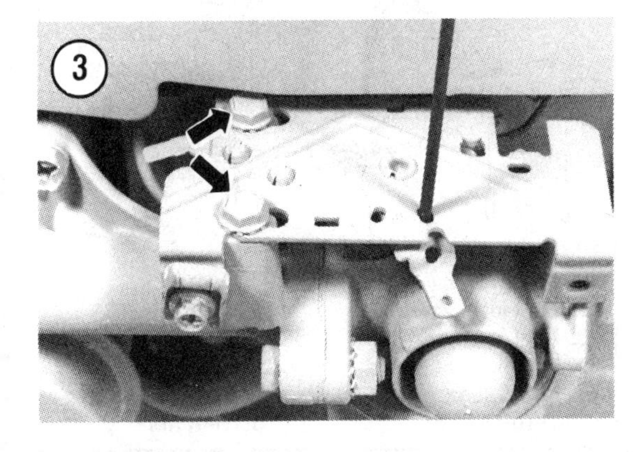

Operation

Refer to **Figure 6** for the following description.

The diaphragm carburetor operates according to the principles outlined in Chapter Two. The diaphragm controls fuel movement past the fuel inlet valve into the fuel chamber above the diaphragm. Note that the fuel inlet valve stem is spring-loaded and rests directly on the rivet head on the diaphragm.

At idle speed, the idle mixture screw controls the amount of fuel flowing to the idle discharge holes adjacent to the throttle plate. Fuel flow from the discharge holes is determined by the position of the throttle plate edge in relation to the holes. At idle, one hole discharges fuel and one hole becomes an air passage. At part throttle, two holes discharge fuel. At full throttle, none of the holes function.

At high speed, the high speed mixture screw controls the amount of fuel flowing to the nozzle. Fuel pressure and vacuum combine to lift the check ball off its seat so fuel can pass into the carburetor bore. At part-throttle and idle engine speeds, the check ball is seated due to insufficient vacuum in the carburetor bore so fuel does not flow from the nozzle. When the choke is closed or during high speed operation, the check ball lifts due to the high vacuum

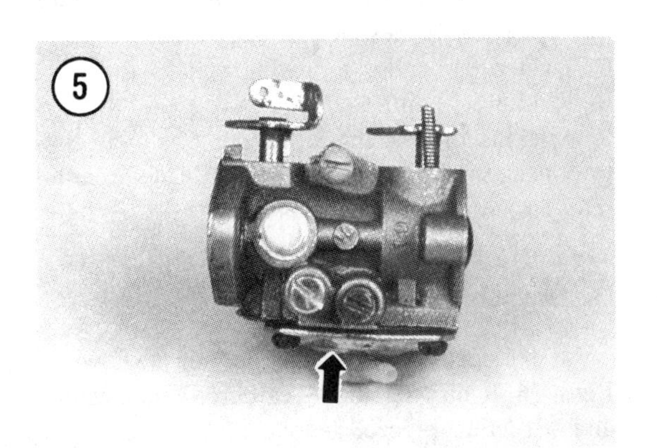

created in the carburetor bore and fuel flows from the nozzle.

Some carburetors may be connected to a primer bulb (**Figure 7**) that is operated when a rich air:fuel mixture is desired to start a cold engine. The primer hose is attached at the vent hole (J, **Figure 6**) and runs to the primer bulb (**Figure 7**). Atmospheric pressure is maintained in the air chamber of the carburetor through the vent hole on the primer bulb, which must be open for normal carburetor operation. The primer bulb is operated by placing a finger over the vent hole and depressing the bulb. The resulting pressure acts against the diaphragm to force additional fuel into the carburetor bore. The diaphragm also opens the fuel inlet valve, but a one-way valve prevents fuel from flowing back to the fuel tank.

Removal/Installation

Note that the following procedure is typical and additional steps may be required to accommodate specific applications.

1. Remove the air cleaner assembly.
2. Remove any metalwork, such as the blower housing or engine control panel, that will prevent removal of the carburetor.
3. Disconnect the fuel line from the carburetor. Close the fuel line valve, if so equipped, or drain the fuel to prevent fuel leakage.
4. Remove the carburetor mounting screws, disconnect control linkage and remove the carburetor from the engine. Note that on some engines it may be necessary to tilt the carburetor so that the linkage can be disconnected.
5. To install the carburetor, reverse the removal Steps 1 through 4. Note that on some engines it may be necessary to tilt the carburetor so that the linkage can be reconnected, in which case the carburetor mounting screws must not yet be installed.
6. Tighten carburetor mounting screws to 48-72 in.-lb. (5.5-8.1 N•m).
7. After installation, check the operation of all control linkage and adjust as required. Adjust the carburetor as outlined in Chapter Five.

Disassembly

An exploded view of the carburetor is shown in **Figure 8**. To disassemble the carburetor for cleaning and inspection, proceed as follows.

1. Unscrew and remove the diaphragm cover, gasket and diaphragm (**Figure 9**).
2. Unscrew the fuel inlet valve and remove the gasket, valve and spring (**Figure 10**).
3. Unscrew the idle mixture screw and high speed mixture screw.
4. Unscrew and remove the throttle plate (12, **Figure 8**).
5. Withdraw the throttle shaft assembly.
6. Unscrew and remove the choke plate (25, **Figure 8**).

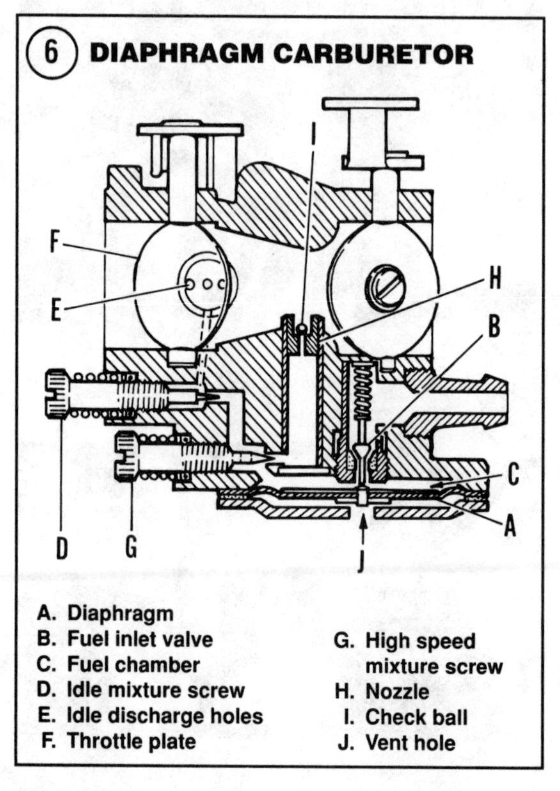

6 DIAPHRAGM CARBURETOR

A. Diaphragm
B. Fuel inlet valve
C. Fuel chamber
D. Idle mixture screw
E. Idle discharge holes
F. Throttle plate
G. High speed mixture screw
H. Nozzle
I. Check ball
J. Vent hole

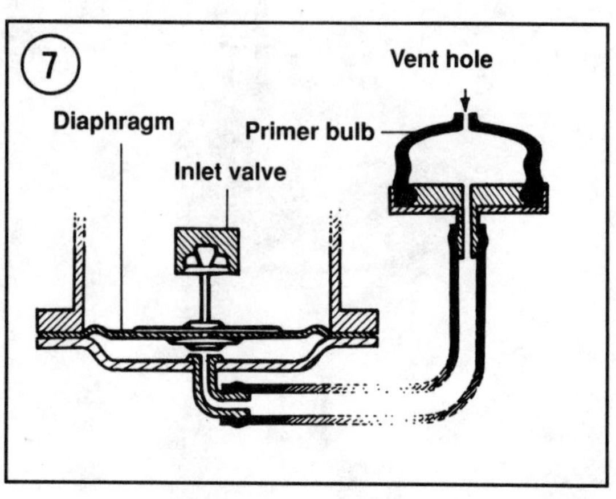

7

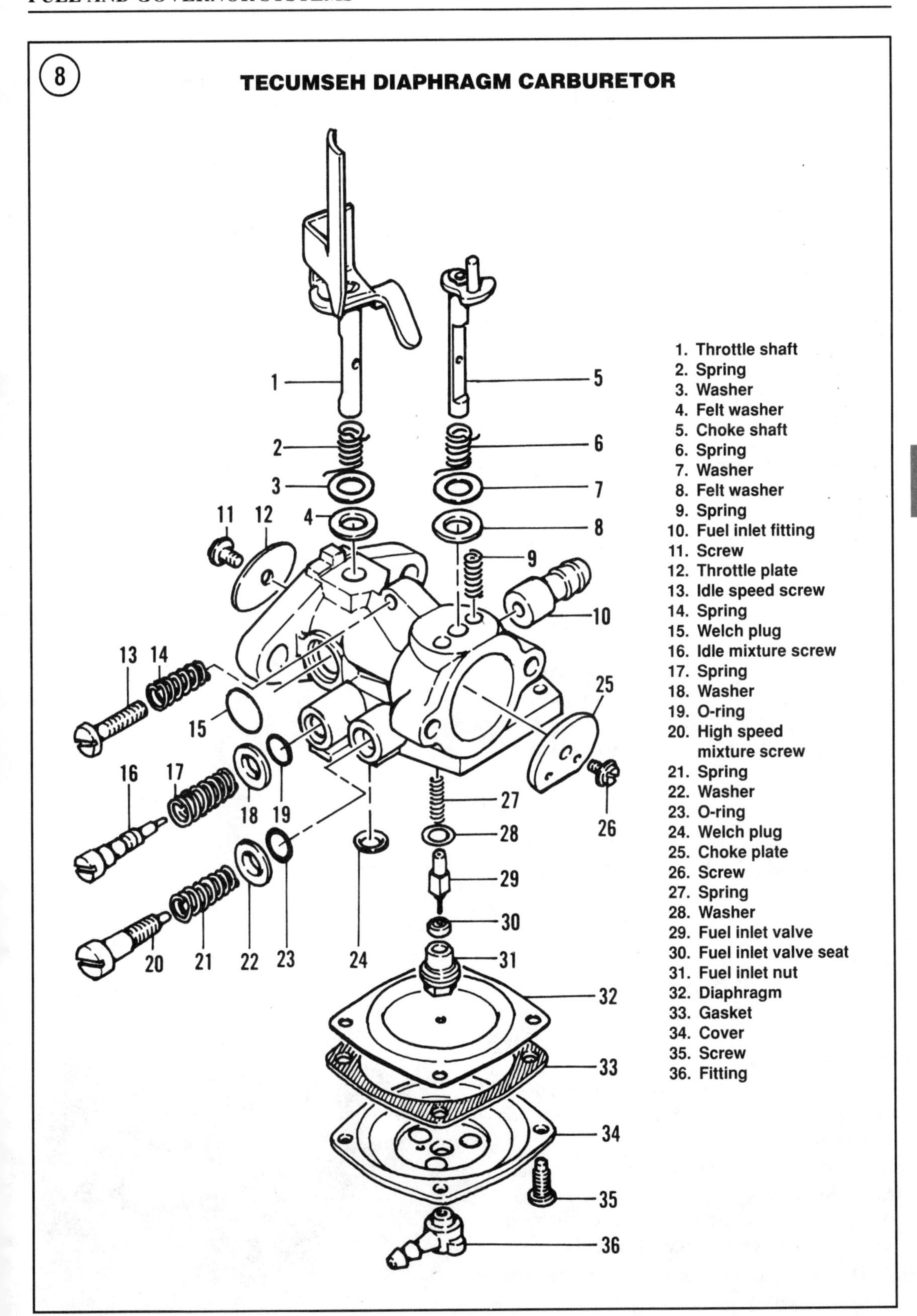
8
TECUMSEH DIAPHRAGM CARBURETOR
1. Throttle shaft
2. Spring
3. Washer
4. Felt washer
5. Choke shaft
6. Spring
7. Washer
8. Felt washer
9. Spring
10. Fuel inlet fitting
11. Screw
12. Throttle plate
13. Idle speed screw
14. Spring
15. Welch plug
16. Idle mixture screw
17. Spring
18. Washer
19. O-ring
20. High speed mixture screw
21. Spring
22. Washer
23. O-ring
24. Welch plug
25. Choke plate
26. Screw
27. Spring
28. Washer
29. Fuel inlet valve
30. Fuel inlet valve seat
31. Fuel inlet nut
32. Diaphragm
33. Gasket
34. Cover
35. Screw
36. Fitting

NOTE
Make note of the position of the choke parts before disassembly so they can be reassembled in their original positions. The choke may operate in either direction, depending on application.

7. Withdraw the choke shaft assembly.
8. Remove the Welch plug (**Figure 11**) by piercing the plug with a sharp chisel or punch, then prying out the plug.

CAUTION
Insert the chisel into the plug only far enough to pierce the plug, otherwise, underlying metal may be damaged.

9. The fuel inlet fitting (10, **Figure 8**) on the side of the carburetor can be removed by simultaneously twisting and pulling. Note the fitting's position before removal so it can be reinstalled in its original position. A filter screen is located behind the fitting on some carburetors.
10. The carburetor should now be ready for soaking in carburetor cleaner. Follow the directions of the cleaner manufacturer. Do not immerse the carburetor longer than 30 minutes. Spray the carburetor with aerosol carburetor cleaner to remove any residue, then use compressed air to blow out passages and dry the carburetor.

CAUTION
Some carburetor parts may be made of plastic and should not be soaked in carburetor cleaner.

Inspection

1. Inspect the fuel inlet valve (**Figure 12**). Note that the fuel inlet valve, valve seat and valve fitting must be replaced as a unit assembly.
2. Inspect the tip of the idle mixture and high speed mixture needles (**Figure 13**) and replace if the tip is bent or grooved.
3. Install the throttle or choke shaft in the carburetor body and check for excessive play between the shaft and body. The body must be replaced if there is excessive play as bushings are not available.
4. Discard the diaphragm if it is torn, creased or otherwise damaged.

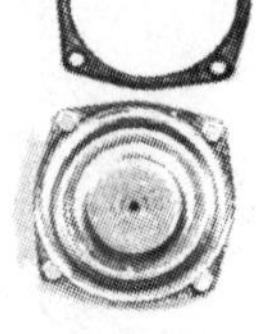

(11)

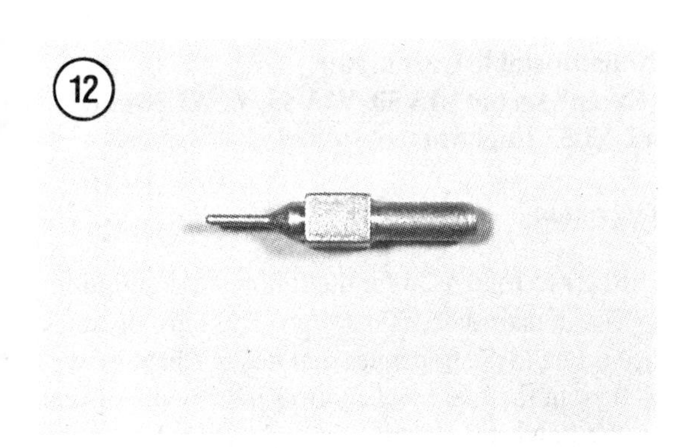

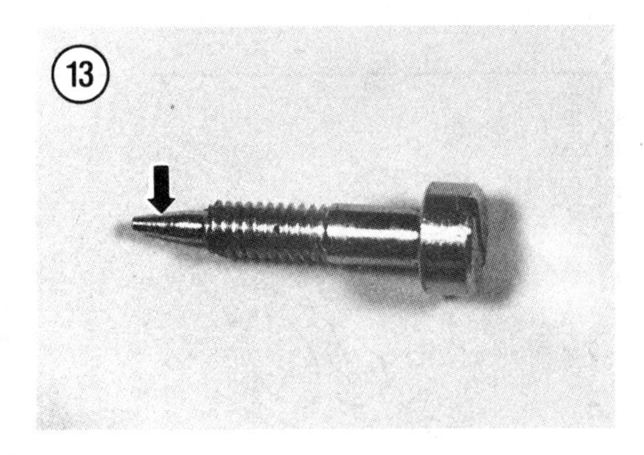

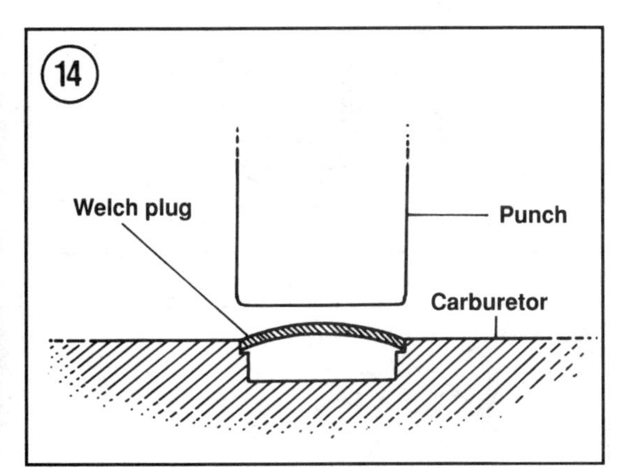

5. Inspect the fuel inlet fitting as some fittings have a nonremovable screen inside that may be dirty or clogged.

Assembly

1. Install the Welch plug with the concave side toward the carburetor (**Figure 14**). Do not indent the plug; the plug should be flat after installation. The installation tool should have a diameter that is the same or larger than the plug.
2. Install the choke plate so the straight edge is towards the diaphragm side of the carburetor (**Figure 15**). If the plate has indentations (dimples), the convex side should be toward the inside of the carburetor when the choke plate is closed. The choke assembly parts should be in the original position noted during disassembly. With the retaining screw just loose, turn the choke shaft so the choke plate is in the closed position, then tighten the retaining screw. Check for binding.
3. Install the throttle plate so the lines are visible when the throttle is closed (**Figure 16**). The lines should be positioned at 12 o'clock and 3 o'clock. With the retaining screw just loose, turn the throttle shaft so the throttle plate is in the closed position, then tighten the retaining screw. Check for binding.
4. The diaphragm must be installed so the rivet head (**Figure 17**) is towards the fuel inlet valve. The assembly order of the diaphragm and gasket is determined by the absence or presence of an "F" stamped on the carburetor bottom flange.
   a. If an "F" is stamped on the flange (**Figure 18**), the diaphragm, gasket and cover are assembled as shown in **Figure 19**.
   b. If there is no "F," the gasket, diaphragm and cover are assembled as shown in **Figure 20**.

6

5. In **Figure 21** note the proper location of spring (A), washer (B) and "O" ring (C) on either mixture screw.

TECUMSEH FLOAT CARBURETORS

Three versions of Tecumseh float type carburetors have been used. Nonadjustable carburetors do not have adjustable mixture screws, but may be equipped with a primer bulb (**Figure 22**). Vector VLV50, VLXL50, VLV55 and VLXL55 engines are equipped with a carburetor that is identified by the finned, extruded aluminum body (**Figure 23**). Some carburetors are adjustable and have an idle mixture screw (A, **Figure 24**). Adjustable carburetors may also have a high speed mixture screw (B, **Figure 24**), although a fixed jet may be used instead of a high speed mixture screw. Refer to the following sections for carburetor operation and service.

17

18

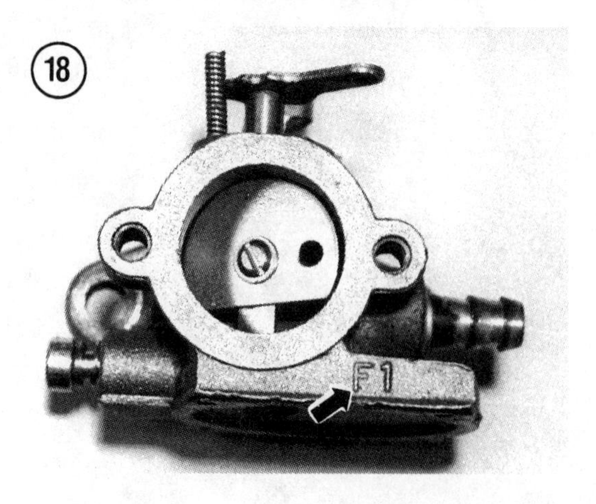

Nonadjustable Carburetor (Except Vector VLV50, VLV55, VLXL50 and VLXL55 Engines)

Operation

Refer to **Figure 25** for the following description.

The nonadjustable float type carburetor operates according to the principles outlined in Chapter Two. A float in the fuel bowl controls the amount of fuel in the bowl. Fuel passes through fuel inlets in the stanchion, through the main jet and into the nozzle.

19

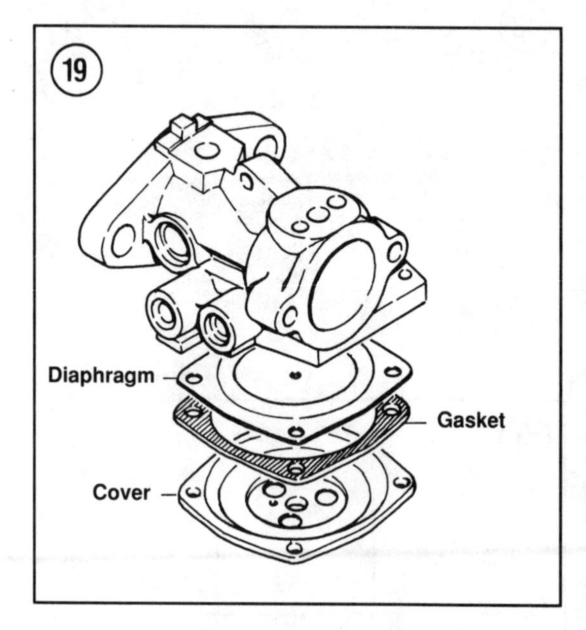

20

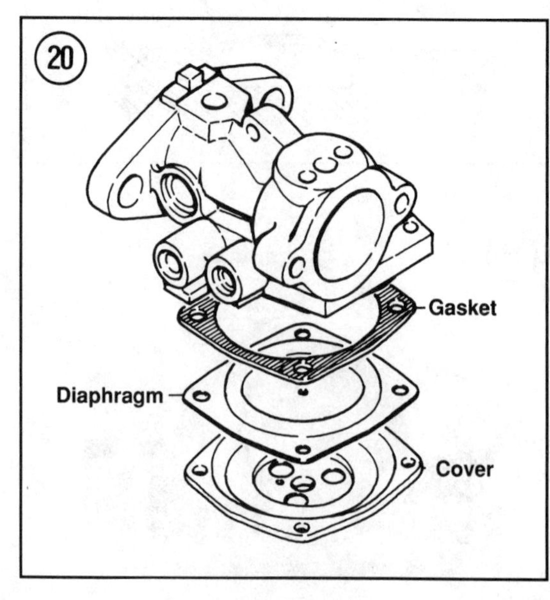

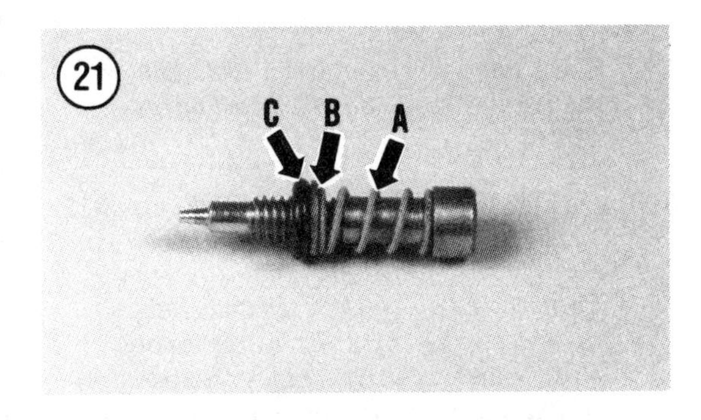

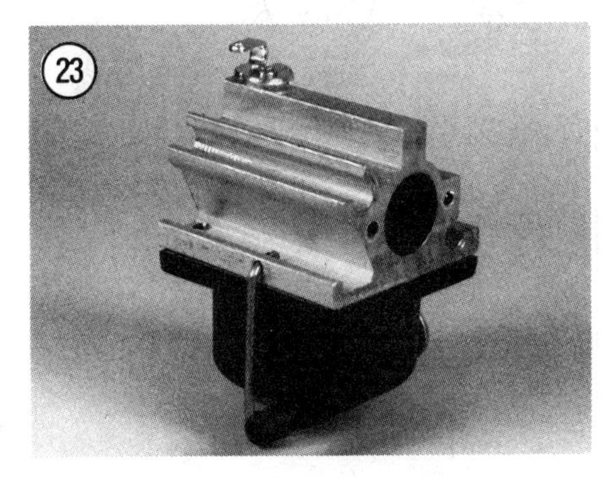

The fuel travels up the nozzle and exits into the carburetor bore. Air flow through the carburetor is controlled by a throttle plate.

To provide a rich mixture for starting, the fuel level below the nozzle is higher when the engine is stopped. This excess fuel enters the carburetor bore when the engine is being started.

Some carburetors are equipped with a primer bulb on the side of the carburetor so additional fuel can be forced into the carburetor bore during starting. When the primer bulb (A, **Figure 26**) is depressed, air inside the bulb is pressurized as well as air in the fuel bowl. Air pressure against the fuel in the fuel bowl forces fuel through the main jet (B, **Figure 26**) and up the nozzle (C, **Figure 26**), thereby enriching the air:fuel mixture for starting.

NOTE
*Some carburetors may be equipped with a primer bulb that is a part of the fuel bowl nut (**Figure 27**).*

Removal/installation

Note that the following procedure is typical and additional steps may be required to accommodate specific applications.

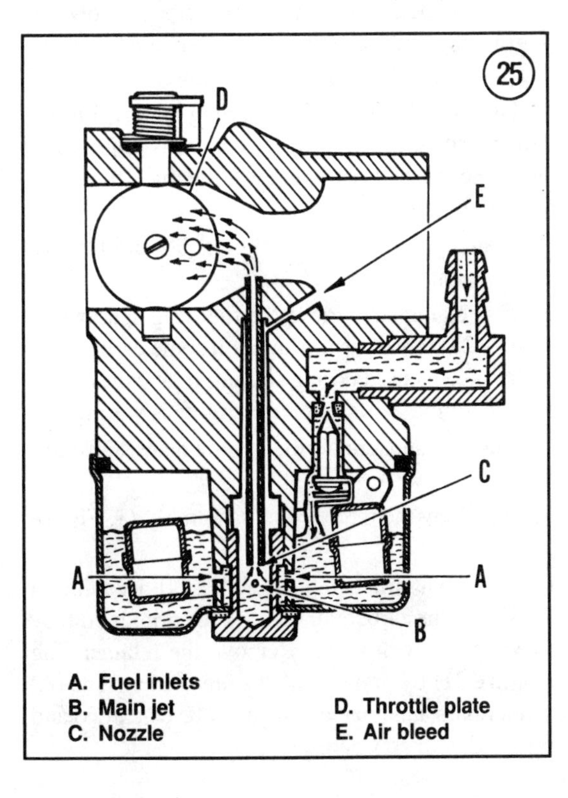

1. Remove the air cleaner assembly.
2. Remove any metalwork, such as the blower housing or engine control panel, that will prevent removal of the carburetor.
3. Disconnect the fuel line from the carburetor. Close the fuel line valve, if so equipped, or drain the fuel to prevent fuel leakage.
4. Remove the carburetor mounting screws, disconnect control linkage and remove the carburetor from the engine. Note that on some engines it may be necessary to tilt the carburetor so that the linkage can be disconnected.
5. To install the carburetor, reverse the removal Steps 1 through 4. Note that on some engines it may be necessary to tilt the carburetor so that the linkage can be reconnected, in which case the carburetor mounting screws must not yet be installed.
6. Tighten carburetor mounting screws to 48-72 in.-lb. (5.5-8.1 N•m).
7. After installation, check the operation of all control linkage and adjust as required. Adjust the carburetor as outlined in Chapter Five.

Overhaul

An exploded view of the carburetor is shown in **Figure 28**. To disassemble the carburetor for cleaning and inspection, proceed as follows.

1. Unscrew the fuel bowl retaining nut (A, **Figure 29**), which may include the primer bulb (**Figure 27**) on some models, and remove the fuel bowl (B, **Figure 29**) and gasket.
2. Dislodge the float pin (**Figure 30**) and remove the float and fuel valve.
3. If so equipped, remove the fuel well spacer (12, **Figure 28**). The spacer is used on some carburetors so less fuel is used during starting if an excessively rich condition exists when the engine is warm.
4. Unscrew and remove the throttle plate (8, **Figure 28**).
5. Withdraw the throttle shaft assembly (1, **Figure 28**).
6. On carburetors equipped with a primer bulb on the side of the carburetor, remove the primer bulb by twisting out with pliers. Remove the retainer ring (**Figure 31**) by prying with a suitable tool. The old primer bulb and retainer ring must be discarded and replaced with new parts.

WARNING
Wear safety eyewear when dislodging the retainer as it can be ejected uncontrolled.

7. Extract the fuel inlet valve seat (A, **Figure 32**).

NOTE
Compressed air can be used to dislodge the inlet valve seat, but extreme care must be exercised. Cover the carburetor so that the seat will not be ejected uncontrolled.

8. Remove the Welch plug (B, **Figure 32**) by piercing the plug with a sharp chisel or punch, then prying out the plug.

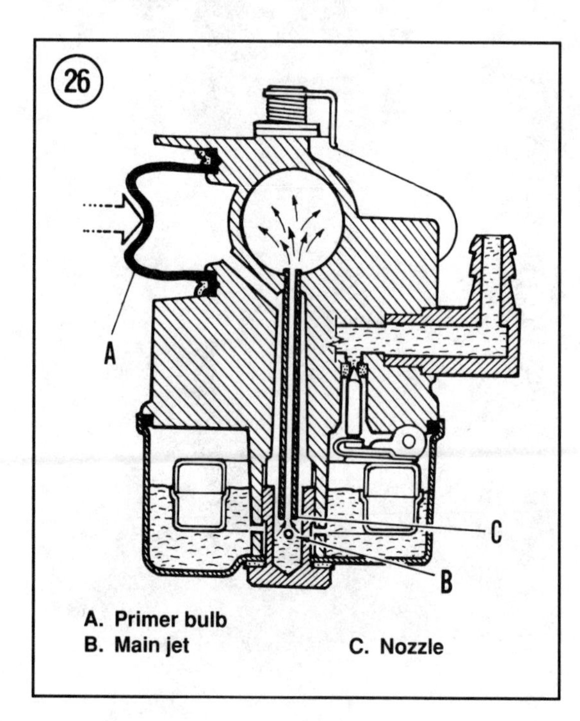

A. Primer bulb
B. Main jet
C. Nozzle

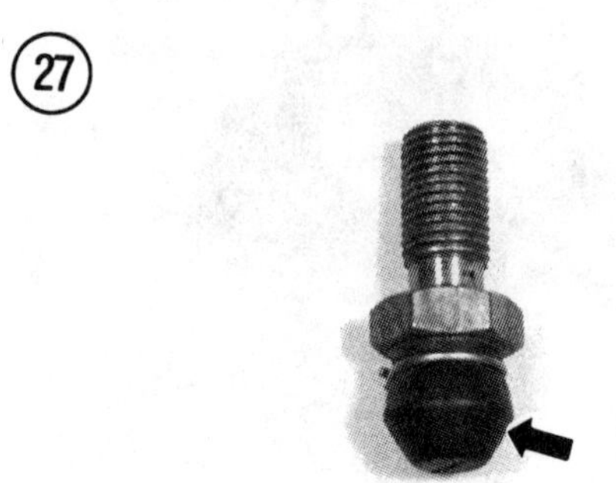

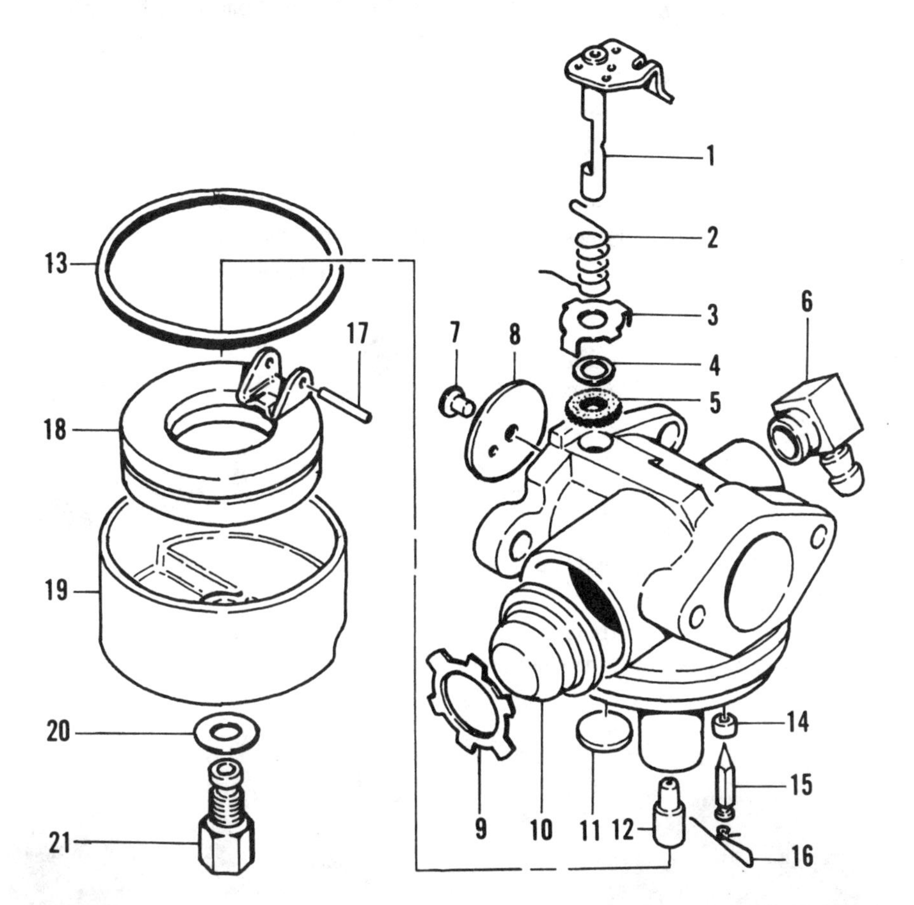

1. Throttle shaft
2. Spring
3. Felt washer retainer
4. Washer
5. Felt washer
6. Fuel inlet fitting
7. Screw
8. Throttle plate
9. Retainer
10. Primer bulb
11. Welch plug
12. Spacer
13. Gasket
14. Fuel inlet seat
15. Fuel inlet valve
16. Clip
17. Float pin
18. Float
19. Fuel bowl
20. Washer
21. Fuel bowl nut

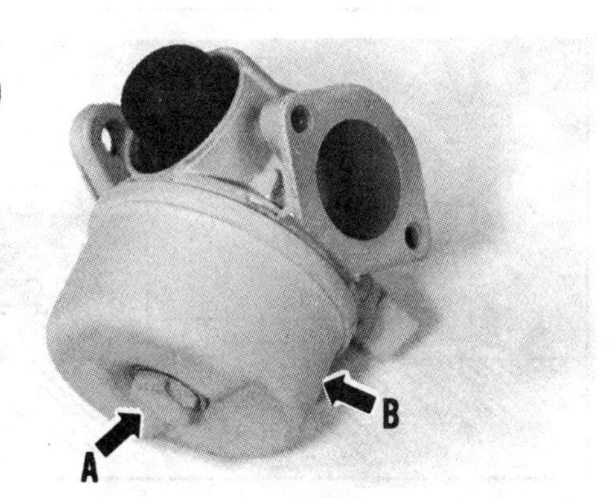

CAUTION
Insert the chisel into the plug only far enough to pierce the plug, otherwise, underlying metal may be damaged.

9. The carburetor should now be ready for soaking in carburetor cleaner. Follow the directions of the cleaner manufacturer. Do not immerse the carburetor longer than 30 minutes. Spray the carburetor with aerosol carburetor cleaner to remove any residue, then use compressed air to blow out passages and dry the carburetor.

CAUTION
Some carburetor parts may be made of plastic and should not be soaked in carburetor cleaner.

Inspection

1. Inspect the fuel inlet valve (**Figure 33**) and replace it if the tip is grooved.
2. Install the throttle shaft in the carburetor body and check for excessive play between the shaft and body. The body must be replaced if there is excessive play as bushings are not available.
3. Replace the float if the float pivot pin holes (A, **Figure 34**) are worn excessively.

Assembly

1. Install the Welch plug with the concave side toward the carburetor (**Figure 35**). Do not indent the plug; the plug should be flat after installation. The

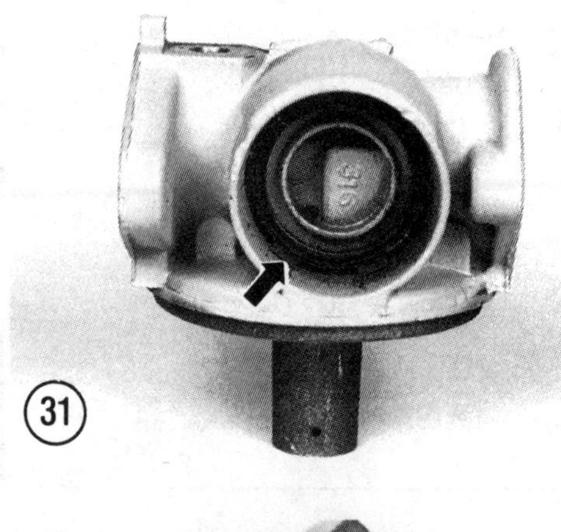

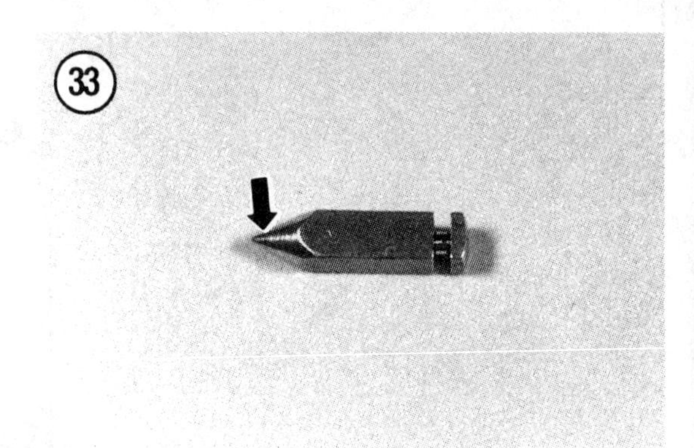

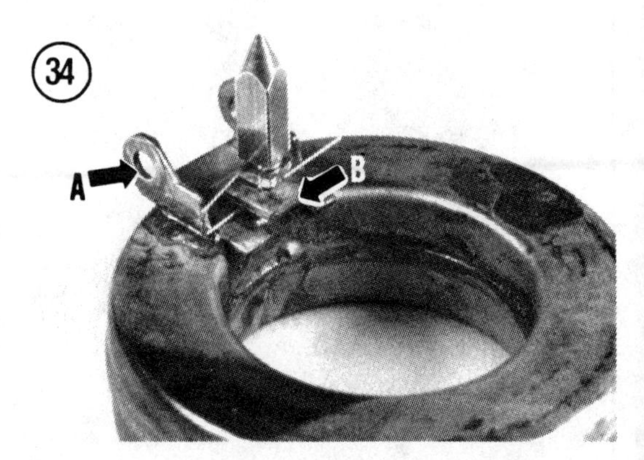

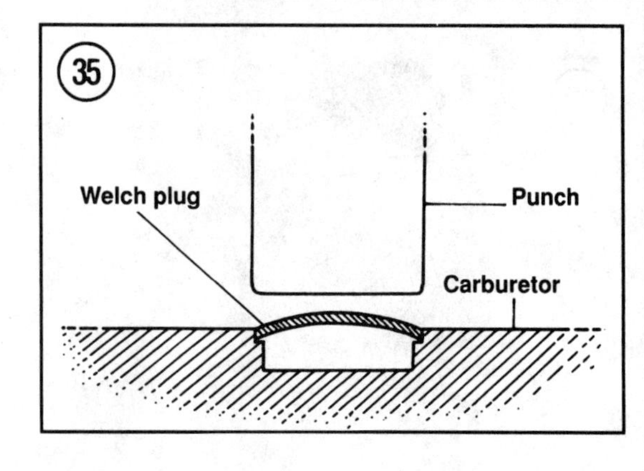

installation tool should have a diameter that is the same or larger than the plug.

2. Install the throttle plate so the line is visible when the throttle is closed (**Figure 36**). With the retaining screw just loose, turn the throttle shaft so the throttle plate is in the closed position, then tighten the retaining screw. Check for binding.

3. Install the fuel valve inlet seat with the flat side out and the grooved side towards the carburetor (**Figure 37**).

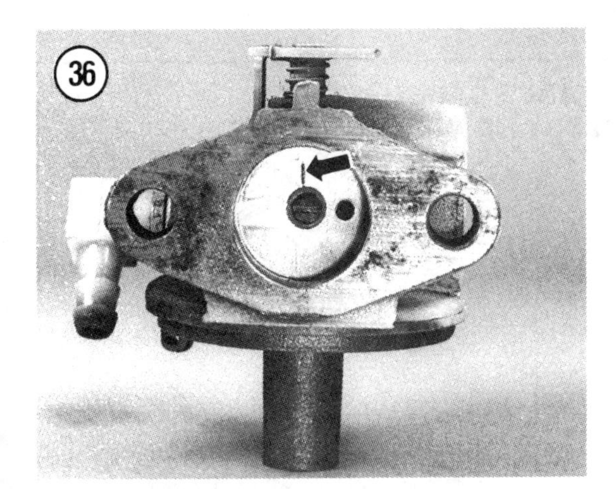

4. The wire clip on the fuel inlet valve must be positioned around the float tab as shown in **Figure 34**.

5A. With the float installed and the carburetor inverted, measure the float level (**Figure 38**). Measure at a point opposite the fuel inlet valve. The float level should be 0.162-0.215 in. (4.11-5.46 mm). Bend the float tab (B, **Figure 34**) to adjust the float level.

5B. Tecumseh has a gauge (Tecumseh part 670253A) that can be used to determine the correct float level. With the carburetor inverted, position gauge 670253A at a 90° angle to the carburetor bore and resting on the nozzle stanchion as shown in **Figure 39**. The toe of the float should not be higher than the first step on the gauge (**Figure 39**), or lower than the second step (**Figure 40**).

6. If removed, reinstall fuel inlet (A, **Figure 41**) by pushing it into the carburetor so the inlet is in its original position. Apply Loctite sealant before installation.

7. Install the fuel bowl so the indented portion (B, **Figure 41**) is on the same side as the fuel inlet (A, **Figure 41**).

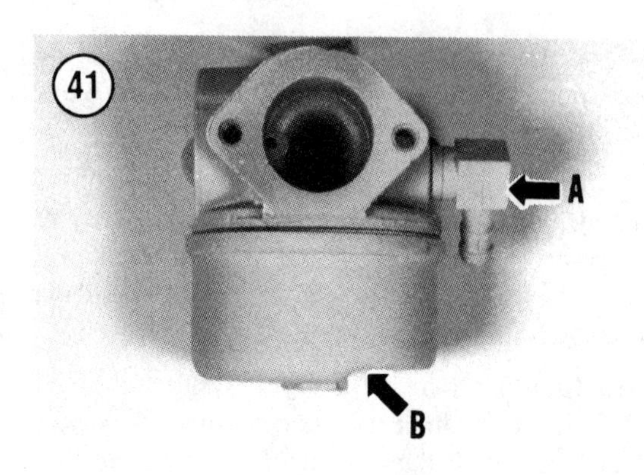

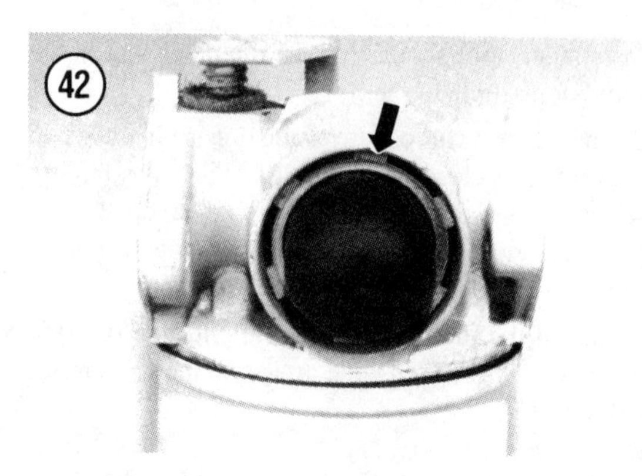

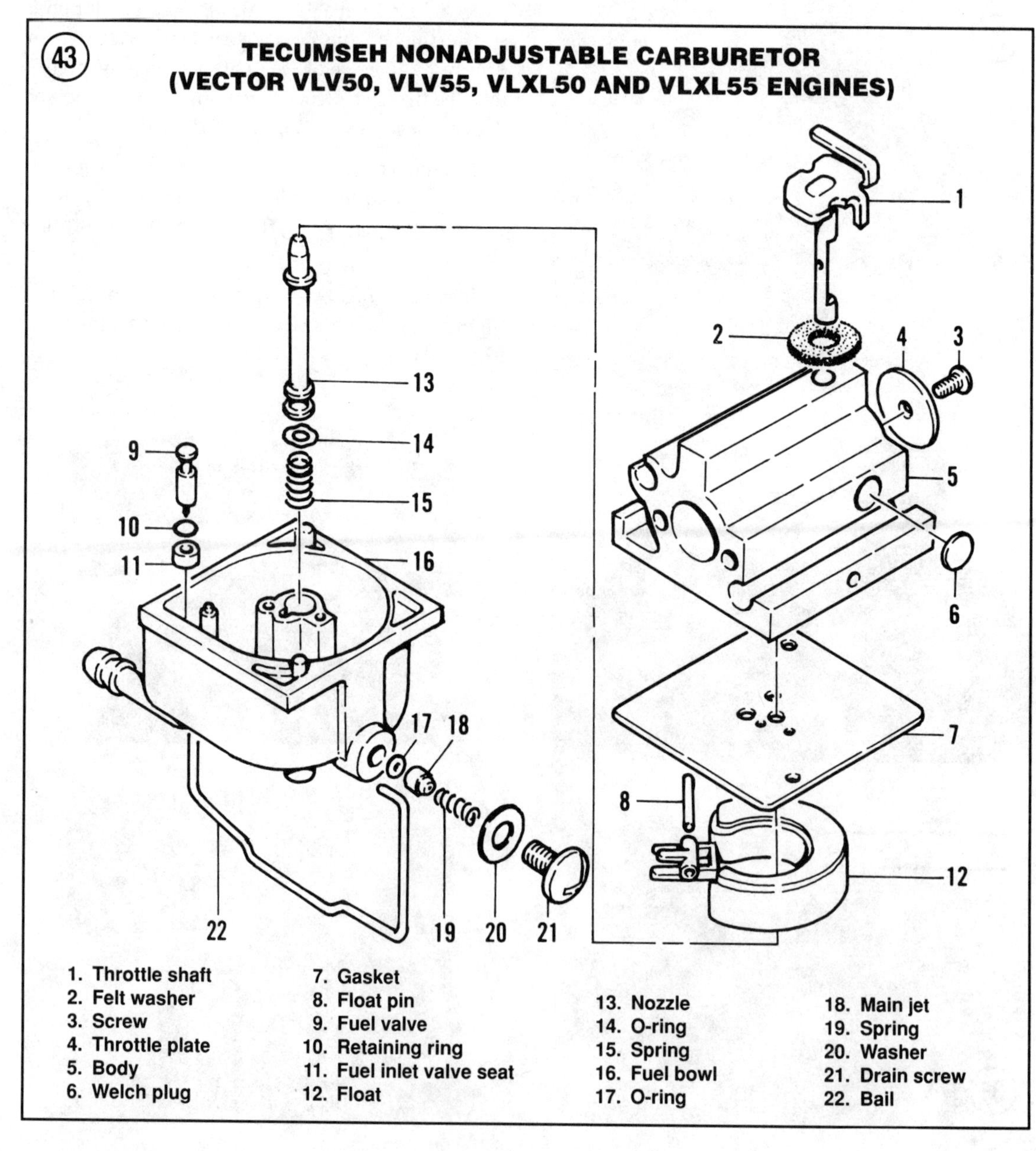

1. Throttle shaft
2. Felt washer
3. Screw
4. Throttle plate
5. Body
6. Welch plug
7. Gasket
8. Float pin
9. Fuel valve
10. Retaining ring
11. Fuel inlet valve seat
12. Float
13. Nozzle
14. O-ring
15. Spring
16. Fuel bowl
17. O-ring
18. Main jet
19. Spring
20. Washer
21. Drain screw
22. Bail

8. To install the primer bulb on carburetors so equipped, place the retainer ring around the bulb as shown in **Figure 42**, then push the bulb and retainer into the carburetor using a 3/4 in. (19 mm) deep-well socket.

Nonadjustable Carburetor (Vector VLV50, VLV55, VLXL50 and VLXL55 Engines)

Operation

The nonadjustable float type carburetor used on Vector VLV50, VLV55, VLXL50 and VLXL55 engines operates according to the principles outlined in Chapter Two. As shown in the exploded view in **Figure 43**, most of the major components are located in the fuel bowl. Some carburetors may be equipped with a removable main jet (18, **Figure 43**), however, only one size is available.

A primer bulb is located on the side of the air cleaner body so additional fuel can be forced into the carburetor bore during starting. When the primer bulb is depressed, air inside the bulb is pressurized as well as air in the fuel bowl. Air pressure against the fuel in the fuel bowl forces fuel and up the carburetor nozzle, thereby enriching the air:fuel mixture for starting.

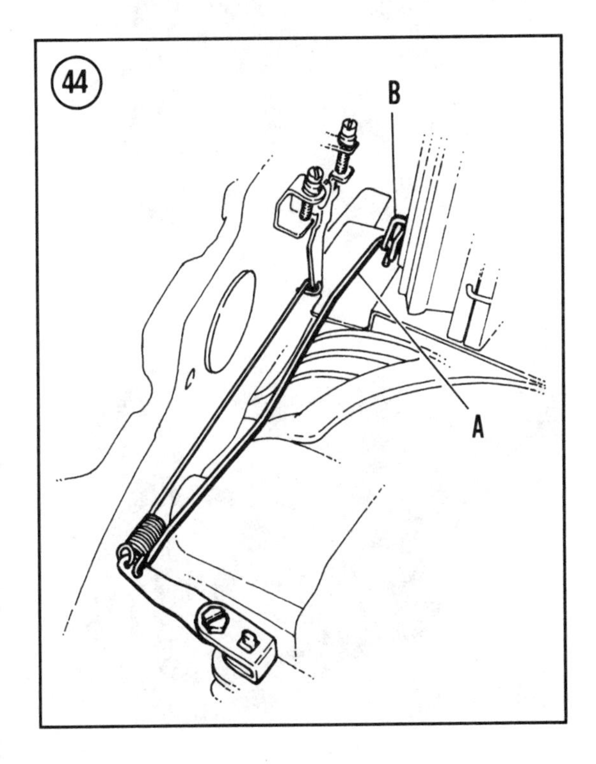

Removal/installation

Because most major carburetor components are contained in the fuel bowl, most service can be performed by removing the fuel bowl without removing the carburetor body. Push the bail that secures the fuel bowl towards the engine to release the fuel bowl.

CAUTION
Exercise care when detaching and reattaching the bail so it is not permanently distorted.

To remove the carburetor, the speed control plate, air cleaner and mounting studs must be removed, and the fuel line and governor link must be detached.

When installing the carburetor, be sure the throttle rod is connected so the end with the long bend (A, **Figure 44**) is attached to the throttle lever (B, **Figure 44**).

Disassembly

An exploded view of the carburetor is shown in **Figure 43**. To disassemble the carburetor for cleaning and inspection, proceed as follows.

1. Detach the fuel bowl retaining bail by pushing it towards the throttle end of the carburetor and remove the fuel bowl.

CAUTION
Exercise care when detaching and reattaching the bail so it is not permanently distorted.

2. Withdraw the nozzle and spring (**Figure 45**) from the fuel bowl. Remove the O-ring if the nozzle is to be immersed in carburetor cleaner.
3. Insert a screwdriver under the float arm (**Figure 46**) and carefully pry the float pin free from the posts in the fuel bowl, then remove the float with the fuel inlet valve.
4. Extract the retaining ring and fuel inlet valve seat (**Figure 47**) from the fuel bowl.

NOTE
Compressed air can be used to dislodge the inlet valve seat, but extreme care must be exercised. Cover the fuel bowl so the seat will not be ejected uncontrolled.

5. Unscrew the drain screw, and on carburetors so equipped, remove the main jet (18, **Figure 43**) assembly.
6. Unscrew and discard the throttle plate retaining screw, then remove the throttle plate and throttle shaft.
7. Remove the Welch plug (**Figure 48**) by piercing the plug with a sharp chisel or punch, then prying out the plug.

CAUTION
Insert the chisel into the plug only far enough to pierce the plug, otherwise, underlying metal may be damaged.

8. If the primer bulb on the side of the air cleaner requires service, remove the primer bulb by twisting out with pliers. Remove the retainer ring (**Figure 49**) by prying with a suitable tool. The old primer bulb and retainer ring must be discarded and replaced with new parts.

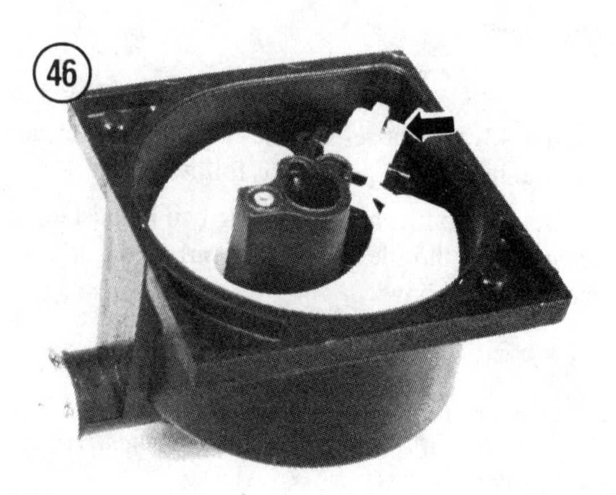
46

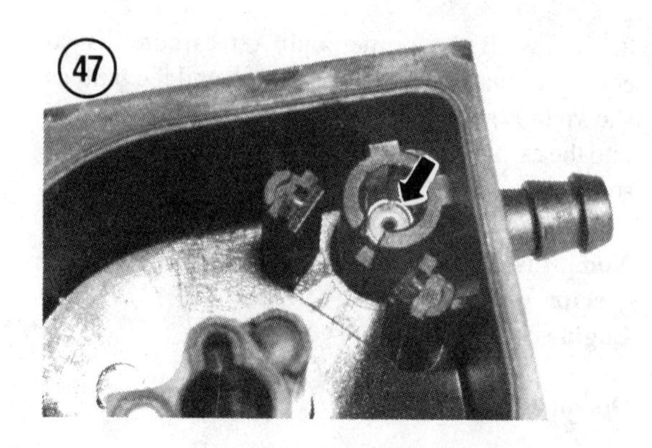
47

48

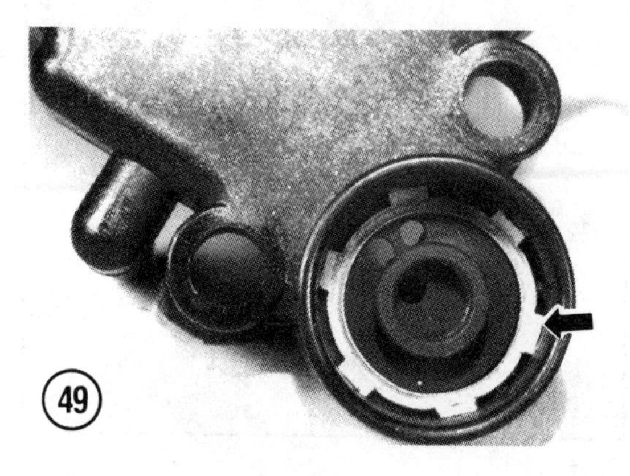
49

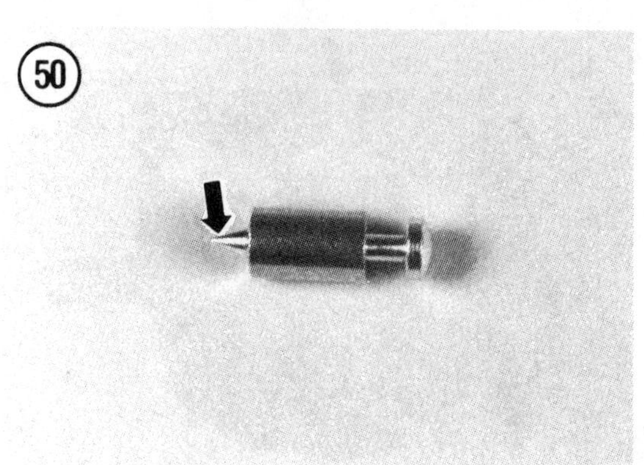
50

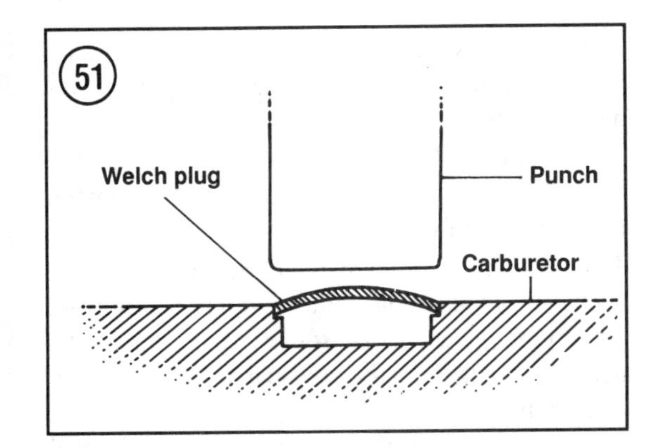

WARNING
Wear safety eyewear when dislodging the retainer as it can be ejected uncontrolled.

9. The carburetor should now be ready for soaking in carburetor cleaner. Follow the directions of the cleaner manufacturer. Do not immerse the carburetor longer than 30 minutes. Spray the carburetor with aerosol carburetor cleaner to remove any residue, then use compressed air to blow out passages and dry the carburetor.

Inspection

1. Inspect the fuel inlet valve (**Figure 50**) and replace it if the tip is grooved.
2. Install the throttle shaft in the carburetor body and check for excessive play between the shaft and body. The body must be replaced if there is excessive play as bushings are not available.
3. If damaged, replace the O-ring on the nozzle.

Assembly

1. Install the Welch plug with the concave side toward the carburetor (**Figure 51**). Do not indent the plug; the plug should be flat after installation. The installation tool should have a diameter that is the same or larger than the plug.
2. Install the throttle plate so the line is visible and toward the top of the carburetor (**Figure 52**). With the retaining screw just loose, turn the throttle shaft so the throttle plate is in the closed position, then tighten the retaining screw. Check for binding.
3. Install the fuel valve inlet seat with the grooved side down and the ridge up (**Figure 53**). Apply a small amount of oil to the outside diameter of the seat to ease insertion. Push the seat in using a tool with a 5/32 in. (4 mm) diameter so the seat bottoms in the bore. Be careful not to scratch the bore. Install the retaining ring and push it down against the valve seat.
4. Apply a small amount of oil to the O-ring on the nozzle to ease insertion of the nozzle, which must be installed as shown in **Figure 45**.
5. The float pin snaps into the fuel bowl posts. The float height is not adjustable.
6. To install the primer bulb, place the retainer ring around the bulb as shown in **Figure 54**, then push

the bulb and retainer into the carburetor using a deep well socket.

Adjustable Carburetor

Operation

The Tecumseh adjustable float carburetor operates according to the principles outlined in Chapter Two. A float (**Figure 55**) in the fuel bowl actuates the fuel inlet valve thereby controlling the amount of fuel in the bowl. When the engine is running, fuel is drawn into the nozzle. A high speed mixture screw below the fuel bowl controls the amount of fuel entering the nozzle. Fuel in the nozzle then exits into the carburetor bore through discharge holes. Fuel in the fuel bowl passes into an idle circuit that routes fuel to discharge holes adjacent to the throttle plate. Fuel delivery from the holes is determined by the position of the throttle plate.

Idle mixture is controlled by the idle mixture screw. On engines with 8 horsepower (6 kW) or more, the idle mixture screw controls fuel flow in the idle circuit. On engines with less than 8 horsepower (6 kW), air for idle speed operation enters through the idle air bleed, and the idle mixture screw controls the amount of air in the idle mixture. Turning the idle mixture screw clockwise on engines with 8 horsepower (6 kW) or more, enriches the idle mixture. Turning the idle mixture screw clockwise on engines with less than 8 horsepower (6 kW), leans the idle mixture.

Air flow through the carburetor is controlled by the throttle plate and choke plate. The main air bleed provides air at atmospheric pressure.

Some carburetors are equipped with a remote primer bulb so additional fuel can be forced into the carburetor bore during starting. When the primer bulb is depressed, air inside the bulb is pressurized as well as air in the fuel bowl. Air pressure against the fuel in the fuel bowl forces fuel up the nozzle, thereby enriching the air:fuel mixture for starting.

Removal/installation

Note that the following procedure is typical and additional steps may be required to accommodate specific applications.

1. Remove the air cleaner assembly.
2. Remove any metalwork, such as the blower housing or engine control panel, that will prevent removal of the carburetor.
3. Disconnect the fuel line from the carburetor. Close the fuel line valve, if so equipped, or drain the fuel to prevent fuel leakage.
4. Remove the carburetor mounting screws, disconnect control linkage and remove the carburetor from the engine. Note that on some engines it may be necessary to tilt the carburetor so that the linkage can be disconnected.
5. To install the carburetor, reverse the removal Steps 1 through 4. Note that on some engines it may

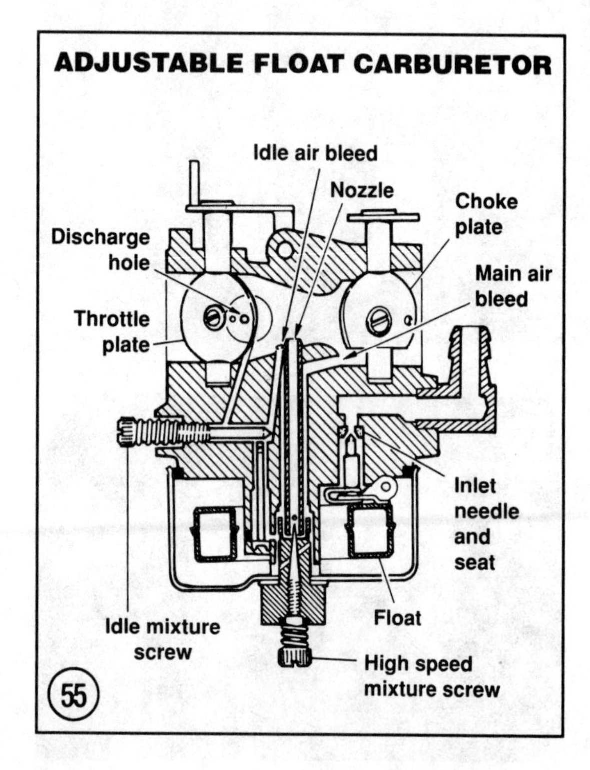

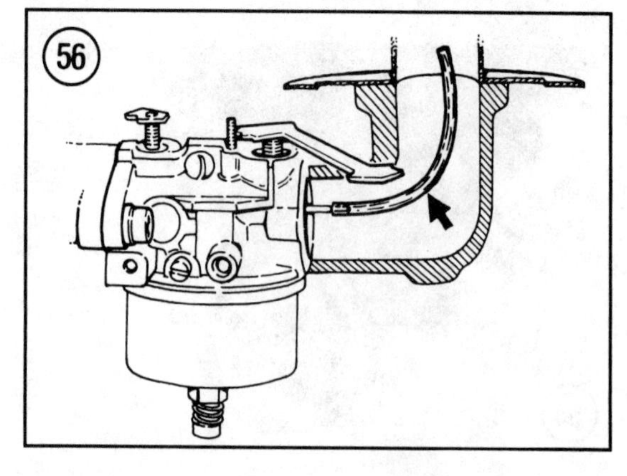

be necessary to tilt the carburetor so that the linkage can be reconnected, in which case the carburetor mounting screws must not yet be installed.

6. Tighten carburetor mounting screws to 48-72 in.-lb. (5.5-8.1 N•m).
7. Some engines are equipped with a tube that extends into the intake passage. See **Figure 56** and **Figure 57**. The tube attaches to a nipple in the carburetor and must be in place for proper carburetor operation.
8. After installation, check the operation of all control linkage and adjust as required. Adjust the carburetor as outlined in Chapter Five.

Disassembly

An exploded view of the carburetor is shown in **Figure 58**. To disassemble the carburetor for cleaning and inspection, proceed as follows.

1. Unscrew the fuel bowl retaining nut, which may include the high speed mixture screw assembly on some carburetors, and remove the fuel bowl and gasket.
2. Dislodge the float pin (**Figure 59**) and remove the float and fuel valve.
3. Unscrew the idle mixture screw.
4. Unscrew and remove the throttle plate.
5. Withdraw the throttle shaft assembly.
6. Unscrew and remove the choke plate.

NOTE
Make note of the position of the choke parts before disassembly so they can be reassembled in their original positions. The choke may operate in either direction depending on application.

7. Withdraw the choke shaft assembly.
8. Extract the fuel inlet valve seat (**Figure 60**).

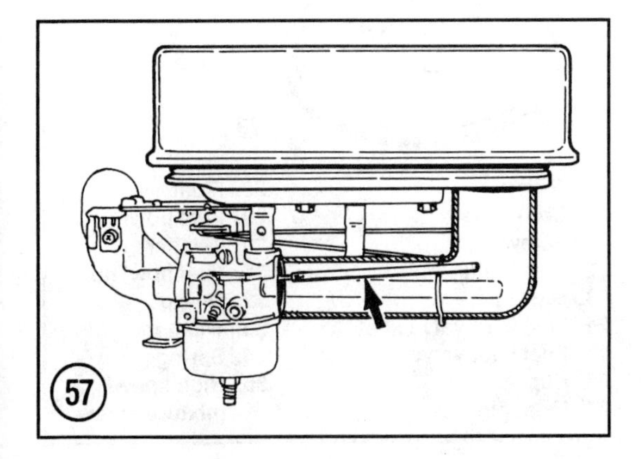

NOTE
Compressed air can be used to dislodge the inlet valve seat, but extreme care must be exercised. Cover the carburetor so that the seat will not be ejected uncontrolled.

9. Remove the Welch plugs (**Figure 61**) by piercing the plug with a sharp chisel or punch, then prying out the plug.

CAUTION
Insert the chisel into the plug only far enough to pierce the plug, otherwise, underlying metal may be damaged.

10. The fuel inlet fitting on the side of the carburetor can be removed by simultaneously twisting and pulling. Note the fitting's position before removal so it can be reinstalled in its original position.

NOTE
*Do not attempt to remove the nozzle (**Figure 62**), as it is pressed into position and movement will affect carburetor operation. Do not remove or loosen any other cup or ball plugs.*

11. The carburetor should now be ready for soaking in carburetor cleaner. Follow the directions of the cleaner manufacturer. Do not immerse the carburetor longer than 30 minutes. Spray the carburetor with aerosol carburetor cleaner to remove any residue, then use compressed air to blow out passages and dry the carburetor.

CAUTION
Some carburetor parts may be made of plastic and should not be soaked in carburetor cleaner.

Inspection

1. Inspect the fuel inlet valve (**Figure 63**) and replace it if the tip is grooved.
2. Inspect the tip of the idle mixture and high speed mixture needles (**Figure 64**) and replace if the tip is bent or grooved.
3. Install the throttle or choke shaft in the carburetor body and check for excessive play between the shaft and body. The body must be replaced if there is excessive play, as bushings are not available.
4. Replace the float if the float pin holes (A, **Figure 65**) are worn excessively.

5. The fuel bowl retaining nut may have one or two holes (**Figure 66**) adjacent to the hex. If a replacement nut is required, the new nut must have the same number of holes as the original nut.

Assembly

1. Install the Welch plugs with the concave side toward the carburetor (**Figure 67**). Do not indent the plug; the plug should be flat after installation. The installation tool should have a diameter that is the same or larger than the plug.

2. Install the fuel valve inlet seat with the flat side up (**Figure 68**) and the grooved side towards the carburetor.

3. Install the choke plate so the straight edge is towards the fuel bowl side of the carburetor (**Figure 69**). The choke assembly parts should be in the

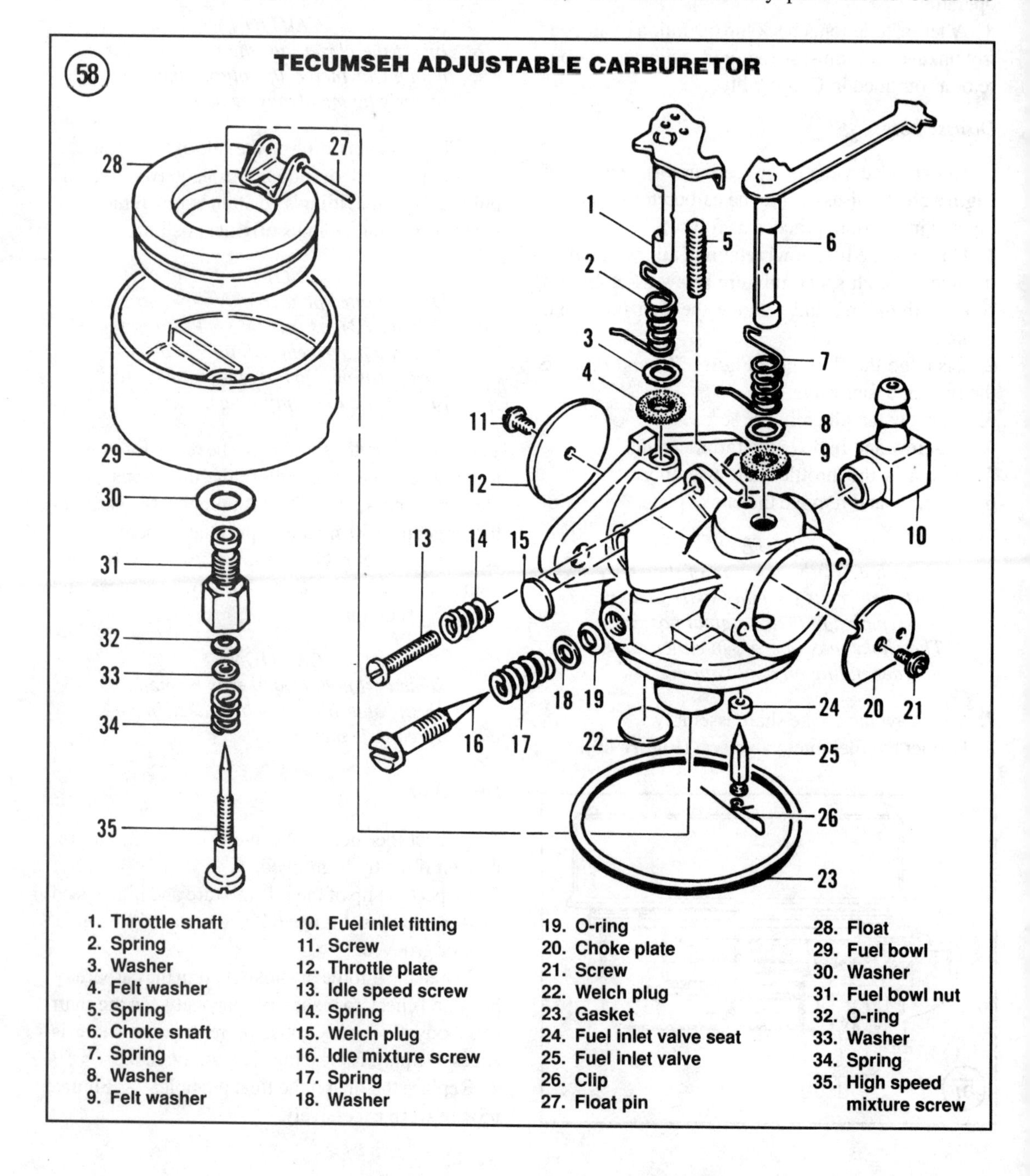

59

60

61

62

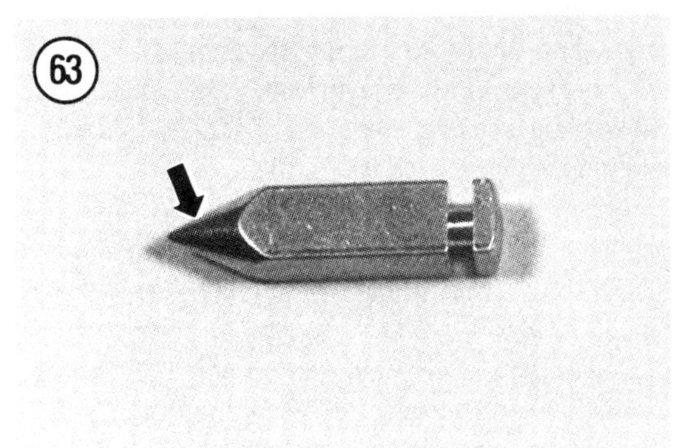
63

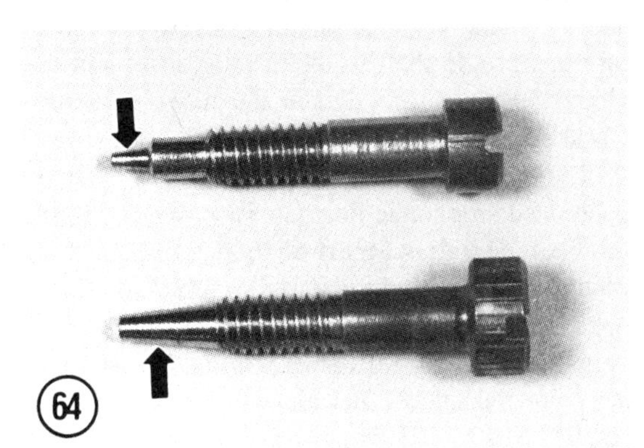
64

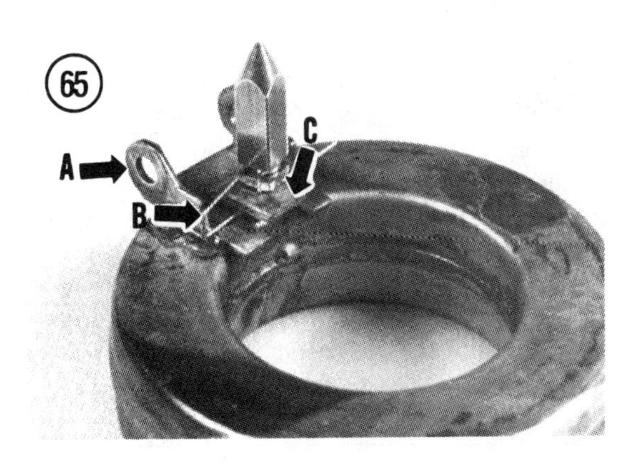
65
A
B
C

66

original position noted during disassembly. With the retaining screw just loose, turn the choke shaft so the choke plate is in the closed position, then tighten the retaining screw. Check for binding.

4. Install the throttle plate so the line or lines are visible when the throttle is closed (**Figure 70**). If there is only one line and the engine has less than 8 horsepower (6 kW), the line should point towards the top of the carburetor. If there is only one line and the engine has 8 horsepower (6 kW) or more, the line should point towards the 3 o'clock position. If there are two lines, they should be positioned at 12 o'clock and 3 o'clock. With the retaining screw just loose, turn the throttle shaft so the throttle plate is in the closed position, then tighten the retaining screw. Check for binding.

5. The wire clip on the fuel inlet valve must be positioned around the float tab as shown in **Figure 65**. The long end (B, **Figure 65**) of the wire clip must point toward the choke end of the carburetor.

6A. With the float installed and the carburetor inverted, measure the float level (**Figure 71**). Measure at a point opposite the fuel inlet valve. The float level should be 0.162-0.215 in. (4.11-5.46 mm). Bend the float tab (C, **Figure 65**) to adjust the float level.

6B. Tecumseh has a gauge (Tecumseh part 670253A) that can be used to determine the correct float level. Position gauge 670253A at a 90° angle to carburetor bore and resting on nozzle stanchion as shown in **Figure 72**. The toe of the float should not be higher than the first step on the gauge (**Figure 72**), or lower than the second step (**Figure 73**).

NOTE
If the carburetor is equipped with a fiber washer between the fuel bowl and stan-

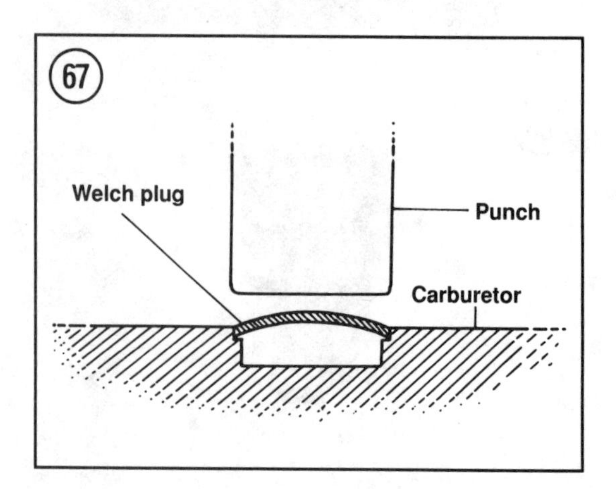

72

73

74

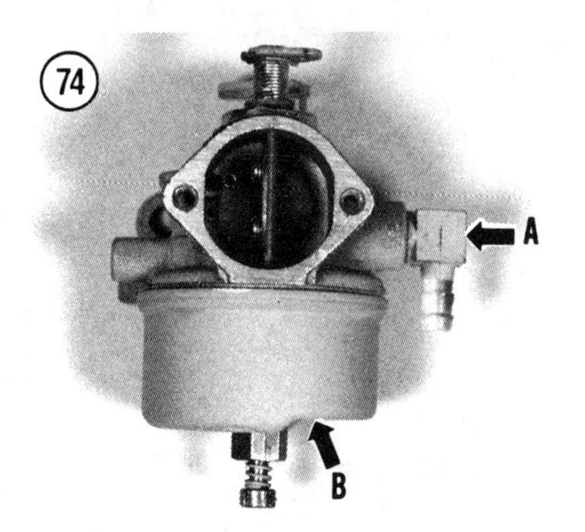

75

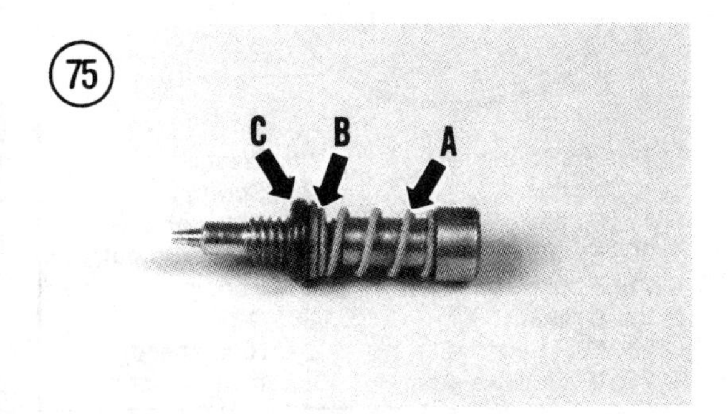

chion, place the washer on the stanchion and measure from the washer rather than the stanchion. If using gauge 670253A, place the gauge on the washer.

7. If removed, reinstall fuel inlet (A, **Figure 74**) by pushing it into the carburetor so the inlet is in its original position. Apply Loctite sealant before installation.
8. Install the fuel bowl so the indented portion (B, **Figure 74**) is on the same side as the fuel inlet (A, **Figure 74**).
9. In **Figure 75** note the proper location of spring (A), washer (B) and O-ring (C) on either mixture screw.

WALBRO LME CARBURETOR

Operation

The Walbro model LME carburetor is an adjustable float type carburetor. The carburetor operates according to the principles outlined in Chapter Two. A float in the fuel bowl actuates the fuel inlet valve thereby controlling the amount of fuel in the bowl. When the engine is running, fuel is drawn into the nozzle past the high speed adjusting screw. Fuel in the nozzle then exits into the carburetor bore through a discharge hole. Fuel is also drawn into the idle circuit and past the idle mixture screw before entering the carburetor bore. Air flow through the carburetor is controlled by throttle and choke plates.

Removal/Installation

Note that the following procedure is typical and additional steps may be required to accommodate specific applications.

1. Remove the air cleaner assembly.
2. Remove any metalwork, such as the blower housing or engine control panel, that will prevent removal of the carburetor.
3. Disconnect the fuel line from the carburetor. Close the fuel line valve, if so equipped, or drain the fuel to prevent fuel leakage.
4. Remove the carburetor mounting screws, disconnect control linkage and remove the carburetor from the engine. Note that on some engines it may be

necessary to tilt the carburetor so that the linkage can be disconnected.

5. To install the carburetor, reverse the removal Steps 1 through 4. Note that on some engines it may be necessary to tilt the carburetor so that the linkage can be reconnected, in which case the carburetor mounting screws must not yet be installed.

6. Tighten carburetor mounting screws to 48-72 in.-lb. (5.5-8.1 N•m).

7. After installation, check the operation of all control linkage and adjust as required. Adjust the carburetor as outlined in Chapter Five.

Disassembly

An exploded view of the carburetor is shown in **Figure 76**. To disassemble the carburetor for cleaning and inspection, refer to **Figure 76** and proceed as follows.

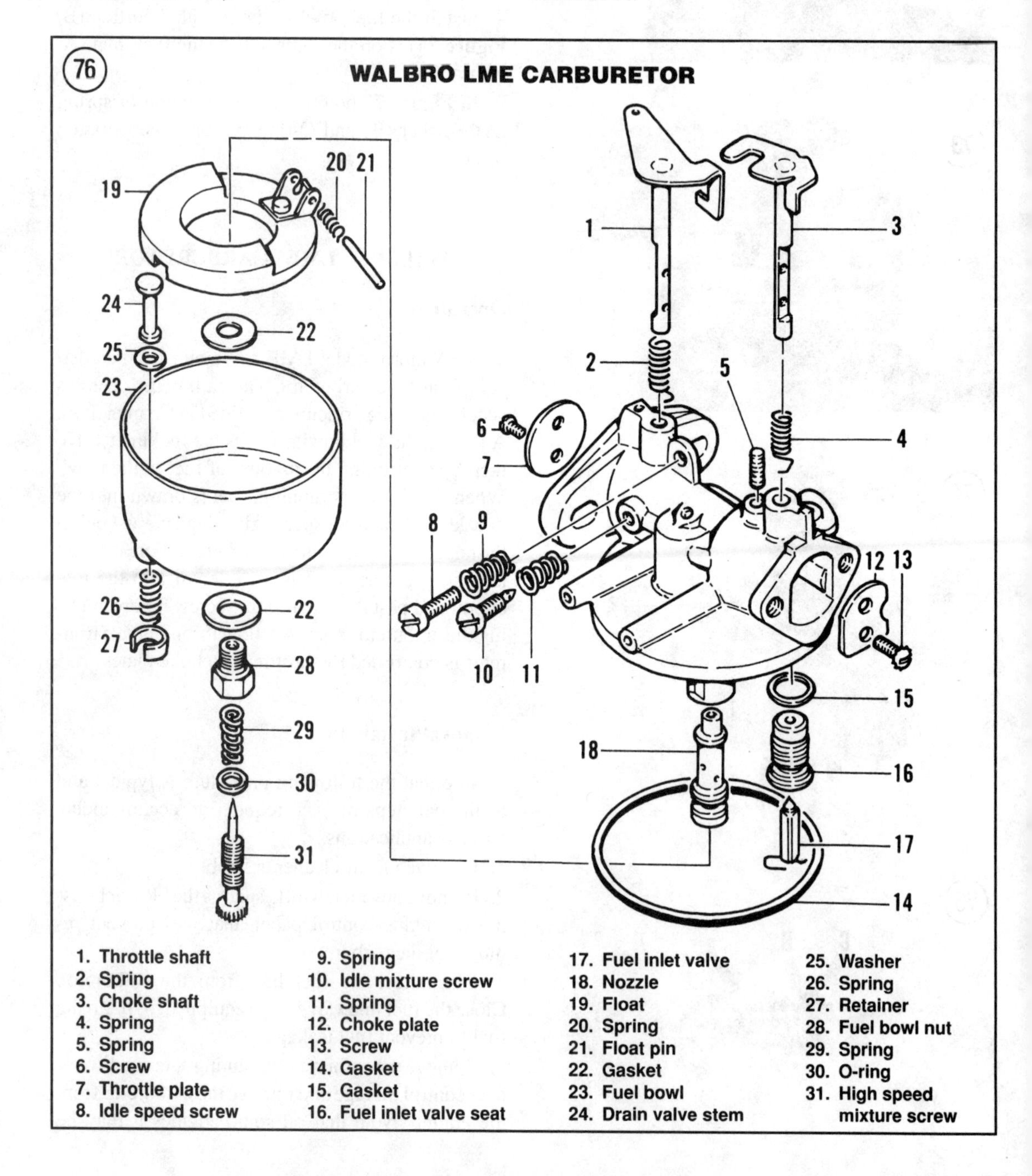

1. Unscrew the fuel bowl retaining nut, which includes the high speed mixture screw assembly, and remove the fuel bowl and gasket.
2. Dislodge the float pin and remove the float and fuel inlet valve.
3. Unscrew the idle mixture screw.
4. Unscrew and remove the throttle plate.
5. Withdraw the throttle shaft assembly.
6. Unscrew and remove the choke plate.
7. Withdraw the choke shaft assembly.
8. Unscrew the fuel inlet valve seat.
9. If the carburetor is excessively dirty or the high speed mixture screw seat on the nozzle is damaged, unscrew the nozzle. Otherwise, the nozzle should not be removed.
10. The carburetor should now be ready for soaking in carburetor cleaner. Follow the directions of the cleaner manufacturer. Do not immerse the carburetor longer than 30 minutes. Spray the carburetor with aerosol carburetor cleaner to remove any residue, then use compressed air to blow out passages and dry the carburetor.

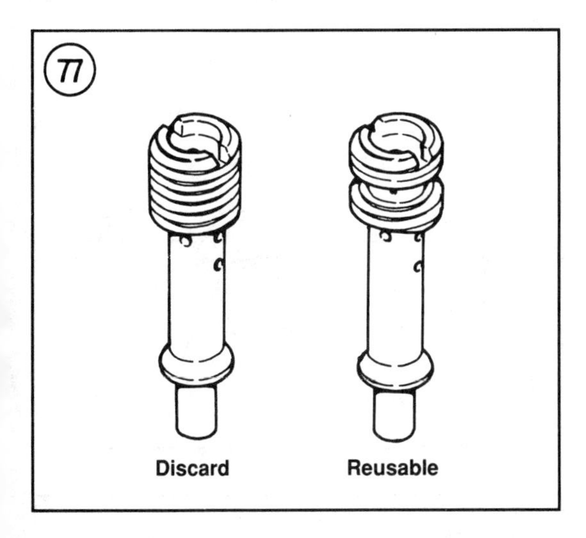

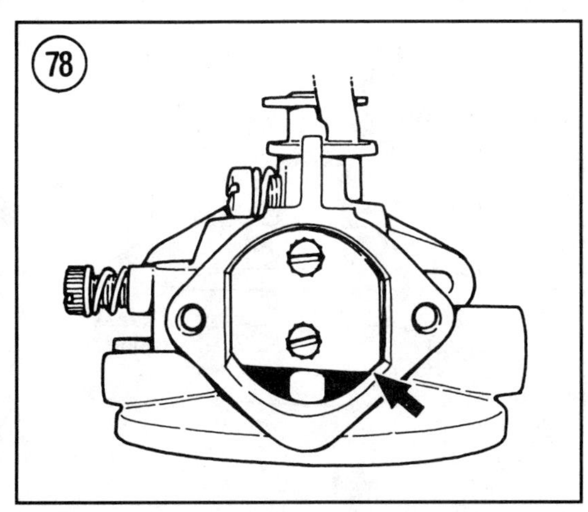

Inspection

1. Inspect the fuel inlet valve and replace it if the tip is grooved.
2. Inspect the tip of the idle mixture and high speed mixture needles and replace if the tip is bent or grooved.
3. Install the throttle or choke shaft in the carburetor body and check for excessive play between the shaft and body. The body must be replaced if there is excessive play as bushings are not available.
4. If the nozzle does not have a groove (**Figure 77**) cut in the threaded portion, discard it. A service replacement nozzle has a groove. A nozzle with a groove can be reused. See **Figure 77**.

Assembly

1. Be sure to install the gasket (15, **Figure 76**) under the fuel inlet valve seat (16, **Figure 76**). Tighten the valve seat to 40-50 in.-lb. (4.5-5.6 N•m).
2. Install the choke plate so the straight edge is towards the fuel bowl side of the carburetor (**Figure 78**). The numbered or lettered side of the choke plate should be on the inside when the choke plate is closed. With the retaining screw just loose, turn the choke shaft so the choke plate is in the closed position, then tighten the retaining screw. Check for binding.
3. Install the throttle plate so the numbers or letters face out when the throttle plate is closed (**Figure 79**). With the retaining screws just loose, turn the throttle shaft so the throttle plate is in the closed position, then tighten the retaining screws. Check for binding.
4. The spring clip on the fuel inlet valve must fit around the tab on the float as shown in **Figure 80**.
5. Assemble the float with the fuel valve and pin (do not install the spring [**Figure 80**]) and check the float level and float drop.
   a. With the float installed and the carburetor inverted, measure the float level as shown in **Figure 81**. Measure at a point opposite the fuel inlet valve. The float level on 8 horsepower (6 kW) engines should be 0.070-0.110 in. (1.78-2.79 mm). The float level on other engines

should be 0.110-0.130 in. (2.79-3.30 mm). Bend the float tab (**Figure 81**) to adjust the float level.

b. With the carburetor upright, the float should drop to a point that the brass plug (**Figure 82**) on the stanchion is fully visible. Push down lightly on the float to bend the tab on the back of the float to adjust the float drop.

6. The ends of float spring (**Figure 80**) must point toward the choke end of the carburetor when the float is installed. The spring fits around the float pin (**Figure 80**). The short spring end must be pushed against spring tension so it fits in the gasket groove of the carburetor body as shown in **Figure 83**. The fuel bowl gasket fits over the spring end.

GOVERNOR SYSTEM

The Tecumseh engines covered in this manual are equipped with a mechanical flyweight governor system. Refer to Chapter Two for a discussion of fundamental governor operating principles.

Refer to **Figure 84** for a view of a typical governor and speed control linkage used on vertical crankshaft engines. A view of a typical governor and speed control linkage used on horizontal crankshaft engines is shown in **Figure 85**.

Observe the following when working on the governor or speed control linkage.

1. Before disconnecting the linkage, mark the linkage so it can be reassembled in its original configuration.

2. Do not stretch the governor spring during removal or installation. Deforming the spring can affect governor operation.

79

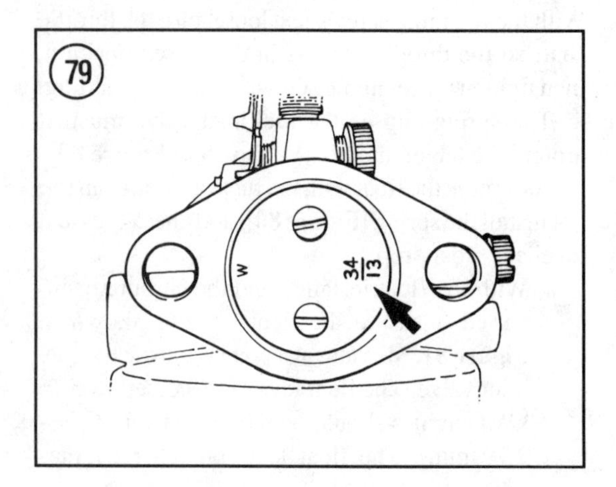

80

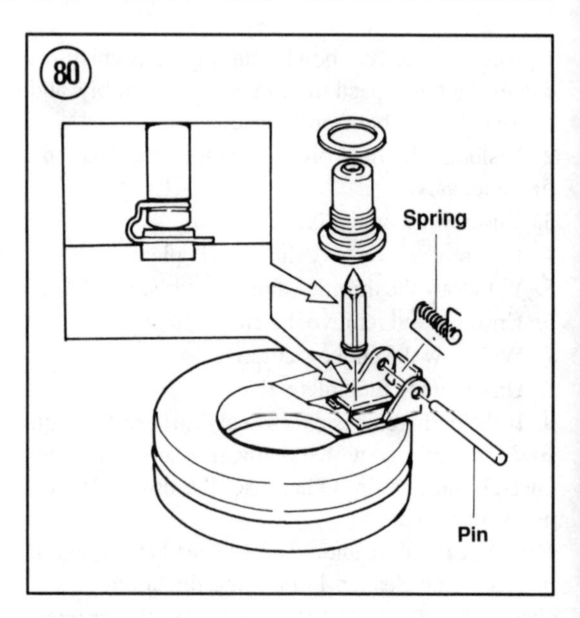

81

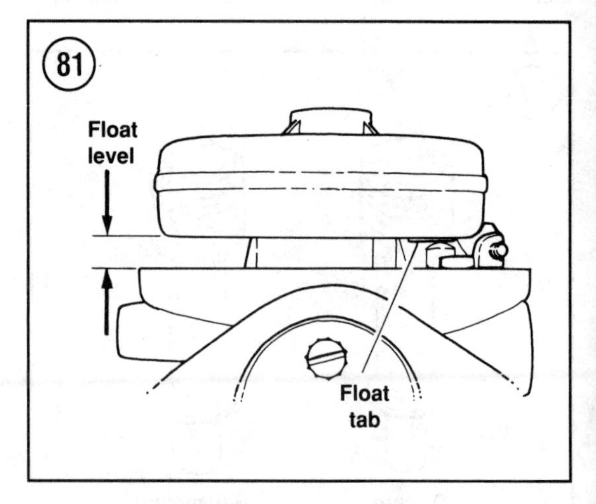

82

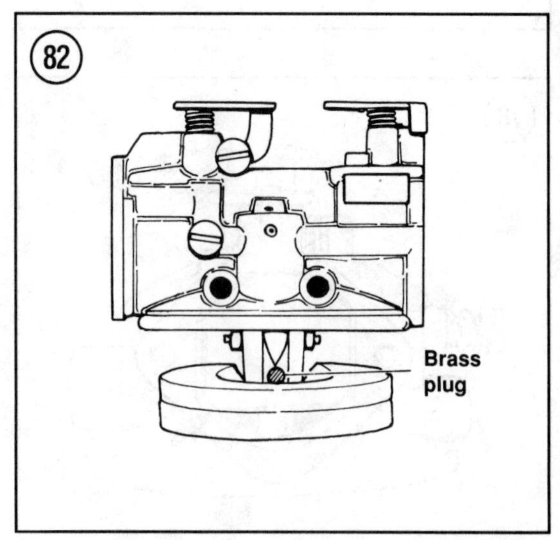

Adjustment

On some engines, the idle speed is governed. Refer to Chapter Five for adjustment of the governed idle speed screw or tab.

On some engines, the maximum governed speed is adjustable. The maximum governed speed is determined by the engine's application and specified by the equipment manufacturer.

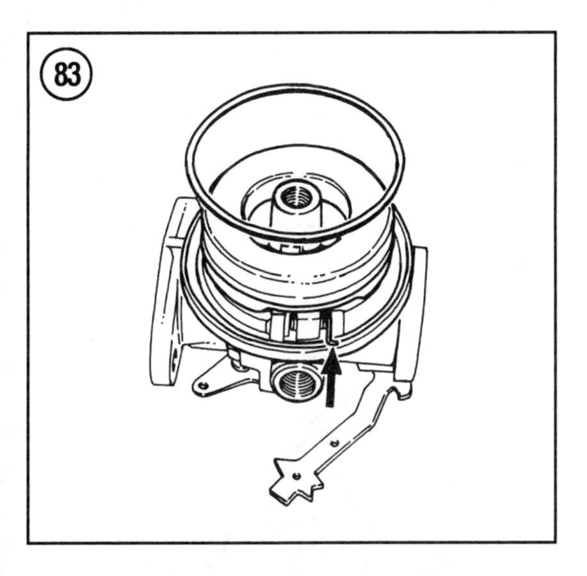

GOVERNOR LINKAGE (VERTICAL CRANKSHAFT)

Speed control
Throttle link
Governor spring
Governor lever

84

CAUTION
Adjusting maximum governed engine speed in excess of the engine speed specified by the manufacturer can cause engine damage as well as damage to engine-driven equipment.

It may be necessary to adjust the governor linkage so a full range of movement is transferred from the internal governor mechanism to the throttle. Before adjusting the linkage, locate the clamp screw on the governor lever. Some examples are shown in **Figure 86** and **Figure 87** for horizontal crankshaft engines and **Figures 88-90** for vertical crankshaft engines.

1. On engines with a horizontal crankshaft (**Figure 86** and **Figure 87**), loosen the clamp screw (A). Move the governor lever (B) so the carburetor throttle plate is in wide open position. Rotate the governor clamp (C) clockwise as far as possible and tighten the clamp screw (A).

2. On engines with a vertical crankshaft (**Figures 88-90**), loosen the clamp screw (A). Move the governor lever (B) so the carburetor throttle plate is in wide open position. On Vector VLV50, VLXL50, VLV55 and VLXL55 engines, rotate the governor clamp (C) clockwise as far as possible and tighten the clamp screw (A). On all other vertical crankshaft engines, rotate the governor clamp (C) counter-

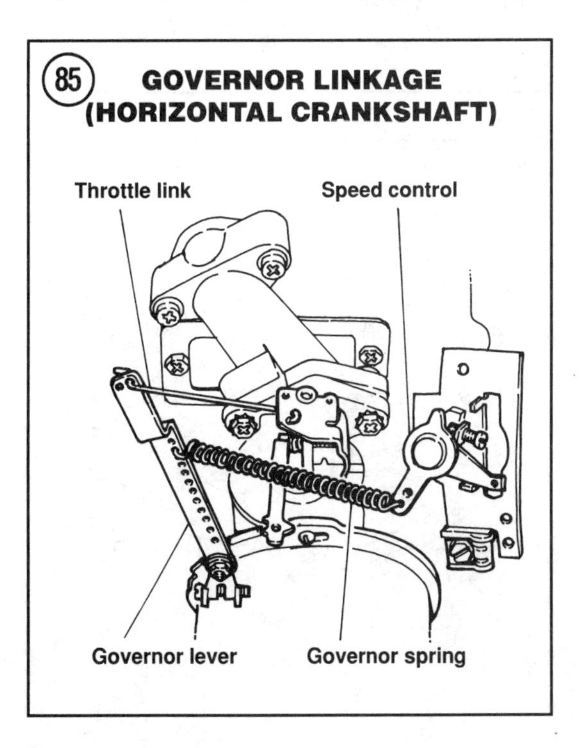

6

clockwise as far as possible and tighten the clamp screw (A).

Troubleshooting

Most governor related problems are caused by incorrect adjustment, linkage connected incorrectly, a stretched governor spring or damaged components.

A simple test can be performed to determine if faulty engine operation is due to the governor system

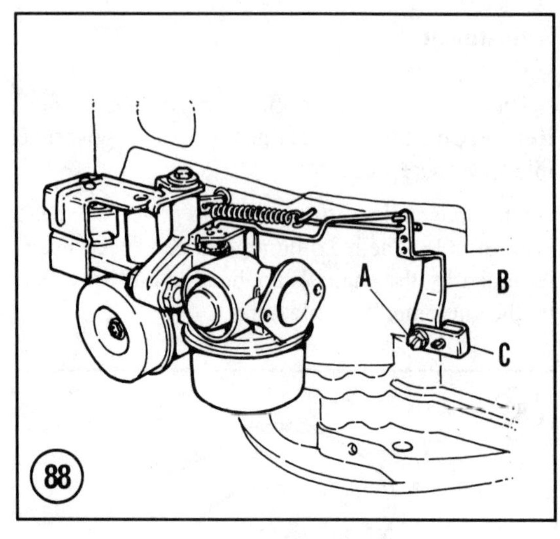

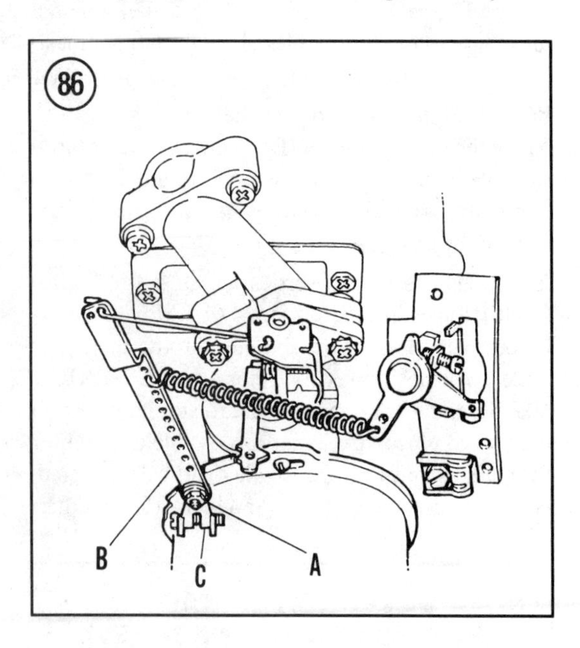

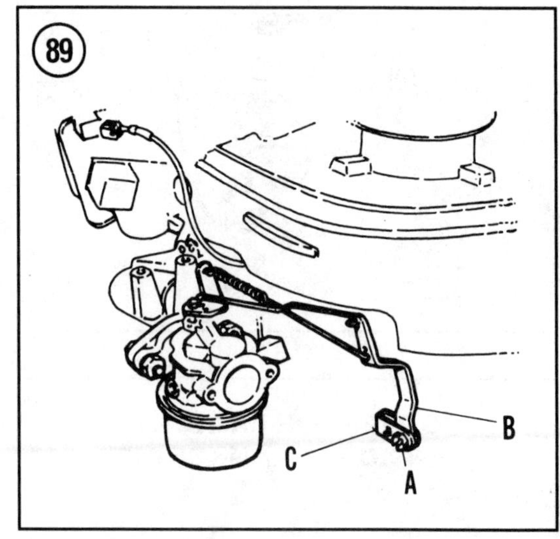

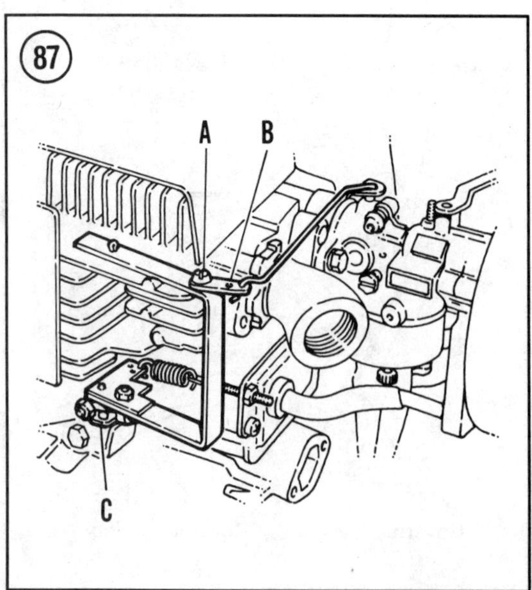

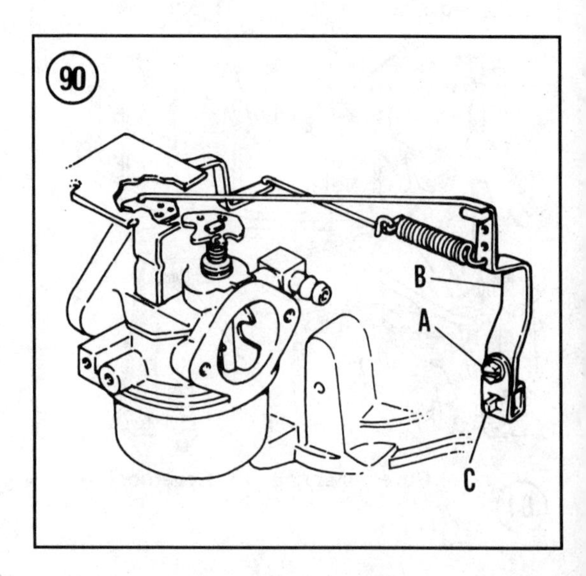

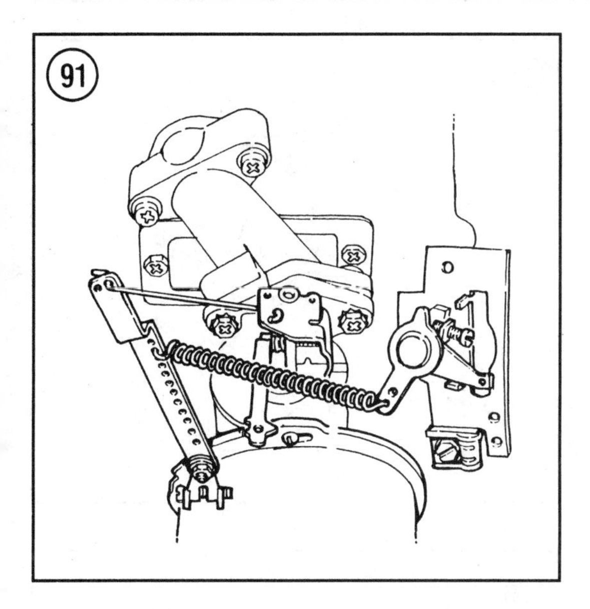

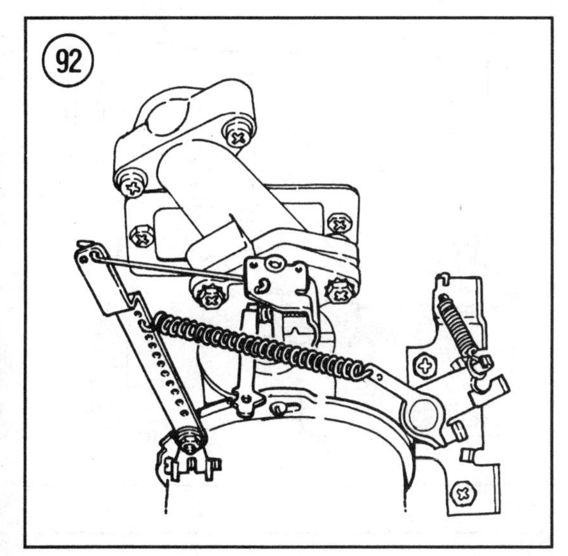

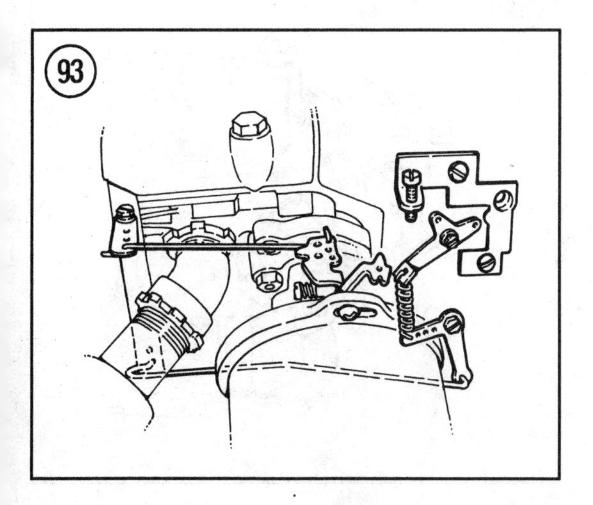

and related linkage. Move the throttle arm or lever on the carburetor through a full range of movement. If the engine operates satisfactorily, then a problem exists in the governor system or speed control linkage. If the engine continues to malfunction, then the problem lies in another area.

When analyzing a suspected problem in the governor system, first check that all linkage is properly connected and no components are damaged. Be sure the governor spring is not stretched. The spring can be easily stretched during removal and installation resulting in abnormal engine operation. A stretched or damaged governor spring should be replaced.

Refer to Chapter Eleven for overhaul information on the internal governor mechanism of mechanical governors.

Some typical linkage configurations are shown in the following illustrations.

a. **Figure 91**—Light frame, horizontal crankshaft.
b. **Figure 92**—Light frame, horizontal crankshaft for recreational vehicle application.
c. **Figure 93**—Medium frame, horizontal crankshaft.
d. **Figure 94**—Horizontal crankshaft for constant speed application.
e. **Figure 95**—Horizontal crankshaft on Snow King engines.
f. **Figure 96**—Medium frame, horizontal crankshaft on Snow King engines.
g. **Figure 97**—Horizontal crankshaft on Snow King engines.

95

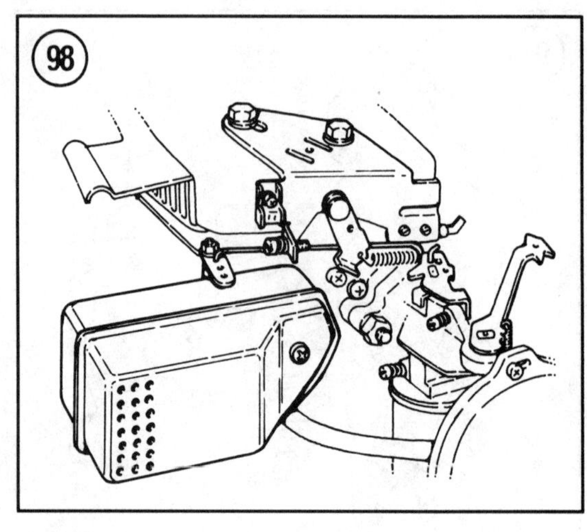
98

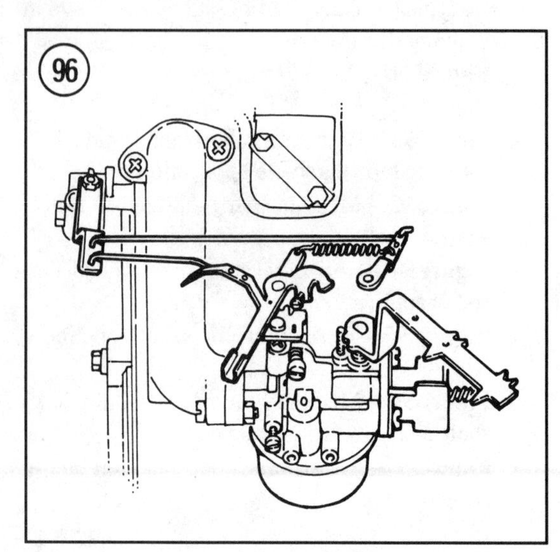
96

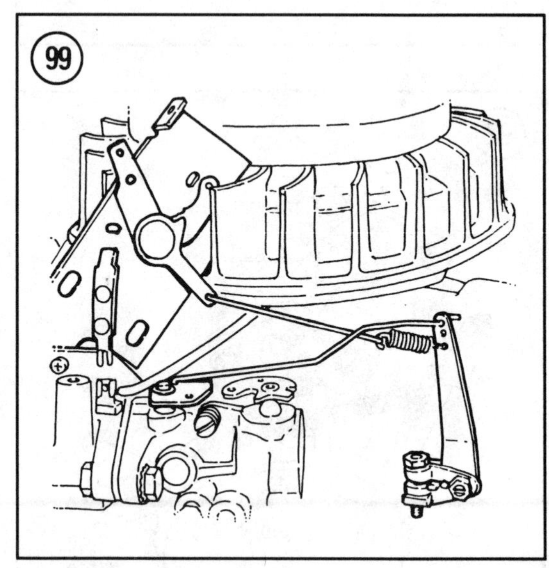
99

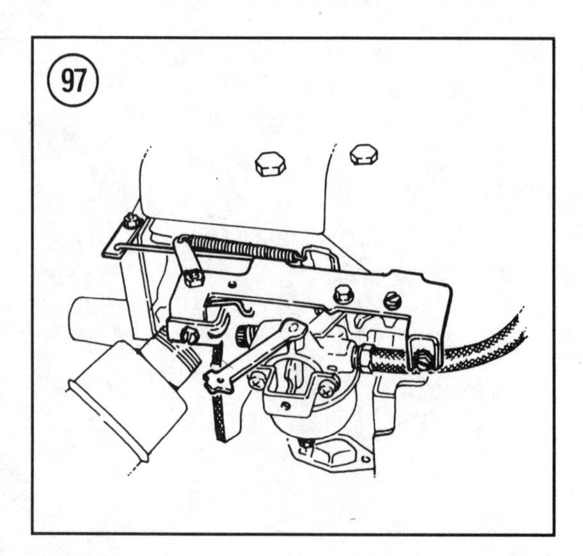
97

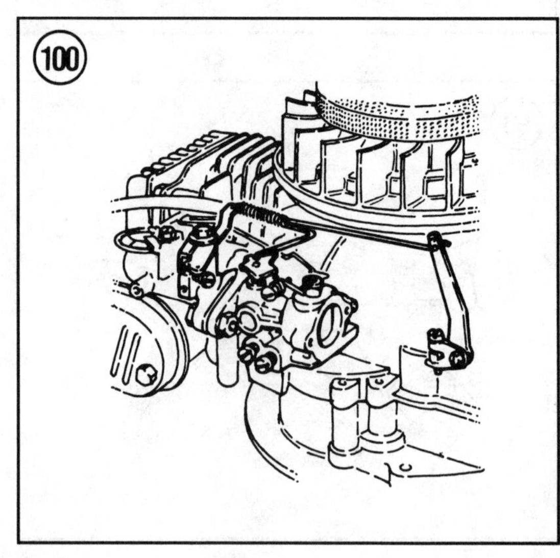
100

h. **Figure 98**—Medium frame, horizontal crankshaft.
i. **Figure 99**—Vertical crankshaft.
j. **Figure 100**—Vertical crankshaft.
k. **Figure 101**—Model TNT100 engine.
l. **Figure 102**—Model TNT100 engine.
m. **Figure 103**—Model TNT120 engine.
n. **Figure 104**—Model TVS engines with a fully adjustable carburetor.

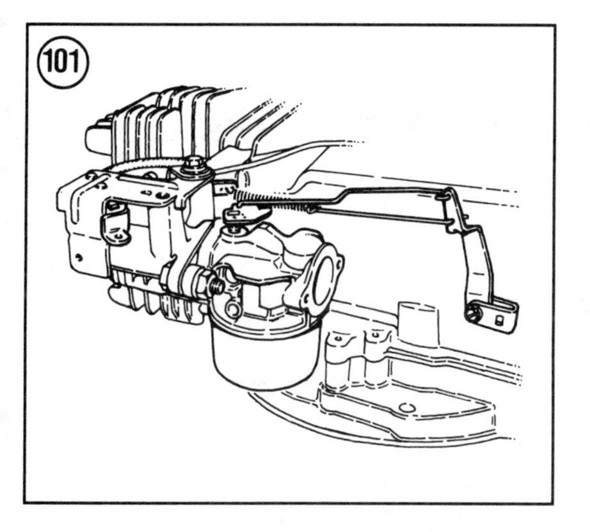

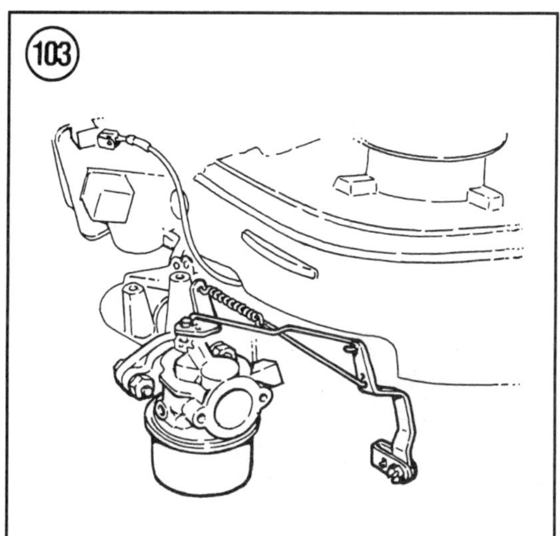

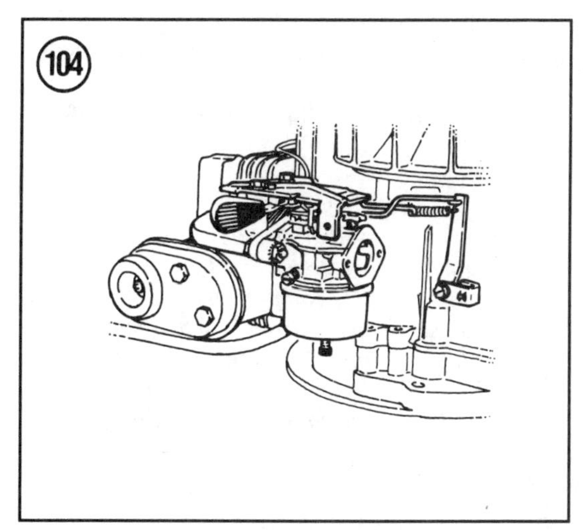

CHAPTER SEVEN

IGNITION SYSTEM AND FLYWHEEL BRAKE

This chapter covers components of the ignition system, except the ignition breaker points and spark plug, which are covered in Chapter Five. Also covered is the flywheel brake used on some later models.

Refer to Chapter Four for troubleshooting information.

CAUTION
The ignition system primary circuit is designed to operate on small electrical current. Ignition components may be damaged if a large current, such as current from a battery, is connected to the ignition. This can happen due to improperly connected wires or a faulty ignition switch.

IGNITION COIL (BREAKER-POINT IGNITION)

Tecumseh engines equipped with breaker points may be equipped with an ignition coil that is mounted behind the flywheel on the stator plate (**Figure 1**), or a round ignition coil that is permanently attached to the laminations (**Figure 2**) may be mounted outside the flywheel adjacent to the rim of the flywheel.

On most electronic ignition systems, the ignition coil and remaining ignition system components are a module that must be serviced as a unit assembly. See the following *Ignition Module/Coil* section for information on the ignition module/coil used on electronic ignition systems.

Flywheel removal is necessary on a breaker-point ignition system, either to gain access to a stator-mounted coil or to disconnect the primary coil wire of an externally mounted coil. See Chapter Eleven for flywheel removal and installation. If the ignition coil is mounted on the stator plate, the coil may be removed after detaching a retainer (**Figure 1**). The retainer may be either a wire or clip. Externally mounted ignition coils can be removed after unscrewing the armature retaining screws.

No test procedures are available for the ignition coil, although some professional shops are equipped with coil testers that can be used to test ignition coils. In most cases, the ignition coil is replaced after tests rule out other components as the cause for an ignition malfunction.

NOTE
Externally mounted ignition coils are no longer available from Tecumseh. An electronic ignition module/coil must be used instead, as well as a different flywheel key. See the ***Flywheel and Key*** *section.*

Whenever the armature retaining screws on an external ignition coil are loosened, the armature air gap must be adjusted. Follow the procedure outlined in Chapter Five.

The high tension wire (spark plug lead) is not available separately from the ignition coil. The spark plug terminal is available separately from small engine, auto and motorcycle dealers and can be attached to the end of the high tension wire.

IGNITION MODULE/COIL

On most Tecumseh electronic ignition systems, the ignition module and ignition coil are a unit assembly (**Figure 3**); neither component is available

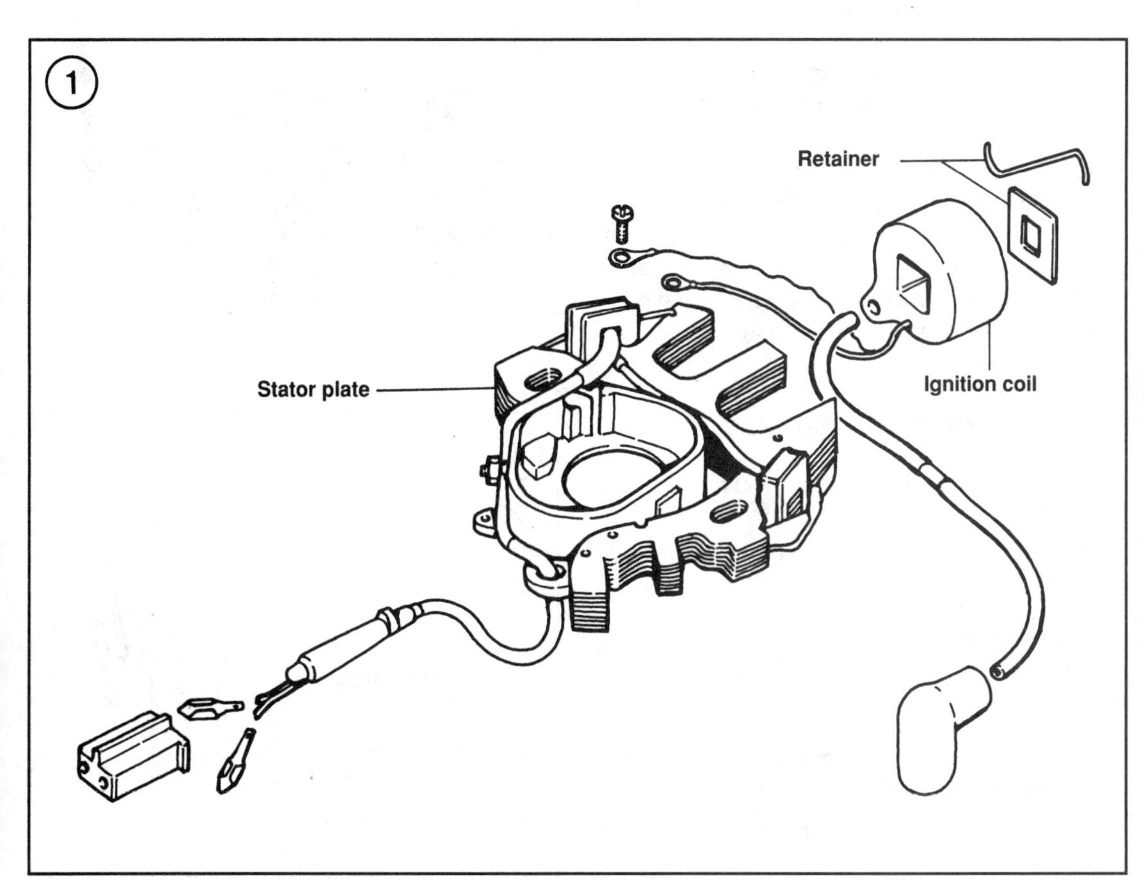

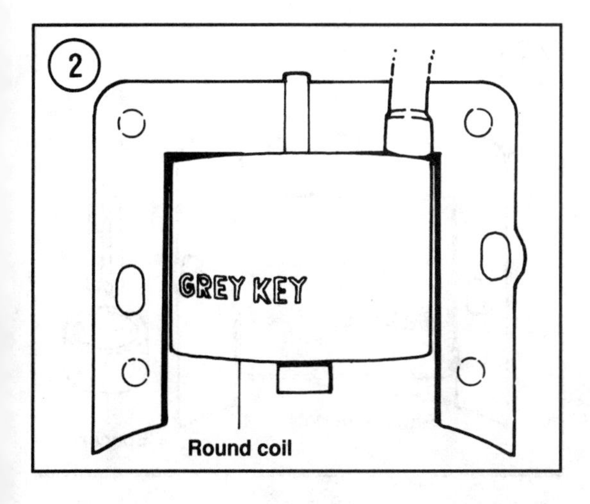

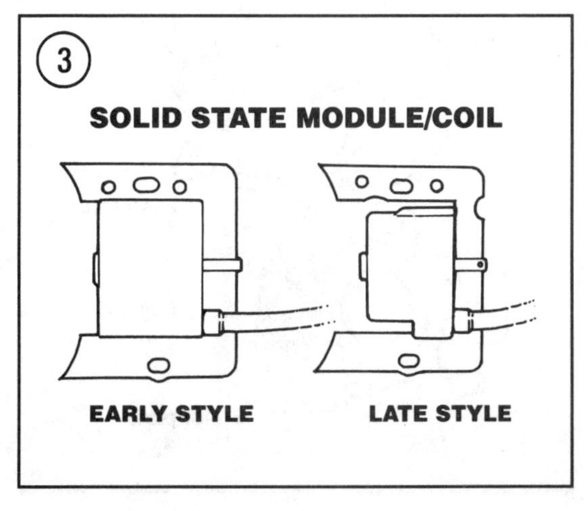

separately. Testing is limited to checking for spark at the spark plug lead. If no spark is produced, and the flywheel and key are in good condition, then the ignition coil/module should be replaced.

Some engines may be equipped with an electronic ignition system (**Figure 4**) that has a separate ignition module and transformer (ignition coil). The ignition module is located on the stator plate, while the transformer is mounted near the flywheel rim. Flywheel removal is necessary to disconnect the primary coil wire. See Chapter Eleven for flywheel removal and installation.

No test procedures are available for the ignition module/coil, although some professional shops are equipped with testers that can be used to test ignition module/coils. In most cases, the ignition module/coil is replaced after tests rule out other components as the cause for an ignition malfunction.

FLYWHEEL AND KEY

Refer to Chapter Eleven for flywheel removal and installation.

Poor ignition performance can be caused by loss of flywheel magnetic strength. A method of checking magnet strength is to suspend a screwdriver with the blade down and approximately 3/4 in. (19 mm) from the flywheel magnet (**Figure 5**). The screwdriver blade should be pulled against the magnet.

On engines originally equipped with a breaker-point ignition system and an external ignition coil,

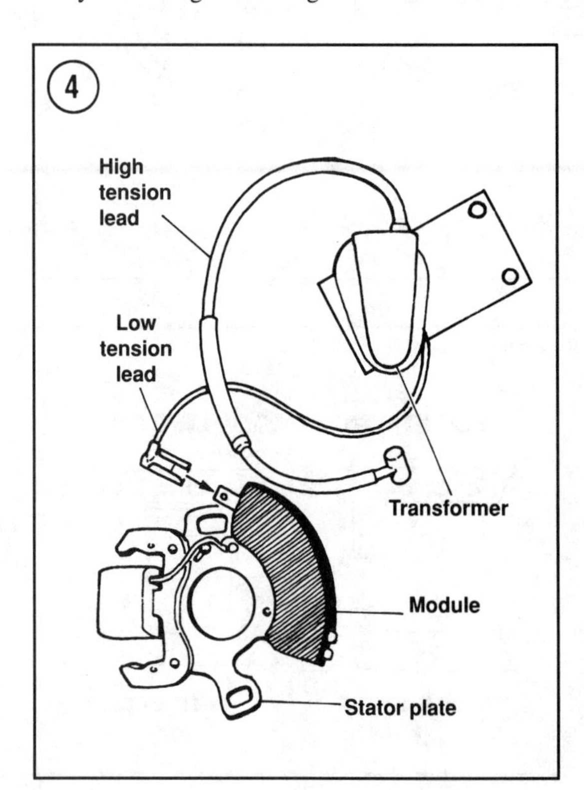

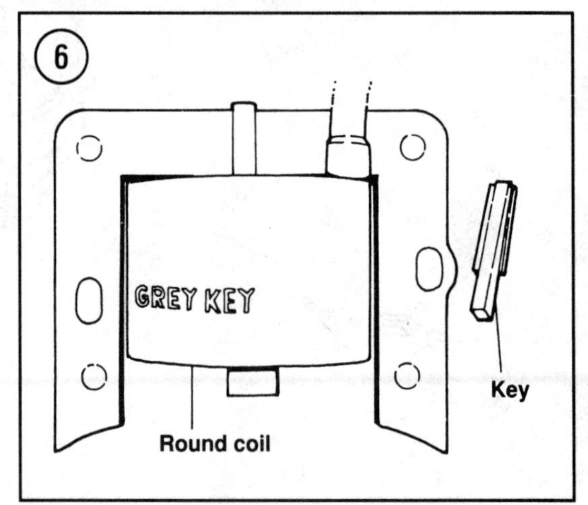

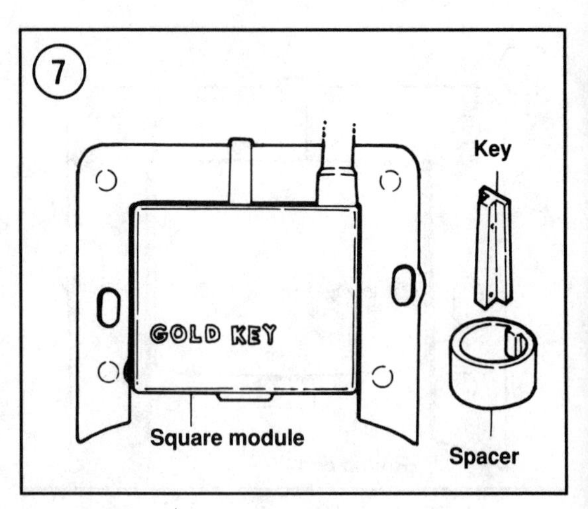

the ignition coil is round and stamped "GREY KEY" as shown in **Figure 6**. The flywheel key is colored grey.

The ignition coil on some engines equipped with an electronic ignition (often used as a replacement for the external coil on the breaker-point ignition system) has a square shape and is stamped "GOLD KEY" as shown in **Figure 7**. A gold colored flywheel key must be used with this ignition coil for proper ignition timing. A spacer must also be installed on the crankshaft.

Install a stepped flywheel key so the stepped end is towards the engine (**Figure 8**). Install a tapered flywheel key so the big end is towards the engine as shown in **Figure 9**. If a spacer is used, install the spacer so the protrusion fits in the keyway and is towards the end of the crankshaft (**Figure 10**).

BREAKER POINT CAM

On engines equipped with a breaker-point type ignition system, a removable breaker point cam is located on the crankshaft under the breaker point cover. A damaged or excessively worn cam should be replaced. Install the cam so the arrow or "TOP" is visible as shown in **Figure 11**.

STOP SWITCH

The engine may be equipped with a stop switch on the engine, or a remotely mounted switch may be located elsewhere on the equipment. The stop switch (often called an ignition switch when mounted on equipment) must stop the ignition when activated. Because a magneto is used on Tecumseh engines, the ignition system is defeated by grounding the primary circuit.

A stop wire runs from the ignition circuit to a variety of stop switches. As an example, note the control panel in **Figure 12** that is used on some engines. A ground wire (A) is connected to the stop switch (B). The engine stops due to grounding the primary ignition circuit when the speed control lever (C) contacts the switch terminal (D). If the engine is equipped with a flywheel brake, the stop switch (**Figure 13** or **Figure 14** on Vector engines) may be a part of the brake mechanism.

Testing a stop switch designed for use with a magneto type ignition system involves using a continuity checker or ohmmeter. The tests determine if the ignition circuit is grounded when the stop switch is actuated. In all cases, the stop switch should show continuity (zero ohms) to engine ground when the switch is in the STOP position and no continuity (infinity) when the switch is in the RUN position. Also check wiring for loose or bad connections.

NOTE
Automotive type ignition switches should not be used as battery voltage may be routed to the ignition circuit, which may damage ignition components.

FLYWHEEL BRAKE

Some later models are equipped with a flywheel brake that simultaneously stops the flywheel and grounds the ignition. The brake should stop the engine within three seconds when the operator releases the mower safety control and the speed control is in high speed position. Engine rotation is stopped by a pad type brake (**Figure 15** or **Figure 16** on Vector engines) that contacts the inside of the flywheel when the operator's handle is released.

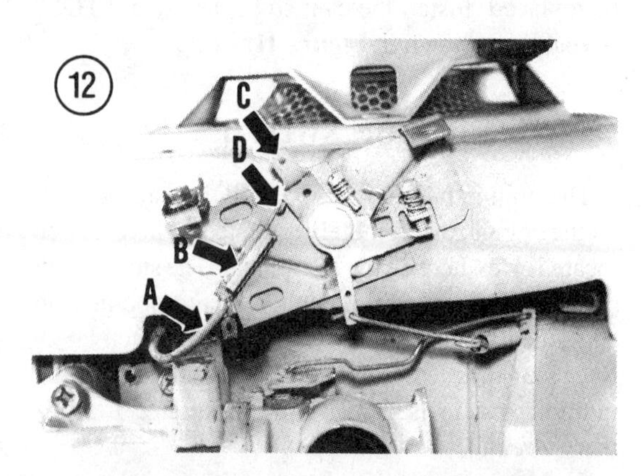

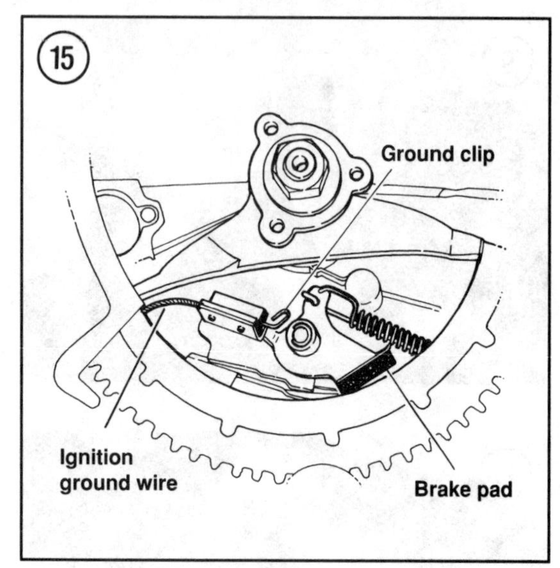

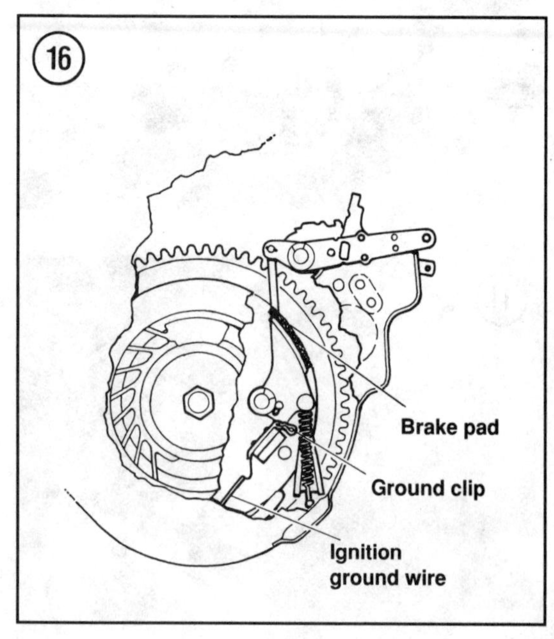

17

On engines with an electric starter, a switch (**Figure 17**) mounted on the flywheel brake bracket or brake lever prevents starter engagement unless the flywheel brake is disengaged.

Disengage Flywheel Brake

To hold the flywheel brake in the disengaged position, except on Vector VLV50, VLXL50, VLV55 and VLXL55 engines, pull the operator's handle or push the lever (A, **Figure 18**) and align the holes in the brake bracket and brake lever. Insert a 7/64 in. (2.75 mm) drill bit (B) or rod through the holes.

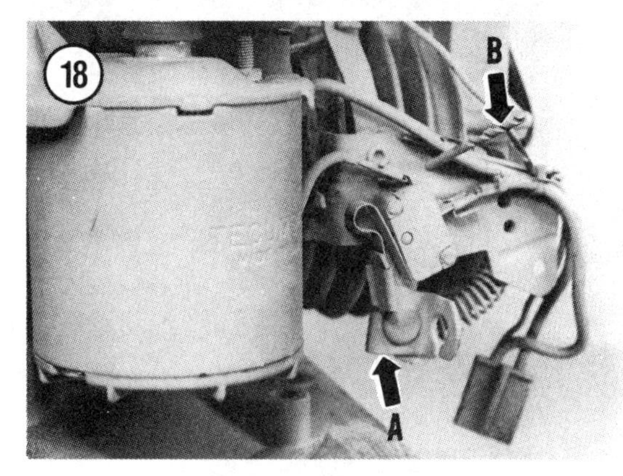

18

To hold the flywheel brake on Vector VLV50, VLXL50, VLV55 and VLXL55 engines in the disengaged position, bend a piece of 3/32 in. (2.5 mm) metal rod as shown in **Figure 19** so there is approximately 1 in. (25 mm) between the bends. Remove the left screw securing the speed control cover. Move the flywheel brake lever to the disengaged position and insert the holder rod into the outer hole of the brake lever and the speed control cover screw hole as shown in **Figure 20**. Tecumseh offers tool 36114 (**Figure 21**) to hold the brake lever in the disengaged position.

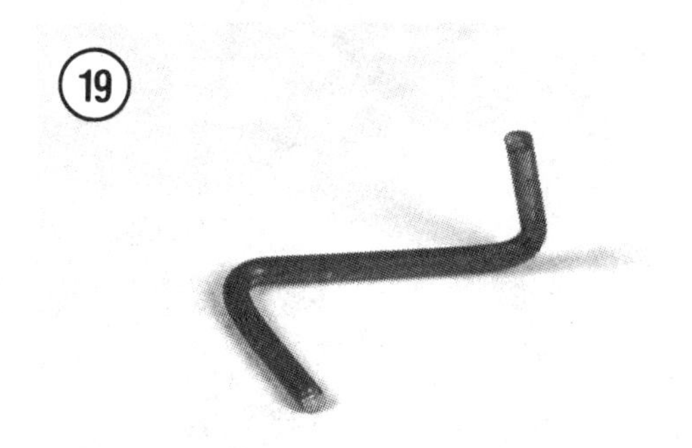
19

Flywheel Brake Removal/Installation (Except Vector VLV50, VLXL50, VLV55 and VLXL55 Engines)

1. Hold the flywheel brake in the disengaged position as outlined in the previous *Disengage Flywheel Brake* section.

2. Remove the flywheel as outlined in Chapter Eleven.

20

21

3. Mark and disconnect wires to the flywheel brake bracket assembly.
4. Remove the brake bracket mounting screws and remove the bracket from the engine.
5. To install the brake assembly, position the brake bracket on the engine and install the mounting screws, but do not tighten the screws.
6. Note that the brake bracket mounting holes are elongated (A, **Figure 22**). Push the bracket downward as far as possible and tighten the mounting screws to 90 in.-lb. (10.17 N•m).
7. Connect the wires leading to the bracket assembly.
8. Reinstall the flywheel. Bc sure that the stop switch wire will not contact the flywheel.
9. Release the flywheel brake.

NOTE
When tightened, the flywheel brake cable clamp screw (B, ***Figure 22****) must not interfere with movement of the brake lever (C). If it does, install a shorter screw.*

Flywheel Brake Removal/Installation (Vector VLV50, VLXL50, VLV55 and VLXL55 Engines)

1. Hold the flywheel brake in the disengaged position as outlined in the previous *Disengage Flywheel Brake* section.

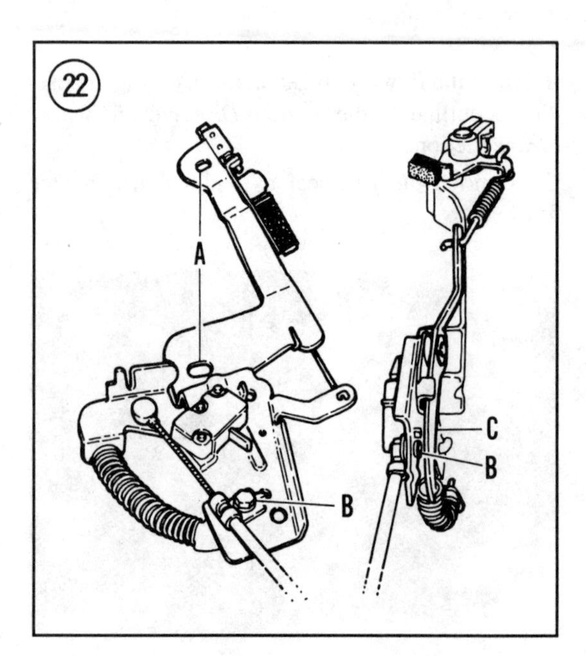

2. Remove the flywheel as outlined in Chapter Eleven.
3. Relieve brake spring pressure by removing the rod or tool holding the brake in the disengaged position.
4. Disconnect the brake spring (A, **Figure 23**) from the brake pad arm (B).
5. Detach the E-ring (C, **Figure 23**) securing the brake pad arm and remove the brake pad arm. Detach the brake link from the arm.

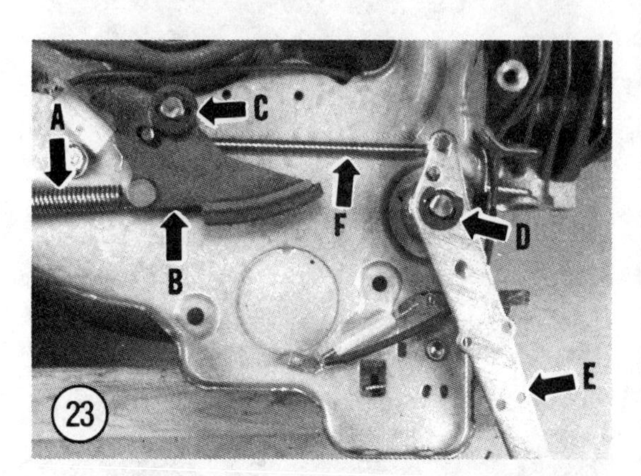

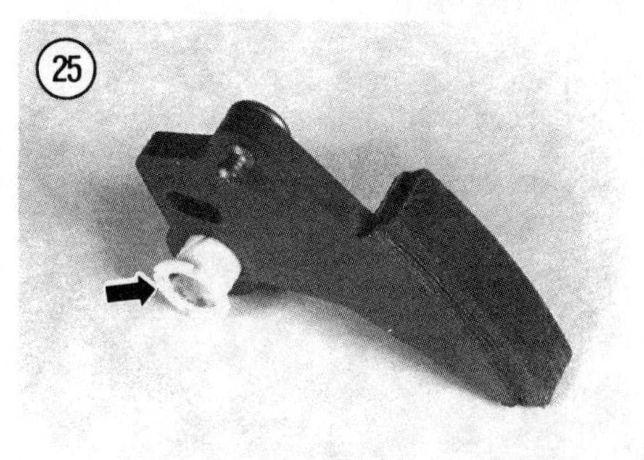

6. On the underside of the baffle plate, mark the hole that holds the brake lever spring end (**Figure 24**).

7. If equipped with an electric starter, mark and disconnect wires to the starter switch on the brake lever.

8. Detach the E-ring (D, **Figure 23**) securing the brake lever (E) and remove the brake lever and spring.

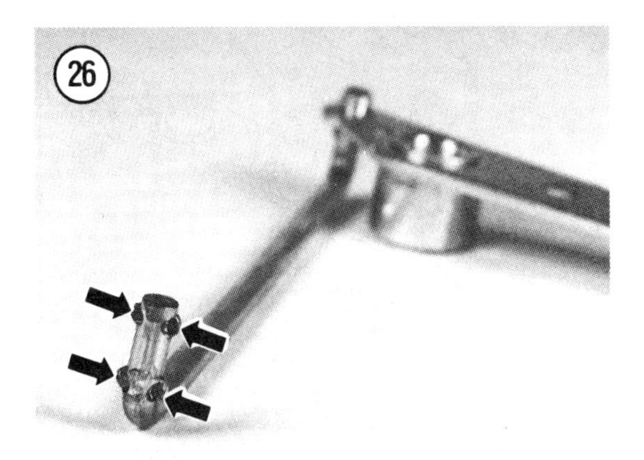

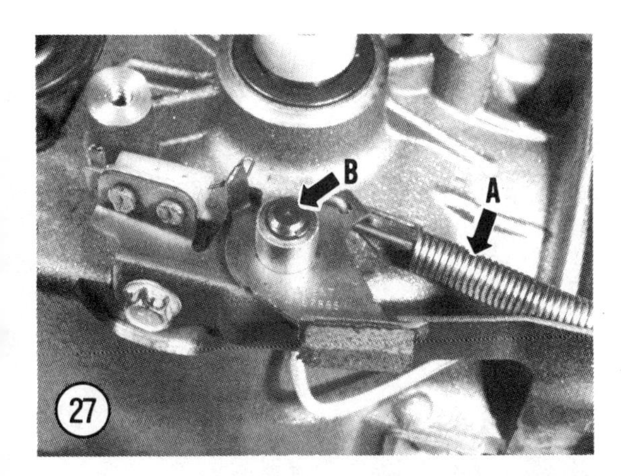

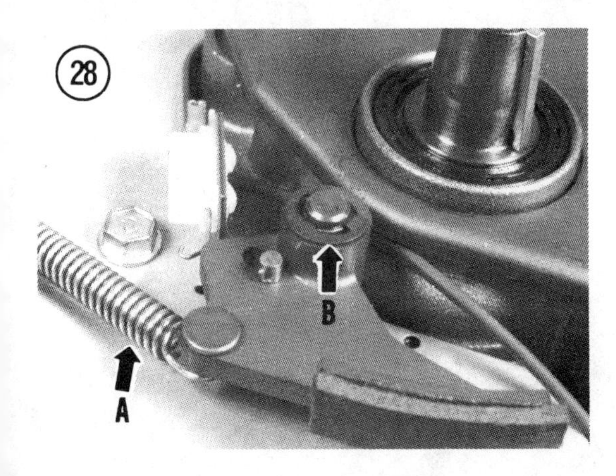

9. The brake pad arm and brake lever ride on renewable plastic bushings. Be sure the bushing is in good condition and install it so the flange is on the bottom side of the brake pad arm (**Figure 25**) or brake lever.

10. Correct installation of the brake link (F, **Figure 23**) is determined by the number of bosses on the link end. The end that attaches to the brake pad arm has four bosses (**Figure 26**), while the brake lever end has three bosses.

11. Install the brake lever, brake link and brake pad arm and secure with the E-rings.

12. The brake lever spring end must be inserted in the original hole in the baffle plate (**Figure 24**).

13. The brake spring (A, **Figure 23**) end with the short hook must be attached to the brake arm (B).

Service Flywheel Brake Pad

The flywheel brake pad is accessible after disengaging the brake as previously outlined and removing the flywheel as outlined in Chapter Eleven. Replace the brake pad and arm if the pad is damaged, contaminated by oil, or worn to a thickness less than 0.060 in. (1.52 mm).

1. Before removing the brake pad arm, relieve spring pressure by removing the rod holding the brake in the disengaged position.

2. Detach the brake spring (A, **Figure 27** or **Figure 28** on Vector engines) from the brake arm.

3. Remove the brake pad arm after detaching the E-ring (B, **Figure 27** or **Figure 28**) on the brake arm post.

4. Attach the new brake pad arm to the brake bracket.

5. On Vector VLV50, VLXL50, VLV55 and VLXL55 engines, note that the brake spring end with the short hook must be attached to the brake pad arm.

Service Electric Starter Switch

1. To check the starter switch (**Figure 29** or **Figure 30** on Vector engines), disconnect the wires to the switch.

2. Connect an ohmmeter or continuity checker to the switch terminals. When the switch button is out, there should be no continuity. When the switch button is in, there should be continuity.

3. To replace the switch, grind off the heads of the mounting rivets, then remove the switch and rivets.
4. Screws are used to secure the new switch. A self-tapping screw provided with the new switch forms threads in the old rivet holes, then the new switch is secured with machine screws. Do not tighten the screws excessively as the switch case may crack.

5. On Vector VLV50, VLXL50, VLV55 and VLXL55 engines, an insulator (**Figure 30**) is located between the switch and brake lever.

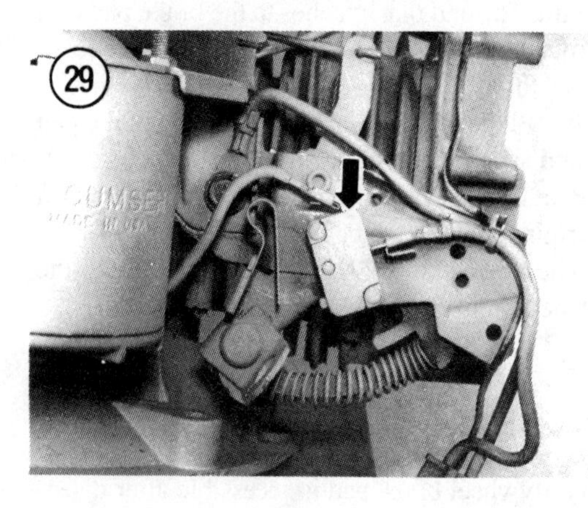

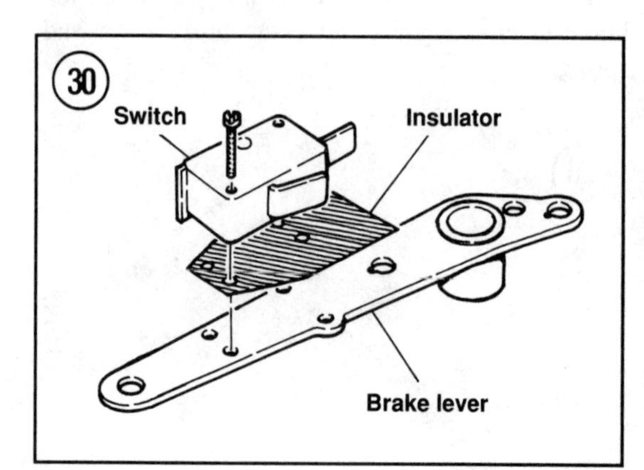

CHAPTER EIGHT

REWIND STARTERS

The rewind starters included in this chapter are divided into sections that cover starters mounted on the blower housing and vertical-pull starters. Refer to the following sections for service information.

NOTE
The starter mounting fasteners must be tight for proper starter operation, including the fasteners securing the blower housing if the starter is attached to the blower housing. Loose fasteners can cause starter drive mechanism damage.

REWIND STARTERS MOUNTED ON BLOWER HOUSING

Most "L" head Tecumseh engines are equipped with a rewind starter that is mounted on the blower housing (**Figure 1**). Refer to **Figure 2** for an illustration showing the various starter types. Early low to medium horsepower engines are commonly equipped with a "teardrop" shaped starter housing. Medium frame engines (HM and VM series) may be equipped with a rewind starter that has a round shape. Some later engines are equipped with a "stylized" rewind starter that is identified by the slots in the starter housing. Refer to the appropriate following section for service information.

EARLY "TEARDROP" SHAPED STARTER

Most early low to medium horsepower Tecumseh engines are equipped with a rewind starter that has a starter housing with a teardrop shape (**Figure 3**) —later teardrop shaped starters are "stylized" and covered in a later section. This is a pawl type starter. When the rope is pulled, one or two pawls extend

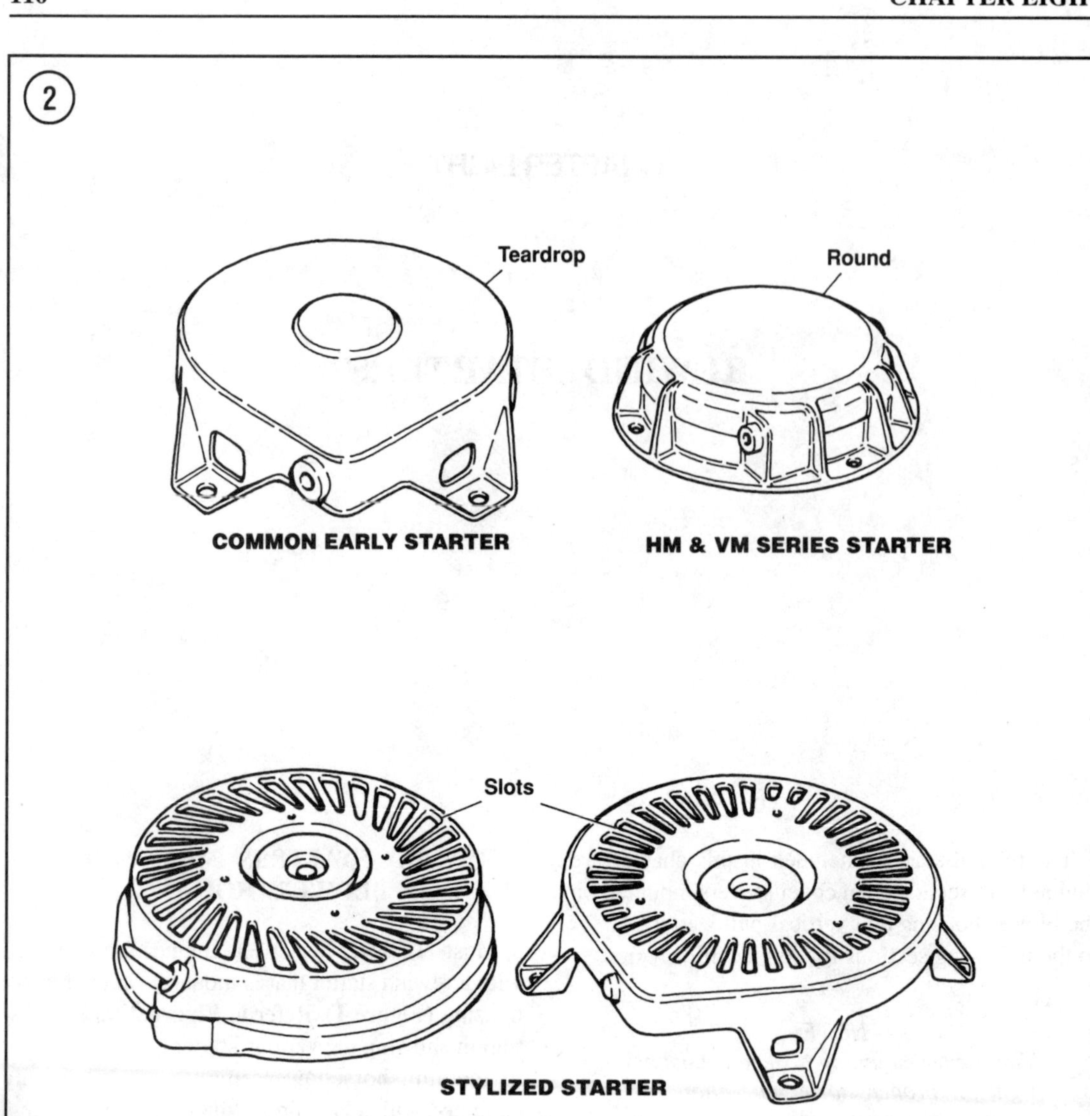
2
Teardrop
Round
COMMON EARLY STARTER
HM & VM SERIES STARTER
Slots
STYLIZED STARTER

3

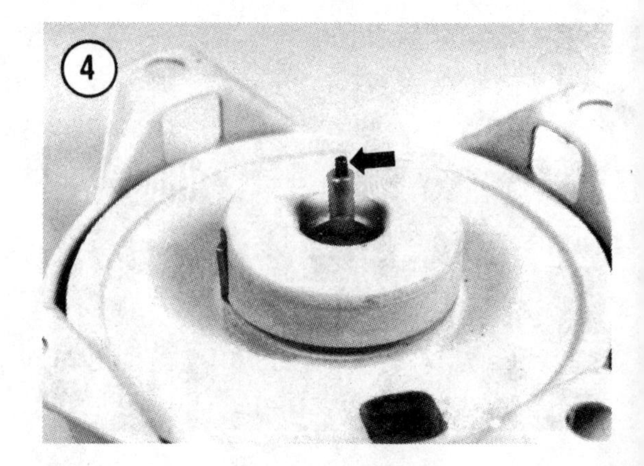
4

outward to engage notches in the starter cup attached to the flywheel.

Major starter problems usually involve a severed rope, a broken rewind spring or poor engagement between the pawl and starter cup. The rope can usually be replaced without completely disassembling the starter, while a broken rewind spring requires disassembling the starter. Poor pawl engagement may be due to worn parts or a loose retainer screw. The following procedures will apply to most starters of this design, but some variations may occur.

Starter Removal/Installation

Starter removal usually only requires removing the starter housing retaining screws. On some models, the starter housing is retained by pop rivets. The blower housing must be removed for starter service if the starter housing is retained by rivets, or the rivets must be removed by drilling. If the rivets are loose, drill out the rivets and mounting holes so suitably sized rivets or bolts can be installed.

On some models, a pin extends from the starter center screw (**Figure 4**) into a bushing in the end of the crankshaft. The pin aligns the starter in the starter cup. When installing the rewind starter on models so equipped, push the pin into the center screw of the starter, place the bushing on the pin, then install the starter.

On all models, check starter operation after installing the starter. If the starter does not operate properly, loosen the starter housing retaining screws, move the starter, retighten the screws and recheck starter operation.

Replace Unbroken Starter Rope

Do not discard the rope before measuring its length and diameter. The same length and diameter of rope must be installed, which is determined by either measuring the old rope or consulting a parts manual. Most starters use a rope that is 54 in. (137 cm) long.

1. To replace an unbroken rope, the starter must be removed as previously outlined.
2. Pull out the rope to its full extended length so the rope end in the pulley is towards the housing rope outlet as shown in **Figure 5**.
3. Hold the pulley by clamping the pulley with lock pliers or a C-clamp (**Figure 5**).
4. Pull out the rope knot in the pulley, untie or cut off the knot, then pull the rope out of the pulley.
5. Pry the rope insert out of the rope handle and detach the rope from the insert and rope handle (**Figure 6**).
6. Measure the rope and obtain a replacement rope of the same length and diameter.
7. If the new rope is made of nylon, melt each rope end with a match to prevent rope from unraveling.
8. Thread the new rope into the rope handle, then attach the rope to the insert as shown in **Figure 6**. Pull the rope and insert into the rope handle.
9. Thread the rope through the housing rope outlet into the pulley and out the pulley rope hole (**Figure 7**).
10. Tie a knot in the inner end of the rope.
11. Hold the rope at the handle end, release the pliers or clamp on the rope pulley and allow the rope to wind slowly onto the pulley.
12. Check starter operation. If the rope handle does not rest snugly against the housing when released, shorten the rope in small increments until it does so.

Replace Broken Starter Rope

Do not discard the rope before measuring its length and diameter. The same length and diameter of rope must be installed, which is determined by either measuring the old rope or consulting a parts manual. Most starters use a rope that is 54 in. (137 cm) long.

1. To replace a broken rope, the starter must be removed as previously outlined.
2. Pull the rope out of the hole in the rope pulley. If the rope pulley must be removed for access to the rope, remove the rope pulley as outlined in the following section for rewind spring removal.
3. Pry the rope insert out of the rope handle and detach the rope from the insert and rope handle (**Figure 6**).
4. Measure the rope and obtain a replacement rope of the same length and diameter.
5. If the new rope is made of nylon, melt each rope end with a match to prevent rope from unraveling.
6. Thread the new rope into the rope handle, then attach the rope to the insert as shown in **Figure 6**. Pull the rope and insert into the rope handle.
7. Rotate the pulley counterclockwise until the rewind spring is tight.
8. Let the pulley unwind until the rope hole in the pulley is aligned with the rope outlet in the housing as shown in **Figure 8**.
9. Hold the pulley by clamping the pulley with lock pliers or a C-clamp (**Figure 8**).
10. Thread the rope through the housing rope outlet into the pulley and out the pulley rope hole (**Figure 7**).
11. Tie a knot in the end of the rope.
12. Hold the rope at the handle end, release the pliers or clamp on the rope pulley and allow the rope to wind slowly onto the pulley.
13. Check starter operation. If the rope handle does not rest snugly against the housing when released, shorten the rope in small increments until it does so.

Service Drive Mechanism or Replace Rewind Spring

Disassembly

To service the drive mechanism or replace the rewind spring, the starter must be removed from the engine as previously outlined.

1. If the rewind spring is under tension, detach the rope from the rope handle. Slowly release the rope and allow it to wind onto the pulley.
2. If the rewind spring is broken, detach the rope from the rope handle. Turn the pulley so the rope is wound onto the pulley.

3A. If equipped with a center retaining screw (**Figure 9**), unscrew the center screw. Note that the drive components will be loose when the center screw is removed.

3B. If equipped with a retaining pin (**Figure 10**), remove the pin by driving the pin out towards the bottom of the starter (**Figure 11**). Note that the drive components will be loose when the center pin is removed. The pin should be discarded and replaced with a new pin.

4. Remove the drive mechanism components.

5. Lift out the rope pulley. The rewind spring is housed on the rope pulley underneath the spring housing. Remove the rope from the pulley.

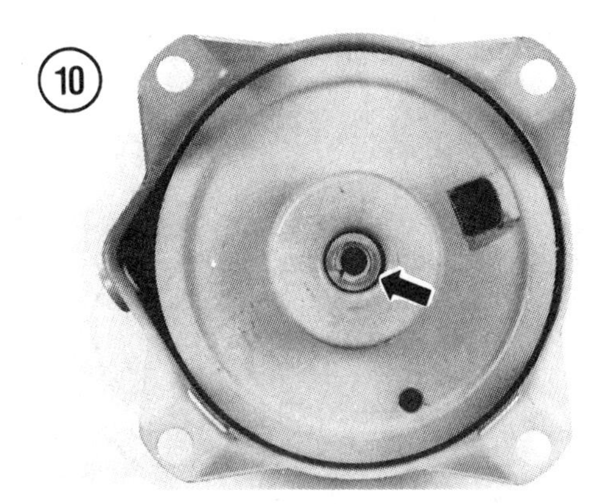

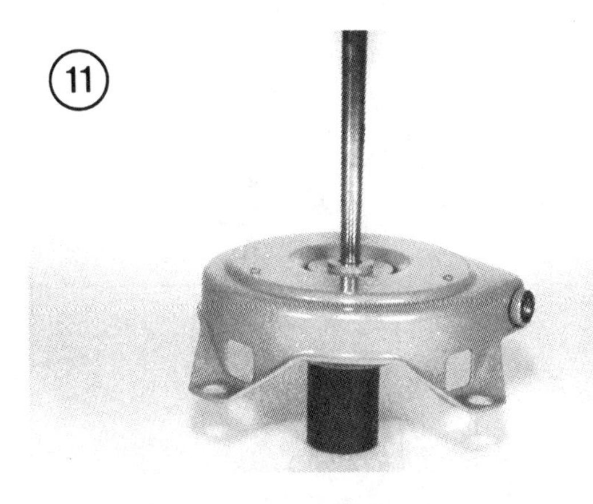

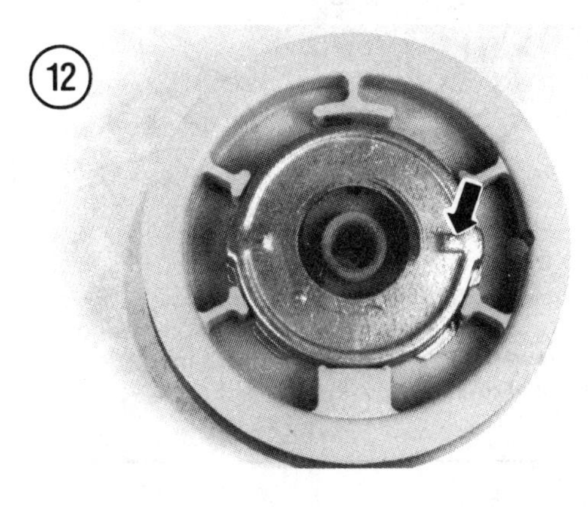

6. If rewind spring removal is necessary, separate the spring housing from the rope pulley. On early models, the spring housing must be rotated clockwise so the flanges on the housing disengage from the pulley lugs.

WARNING
The rewind spring is sharp and can uncoil uncontrolled. Safety eyewear and gloves should be worn when working on or around the rewind spring.

NOTE
The rewind spring is available only as a unit assembly with the spring housing. Do not attempt to separate the spring housing and rewind spring.

Inspection

1. Inspect the drive mechanism and replace any parts which are damaged or excessively worn.

2. Inspect the starter cup on the flywheel for wear or damage that may cause faulty starter engagement.

3. Inspect the pulley for cracks and other damage as well as excessive wear.

4. If a new retaining screw (**Figure 9**) is required on models so equipped, note that early models are equipped with a 10-32 screw while later models are equipped with a 12-28 screw. Only the 12-28 screw is available from Tecumseh. To install the larger screw, enlarge the screw hole using a 13/64 in. drill bit.

Assembly

1. If rewind spring was removed, apply a light coat of grease on the spring. Install the rewind spring and spring housing on the rope pulley (**Figure 12**). On some models, the spring housing must be rotated counterclockwise so the flanges on the housing are forced into the lugs on the pulley.

2. Apply a light coat of grease to the contact surfaces of the starter housing, rope pulley and rope pulley shaft, then install the rope pulley in the starter housing.

3. Install the pawl spring in the recess in the pulley so the long end points up as shown in **Figure 13**.

4. Install the pawl so the spring end forces the pawl toward the center of the pulley (**Figure 14**).

5A. If equipped with a retaining screw (**Figure 9**), proceed as follows:

a. Install the brake spring (**Figure 15**).
b. Install the retainer and retaining screw. Tighten the screw to 65-75 in.-lb. (7.34-8.47 N•m).

NOTE

*On Snow King engines, refer to **Figure 16** for installation of the drive mechanism. Note that the pawl (3) is secured by a retainer (4) and screw (5). The drawing in the inset shows the components assembled.*

5B. If equipped with a retaining pin (**Figure 10**), proceed as follows:

a. Install the plastic washer (**Figure 17**).
b. Install the retainer and brake spring (**Figure 18**). Install the spring so the large end of the spring contacts the retainer.
c. Install the washer and a new retaining pin (**Figure 19**). Drive the pin in until it bottoms against the shoulder in the starter housing or the end of the pin is 1/4 in. (6.4 mm) from the top of the starter housing.

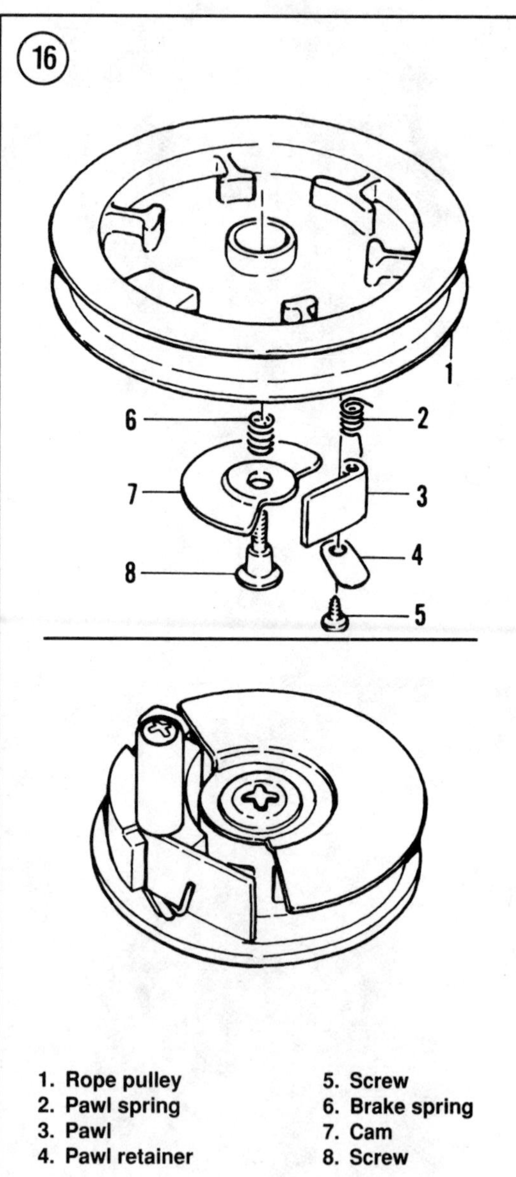

1. Rope pulley
2. Pawl spring
3. Pawl
4. Pawl retainer
5. Screw
6. Brake spring
7. Cam
8. Screw

NOTE
On engines with an alternator mounted under the rewind starter, refer to ***Figure 20*** *for installation of the drive mechanism. Note that the pawl (3) is secured by a retainer (4) and screw (5). The drawing in the inset shows the components assembled.*

6. Install the rope as outlined in the previous section covering replacement of a broken rope.

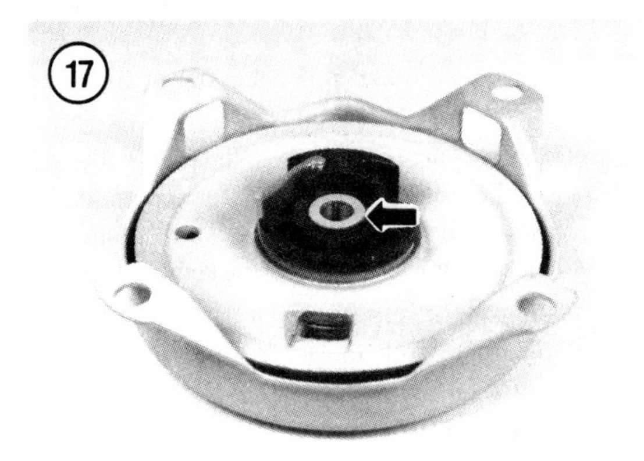

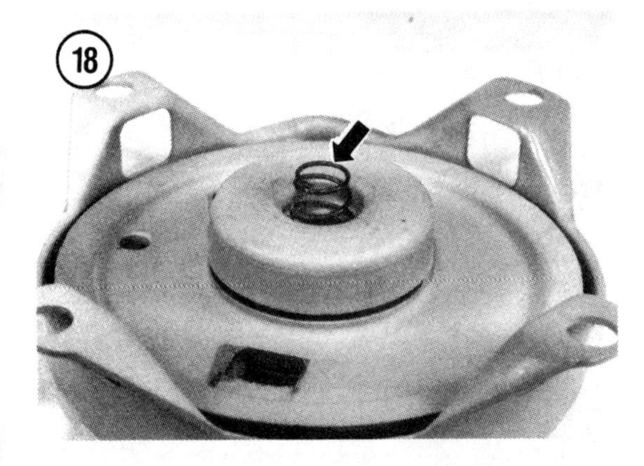

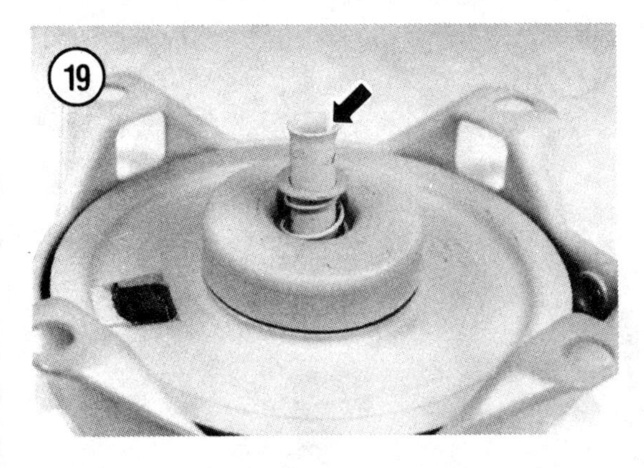

STARTER USED ON HM AND VM SERIES ENGINES

The rewind starter used on some HM and VM series engines is identified by the round shape of the starter housing (**Figure 21**). This is a pawl type starter. When the rope is pulled, two pawls extend outward to engage notches in the starter cup attached to the flywheel.

Major starter problems usually involve a severed rope, a broken rewind spring or poor engagement between the pawls and starter cup. The rope can

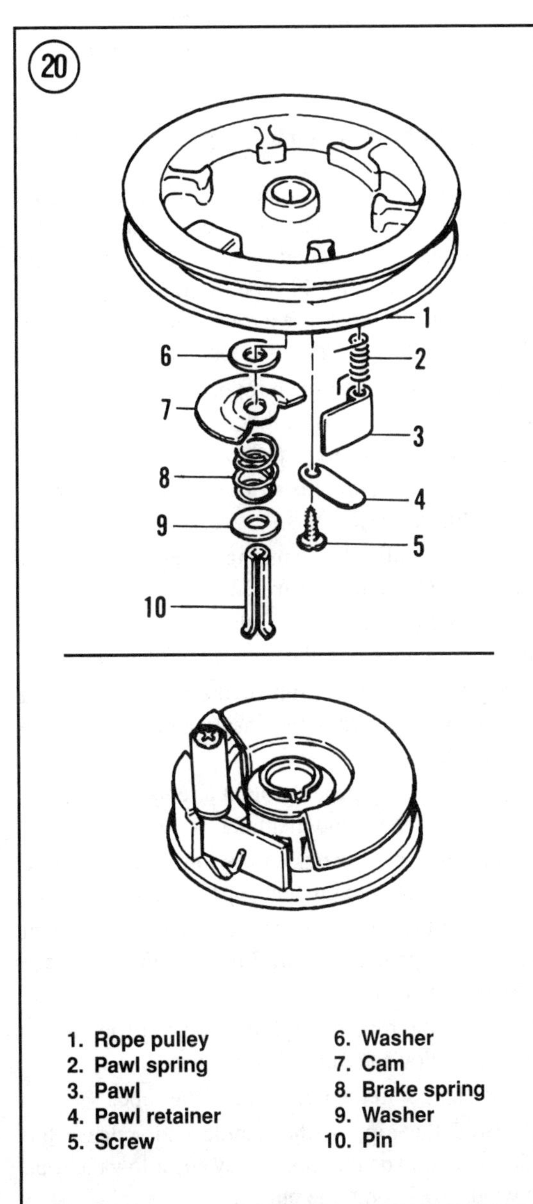

usually be replaced without completely disassembling the starter, while a broken rewind spring requires disassembling the starter. Poor pawl engagement may be due to worn parts or a loose retainer screw.

Starter Removal/Installation

Starter removal usually only requires removing the starter housing retaining screws.

Check starter operation after installing the starter. If the starter does not operate properly, loosen the starter housing retaining screws, move the starter, retighten the screws and recheck starter operation.

Replace Unbroken Starter Rope

Do not discard the rope before measuring its length and diameter. The same length and diameter of rope must be installed, which is determined by either measuring the old rope or consulting a parts manual. Most starters use a rope that is either 68 in. (172 cm) or 114 in. (290 cm) long.

1. To replace an unbroken rope, the starter must be removed as previously outlined.
2. Pull out the rope to its full extended length so the rope end in the pulley is towards the housing rope outlet (**Figure 22**).
3. Hold the pulley by clamping the pulley with lock pliers or a C-clamp (**Figure 22**).
4. Pull out the rope knot in the pulley, untie or cut off the knot, then pull the rope out of the pulley.
5. Pry the rope insert out of the rope handle and detach the rope from the insert and rope handle (**Figure 23**).
6. Measure the rope and obtain a replacement rope of the same length and diameter.
7. If the new rope is made of nylon, melt each rope end with a match to prevent rope from unraveling.
8. Thread the new rope into the rope handle, then attach the rope to the insert. Pull the rope and insert into the rope handle.
9. Thread the rope through the housing rope outlet into the pulley and out the pulley rope hole.
10. Tie a knot in the inner end of the rope.
11. Hold the rope at the handle end, release the pliers or clamp on the rope pulley and allow the rope to wind slowly onto the pulley.
12. Check starter operation. If the rope handle does not rest snugly against the housing when released, shorten the rope in small increments until it does so.

Replace Broken Starter Rope

Do not discard the rope before measuring its length and diameter. The same length and diameter of rope must be installed, which is determined by either measuring the old rope or consulting a parts manual. Most starters use a rope that is either 68 in. (172 cm) or 114 in. (290 cm) long.

1. To replace a broken rope, the starter must be removed as previously outlined.
2. Pull the rope out of the hole in the rope pulley. If the rope pulley must be removed for access to the rope, remove the rope pulley as outlined in the following section for rewind spring removal.
3. Pry the rope insert out of the rope handle and detach the rope from the insert and rope handle (**Figure 23**).
4. Measure the rope and obtain a replacement rope of the same length and diameter.

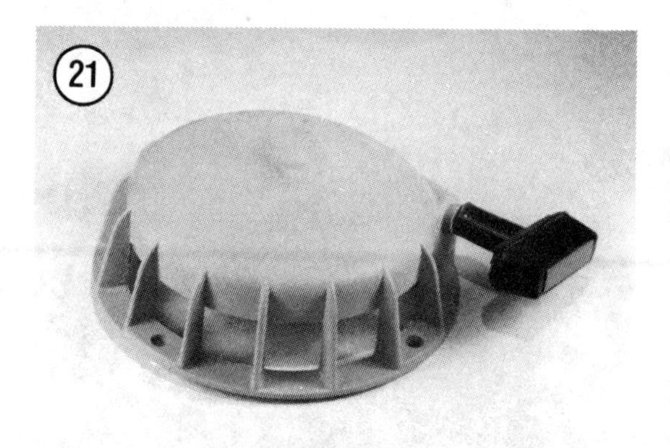

5. If the new rope is made of nylon, melt each rope end with a match to prevent rope from unraveling.
6. Thread the new rope into the rope handle, then attach the rope to the insert. Pull the rope and insert into the rope handle.
7. Rotate the pulley counterclockwise until the rewind spring is tight.
8. Let the pulley unwind until the rope hole in the pulley is aligned with the rope outlet in the housing.
9. Hold the pulley by clamping the pulley with lock pliers or a C-clamp.

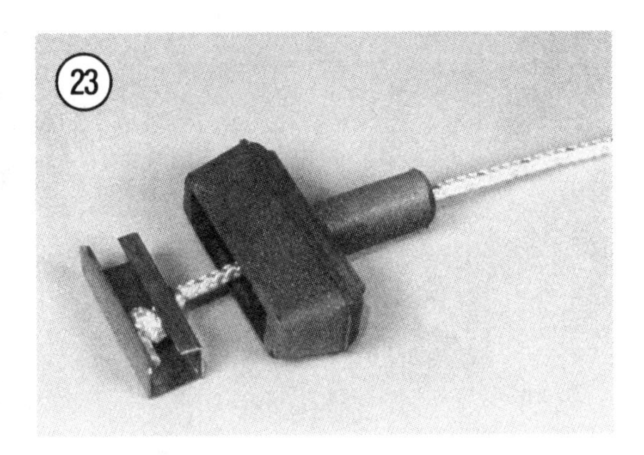

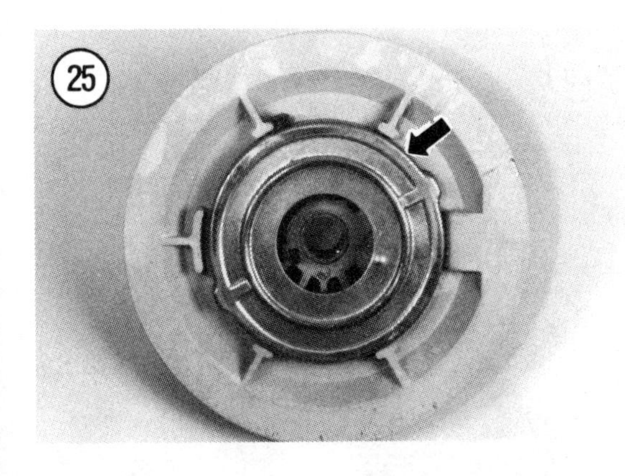

10. Thread the rope through the housing rope outlet into the pulley and out the pulley rope hole.
11. Tie a knot in the end of the rope.
12. Hold the rope at the handle end, release the pliers or clamp on the rope pulley and allow the rope to wind slowly onto the pulley.
13. Check starter operation. If the rope handle does not rest snugly against the housing when released, shorten the rope in small increments until it does so.

Service Drive Mechanism or Replace Rewind Spring

Disassembly

1. To service the drive mechanism or replace the rewind spring, the starter must be removed from the engine as previously outlined.
2. If the rewind spring is under tension, detach the rope from the rope handle. Slowly release the rope and allow it to wind onto the pulley.
3. If the rewind spring is broken, detach the rope from the rope handle. Turn the pulley so the rope is wound onto the pulley.
4. Unscrew the center screw (**Figure 24**).

NOTE
Some models are equipped with an E-ring in place of a center screw. All other components are the same.

5. Remove the drive mechanism components.
6. Lift out the rope pulley. The rewind spring is located in the spring housing (**Figure 25**). Remove the rope from the pulley.
7. If rewind spring removal is necessary, separate the spring housing from the rope pulley. Rotate the spring housing clockwise so the flanges on the housing disengage from the pulley lugs.

WARNING
The rewind spring is sharp and can uncoil uncontrolled. Safety eyewear and gloves should be worn when working on or around the rewind spring.

NOTE
The rewind spring is available only as a unit assembly with the spring housing. Do not attempt to separate the spring housing and rewind spring.

Inspection

1. Inspect the drive mechanism and replace any parts which are damaged or excessively worn.
2. Inspect the starter cup on the flywheel for wear or damage that may cause faulty starter engagement.
3. Inspect the pulley for cracks and other damage as well as excessive wear.

Assembly

1. If rewind spring was removed, apply a light coat of grease on the spring.
2. Install the rewind spring and spring housing on the rope pulley (**Figure 25**). On some models, the spring housing must be rotated counterclockwise so the flanges on the housing are forced into the lugs on the pulley.
3. Apply a light coat of grease to the contact surfaces of the starter housing, rope pulley and rope pulley shaft, then install the rope pulley in the starter housing. The inner spring end must engage the notch near the starter post (**Figure 26**).
4. Install the washer (**Figure 27**) and brake spring.
5. Install the cam (**Figure 28**). The ends of the cam should be down and between the pawls. Install the retaining screw or E-ring. If equipped with a screw, tighten it to 115-135 in.-lb. (13.00-15.25 N•m).
6. Install the pawl springs on the posts of the pulley so the long end of the spring is up as shown in **Figure 29**.
7. Install the pawls so the spring end forces the pawl towards the center of the pulley (**Figure 24**), then install the E-ring.
8. Install the rope as outlined in the previous section covering replacement of a broken rope.

STYLIZED REWIND STARTER

Later Tecumseh engines, including Vector VLV50, VLXL50, VLV55 and VLXL55 engines, may be equipped with a "stylized" starter. This type starter is identified by the slots in the top of the starter housing. Refer to **Figure 30** for an illustration of two versions of the stylized starter.

This is a pawl type starter. When the rope is pulled, two pawls extend outward to engage notches in the starter cup attached to the flywheel.

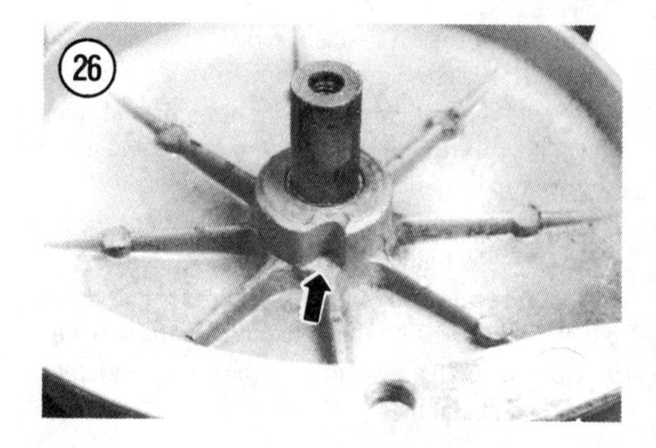

(30)

(31)

(32)

(33)

Major starter problems usually involve a severed rope, a broken rewind spring or poor engagement between the pawls and starter cup. The rope can usually be replaced without completely disassembling the starter, while a broken rewind spring requires disassembling the starter. Poor pawl engagement may be due to worn parts.

Starter Removal/Installation

Starter removal usually only requires removing the starter housing retaining screws. The retaining screws are located either at the ends of the starter housing legs, or on starters with a round starter housing, the retaining screws are located on the outside of the starter housing. On some models, a trim ring surrounds the starter. The trim ring is secured by screws (**Figure 31**) or it snaps into the blower housing (**Figure 32**) and is removed by prying up.

NOTE
*A screen fits inside the starter on some models (**Figure 33**). The screen must be in place when the engine operates to keep out debris.*

On all models, check starter operation after installing the starter. If a leg-type starter does not operate properly, loosen the starter housing retaining screws, move the starter, retighten the screws and recheck starter operation.

Replace Unbroken Starter Rope

Do not discard the rope before measuring its length and diameter. The same length and diameter of rope must be installed, which is determined by either measuring the old rope or consulting a parts manual. Typical rope lengths are 69 in. (175 cm), 98 in. (249 cm) and 114 in. (290 cm).

1. To replace an unbroken rope, remove the rewind starter as previously outlined.
2. Pull out the rope to its full extended length so the rope end in the pulley is towards the housing rope outlet, then hold the pulley by installing a restraining device (such as a tie-wrap) around the starter housing and a strut (**Figure 34**) on the rope pulley so the spring cannot rewind.
3. Pull out the rope knot in the pulley, untie or cut off the knot, then pull the rope out of the pulley.

8

4. Detach the rope from the rope handle. On some models, the rope is retained in the rope handle by a staple (**Figure 35**). Pry out the staple to release the rope.
5. Measure the rope and obtain a replacement rope of the same length and diameter. If the new rope is made of nylon, melt each rope end with a match to prevent rope from unraveling.
6. Tie a knot in one end of the rope.
7. Thread the rope through the pulley rope hole and out the rope outlet in the housing.

NOTE
It may be helpful to hook a thin wire into the rope end so the wire can be used to guide and pull the rope through the holes.

8. Attach the rope handle to the rope. If the rope was originally retained by a staple (**Figure 35**), use a knot tied in the rope end to keep the rope in the handle.
9. Hold the pulley, release the restraining device on the rope pulley and slowly allow the rope to wind onto the pulley.
10. Check starter operation.

Replace a Broken Starter Rope

Do not discard the rope before measuring its total length. The same length of rope must be installed, which is determined by either measuring the old rope or consulting a parts manual. Typical rope lengths are 69 in. (175 cm), 98 in. (249 cm) and 114 in. (290 cm).

1. To replace a broken rope, remove the starter as previously outlined.
2. Pull the rope out of the hole in the rope pulley. If the rope pulley must be removed for access to the rope, remove the rope pulley as outlined in the following procedure for rewind spring removal.
3. Detach the rope from the rope handle. On some models, the rope is retained in the rope handle by a staple (**Figure 35**). Pry out the staple to release the rope.
4. Measure the rope and obtain a replacement rope of the same length and diameter. If the new rope is made of nylon, melt each rope end with a match to prevent rope from unraveling.
5. Rotate the rope pulley counterclockwise as far as possible, then allow the pulley to turn clockwise so the rope hole in the pulley and the rope outlet in the housing are aligned. Hold the pulley by installing a restraining device (such as a tie-wrap) around the starter housing and a strut (**Figure 36**) on the rope pulley so the spring cannot rewind.
6. Tie a knot in one end of the rope.
7. Thread the rope through the pulley rope hole and out the rope outlet in the housing.

NOTE
It may be helpful to hook a thin wire into the rope end so the wire can be used to

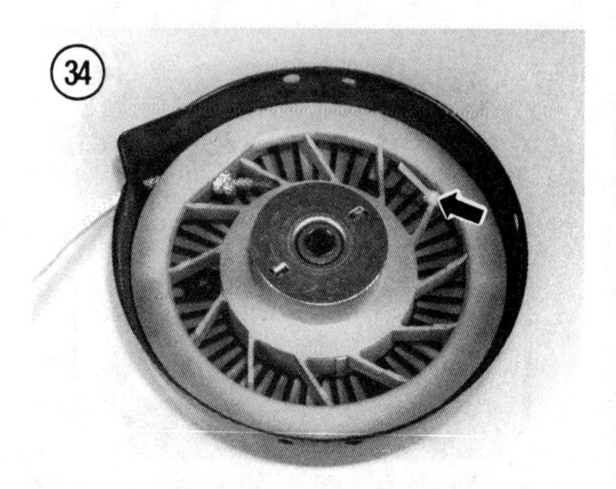
34

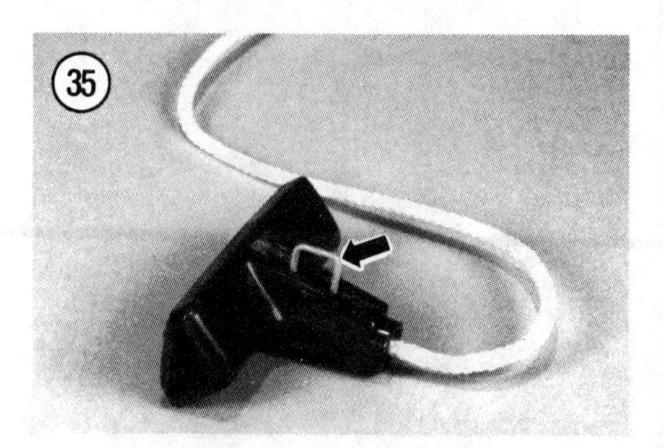
35

36

guide and pull the rope through the holes.

8. Attach the rope handle to the rope. If the rope was originally retained by a staple (**Figure 35**), use a knot tied in the rope end to keep the rope in the handle.
9. Hold the pulley, release the restraining device on the rope pulley and slowly allow the rope to wind onto the pulley.
10. Check starter operation.

37

38

39

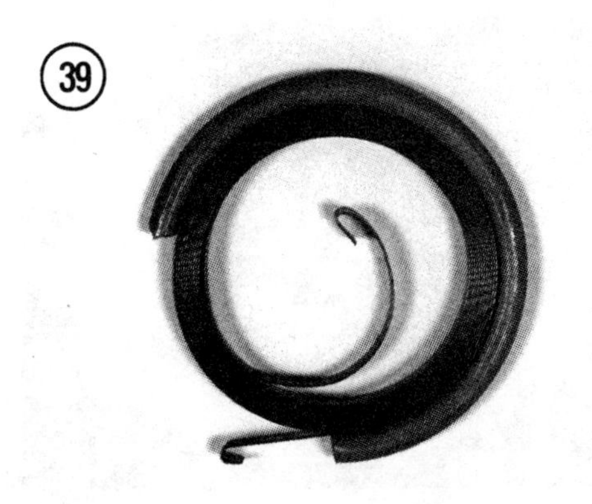

Service Drive Mechanism or Replace Rewind Spring

Disassembly

1. To service drive mechanism or replace the rewind spring, the starter must be removed from the engine as previously outlined.
2. Detach the rope handle and allow the rope to wind into the starter.
3. Place a round support such as a piece of pipe or deep-well socket under the retainer and drive out the center pin using a 5/16 in. (8 mm) rod or punch as shown in **Figure 37**. Note that the drive components will be loose when the center pin is removed.
4. Remove the drive mechanism components.
5. Lift out the rope pulley. Remove the rope from the pulley.
6. The rewind spring is contained under a cover (**Figure 38**) in the rope pulley. On some models the rewind spring is available separately, while on other models the rewind spring is available only as a unit with the pulley. If the rewind spring is not broken, do not remove the cover.

WARNING
The rewind spring is sharp and can uncoil uncontrolled. Safety eyewear and gloves should be worn when working on or around the rewind spring.

Inspection

1. Inspect the drive mechanism and replace any parts which are damaged or excessively worn.
2. Inspect the starter cup on the flywheel for wear or damage that may cause faulty starter engagement.
3. Inspect the pulley for cracks and other damage as well as excessive wear.

Assembly

1. On models with a renewable rewind spring, the new spring is contained in a holder (**Figure 39**). The new spring is installed by placing the holder with the spring on the rope pulley and pushing the spring into the spring cavity in the pulley. Be sure the outer spring end points in the direction shown in **Figure 39** and fits in the groove in the pulley as shown in **Figure 40**.

2. Place the rope pulley in the starter housing so the inner spring end is aligned with the end of the flange adjacent to the starter post (**Figure 41**).
3. Rotate the pulley counterclockwise so the inner spring end engages the flange.
4. Install the pawl return springs in the wells in the pulley hub. Note that the straight spring end fits in a slot while the angled end points up as shown in **Figure 42**.
5. Install the pawls as shown in **Figure 43** so the angled spring end forces the pawl toward the center of the pulley.
6. Place the gray plastic washer on the pulley hub (**Figure 44**).
7. Place the retainer on the pulley hub so the tabs (**Figure 45**) are adjacent to the ends of the pawls.
8. Place the conical spring and steel washer on the retainer (**Figure 46**). The large end of the spring should contact the retainer.
9. Insert the center pin in the starter (**Figure 47**), then drive the pin into the starter until the pin is 1/8 in. (3.2 mm) below the top of the starter housing.

CAUTION
Driving the pin in too far will damage the retainer and cause malfunctioning.

10. Install the rope as outlined in the previous section covering replacement of a broken rope.

VERTICAL-PULL REWIND STARTERS

Two types of vertical-pull starters have been used. The types are identified by the manner of gear engagement, either horizontal or vertical.

Horizontal Engagement Gear Starter

When the rope handle is pulled, the starter gear moves horizontally to engage the gear teeth on the flywheel. An exploded view of the starter is shown in **Figure 48**.

Starter removal/installation

Refer to **Figure 48** for the following procedure.

The starter can be removed after unscrewing the starter bracket (**Figure 48**) retaining screws.

After installing the starter assembly, the clearance between the teeth on gear and the teeth on the

40

41

42

43

flywheel should be checked when the starter is operated. When the teeth are fully engaged, there should be at least 1/16 in. (1.6 mm) clearance from the top of a gear tooth to the base of the opposite gear teeth. Remove the spark plug wire, operate the starter several times and check gear engagement. Insufficient gear tooth clearance could cause the starter gear to hang up on the flywheel gear when the engine starts, which could damage the starter.

Disassembly

Refer to **Figure 48** for the following procedure.

Do not discard the rope before measuring its length and diameter. The same length and diameter of rope must be installed, which is determined by either measuring the old rope or consulting a parts manual. Most starters use a rope that is 61 in. (155 cm) long.

1. Detach the rope from insert (**Figure 48**) and handle, then allow the rope to wind onto the pulley to relieve spring tension.
2. Unscrew the screws securing spring cover (**Figure 48**) and carefully remove the cover without disturbing the rewind spring.
3. Remove the rewind spring.

WARNING
The rewind spring is sharp and can uncoil uncontrolled. Safety eyewear and gloves should be worn when working on or around the rewind spring.

4. Remove hub screw (**Figure 48**) and spring anchor, then withdraw the pulley and gear assembly.
5. Remove snap ring (**Figure 48**), washer, brake spring and gear from pulley.

Assembly

Refer to **Figure 48** for the following procedure.

When assembling the starter, keep the following points in mind: Do not lubricate the brake spring. Do not lubricate the helix in the gear or on the pulley shaft. Lubricate only the edges of the rewind spring and the shaft on the mounting bracket.

1. If removed, install rope guide (**Figure 48**) on the starter bracket so the dimple on the guide fits in the depression on the bracket.
2. Install the rope on the pulley. Wrap the rope around the pulley in a counterclockwise direction

8

when viewing the pulley from the rewind spring side of the pulley.

3. Place the gear on the pulley. Do not lubricate the helix in the gear or on the pulley shaft.

4. Install brake spring (**Figure 48**) in the groove of gear. The bent end of the brake spring should point away from the gear. The brake spring should fit snugly in the groove. Do not lubricate the brake spring.

5. Install washer (**Figure 48**) and snap ring on the pulley shaft.

6. Lightly lubricate the shaft on the starter bracket then install the pulley and gear assembly on the bracket shaft. The closed end of brake spring must fit around tab (A, **Figure 48**) on the bracket.

7. Install spring anchor (**Figure 48**) and screw. Tighten the screw to 44-55 in.-lb. (5.08-6.21 N•m).

8. Install the rewind spring. The spring coils should wind in a clockwise direction from the outer end. A new spring is contained in a holder that allows spring installation by pushing the spring from the holder into the spring cavity on the pulley.

9. Pass the outer end of the rope through rope bracket (**Figure 48**) and install the rope handle and insert.

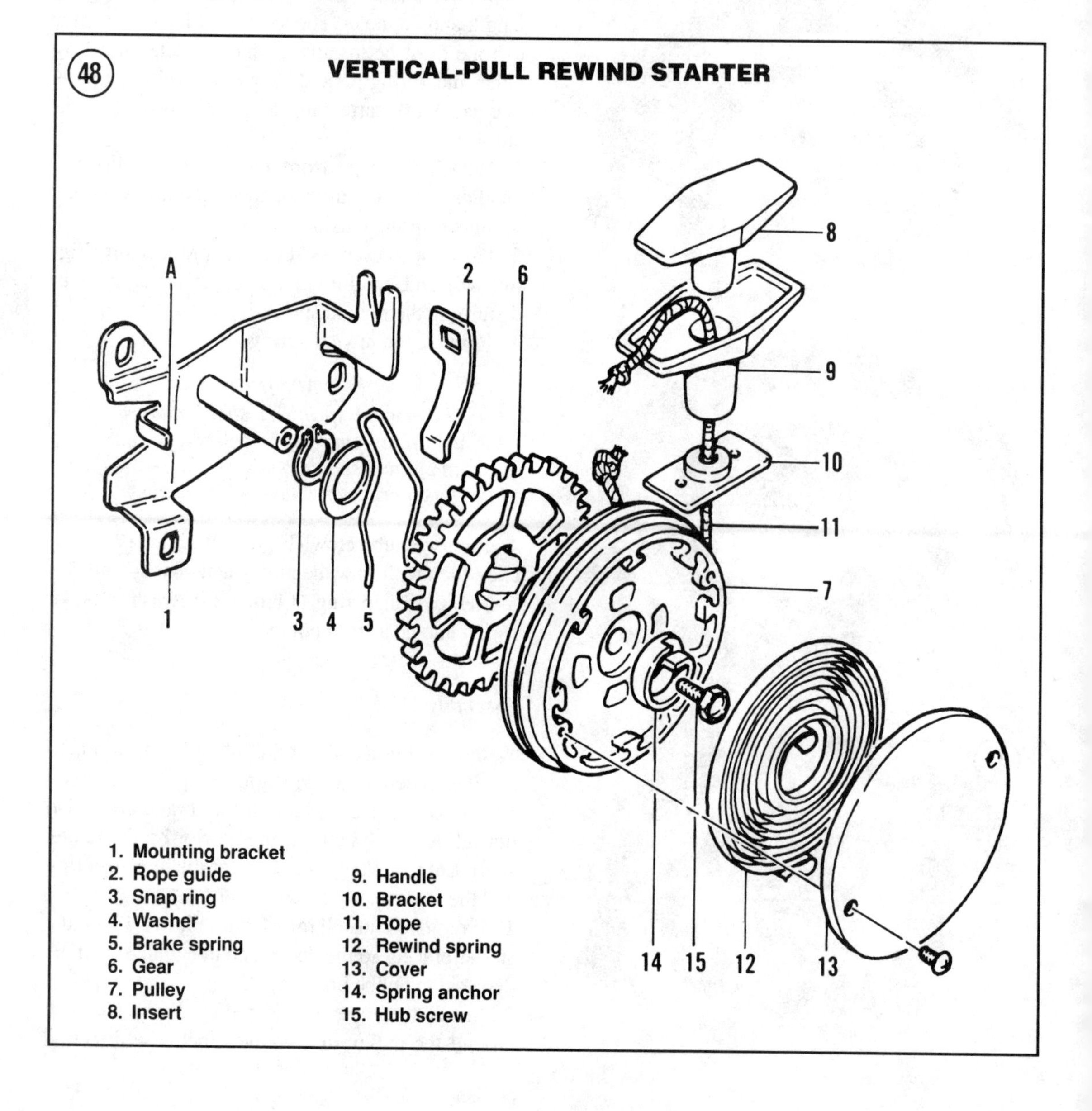

10. Pull a portion of the rope out past the rope guide and wrap any excess rope around the pulley, then turn the pulley 2 to 2 1/2 turns against spring tension to preload the rewind spring.

11. Pull the rope handle and check starter operation.

Vertical Engagement Gear Starter

When the rope handle is pulled, the starter gear moves vertically to engage the gear teeth on the flywheel. An exploded view of the starter is shown in **Figure 49**.

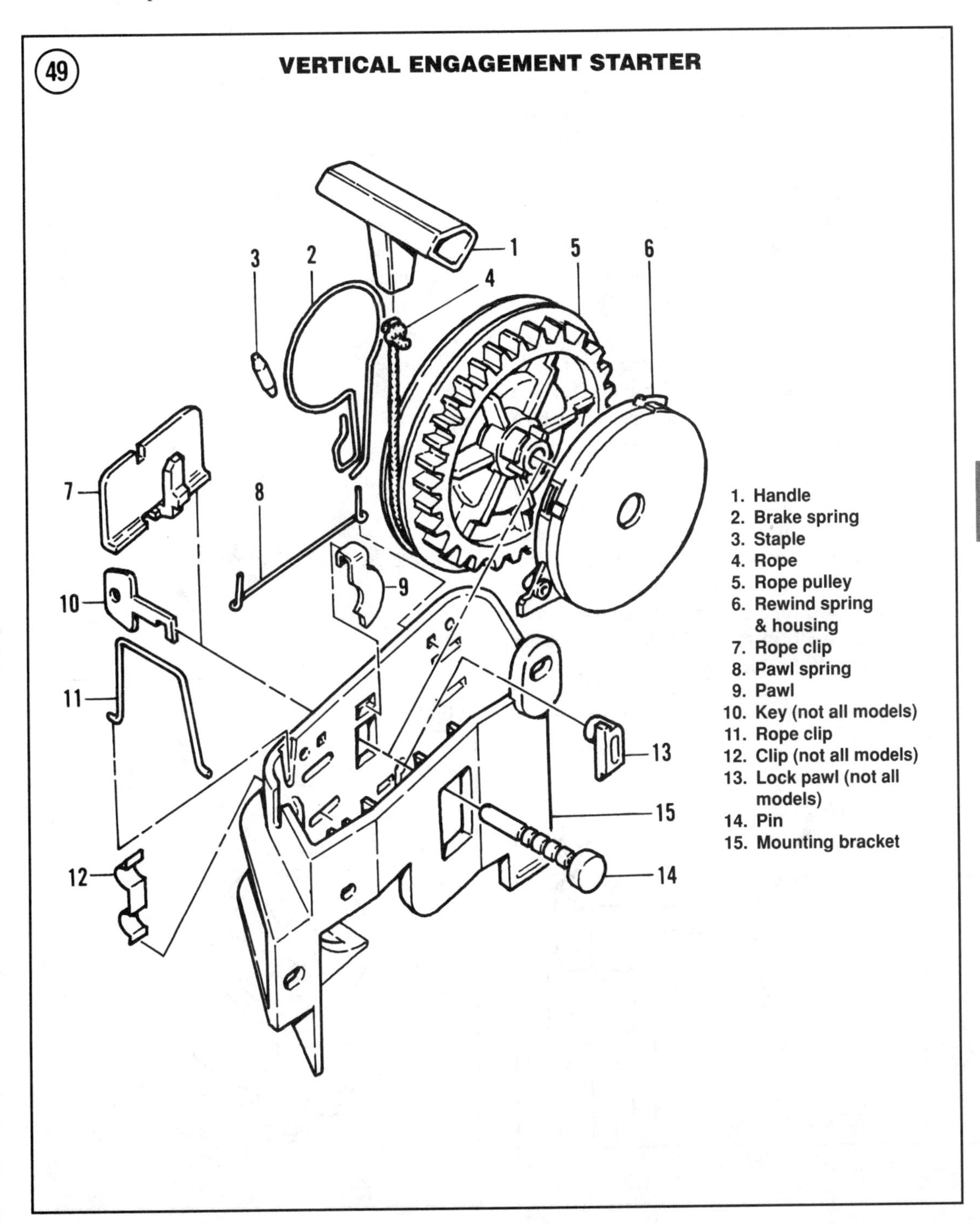

Starter removal/installation

Starter removal and installation is accomplished by removing or installing the mounting bracket (15, **Figure 49**) retaining screws.

Replace starter rope

If the starter bracket has a "V" notch (**Figure 50**), then the inner end of the rope is accessible and the rope can be replaced without disassembling the starter.

Do not discard the rope before measuring its length and diameter. The same length and diameter of rope must be installed, which is determined by either measuring the old rope or consulting a parts manual. Typical rope lengths are 65 in. (165 cm) and 98 in. (249 cm).

1. If the rope is unbroken, remove the rope handle and let the rope wind onto the rope pulley. On some models, the rope is retained in the rope handle by a staple (**Figure 51**). Pry out the staple to release the rope.
2. Note in **Figure 52** that the inner end of the rope was originally retained by a staple, while the inner end on replacement ropes is inserted through a hole in the pulley and knotted. Rotate the pulley so that either the stapled rope end or knotted rope end is visible (**Figure 50**) in the "V" notch. Pry out the staple or untie the knot and pull out the rope.

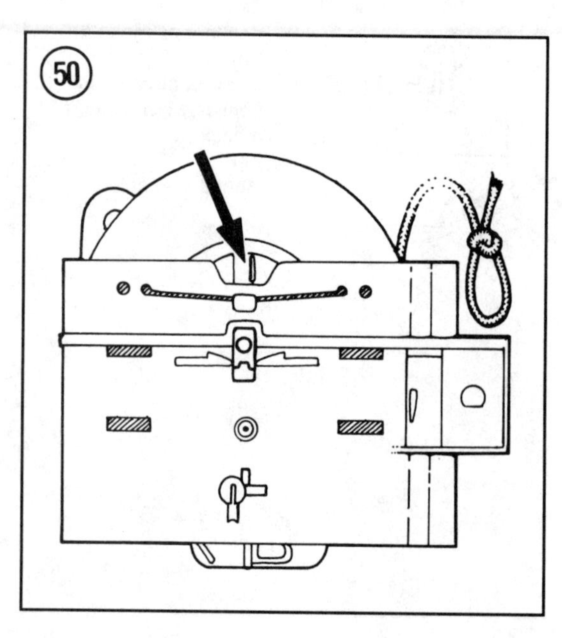

3. Rotate the pulley counterclockwise until the rewind spring is tight, then allow the pulley to turn clockwise until the rope hole in the pulley is visible in the "V" notch.
4. Route the new rope through the pulley hole and tie a knot in the rope end (**Figure 52**). Pull the knot into pulley cavity so rope end does not protrude.
5. Allow the pulley to wind the rope onto the pulley.
6. Attach the rope handle to the rope end.

Disassembly

Do not discard the rope before measuring its length and diameter. The same length and diameter of rope must be installed, which is determined by either measuring the old rope or consulting a parts

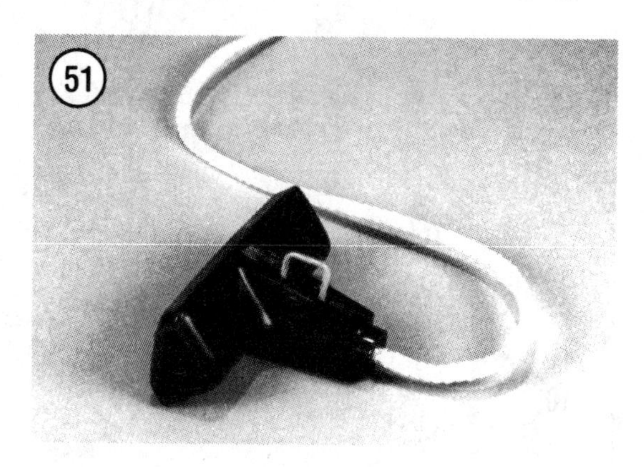

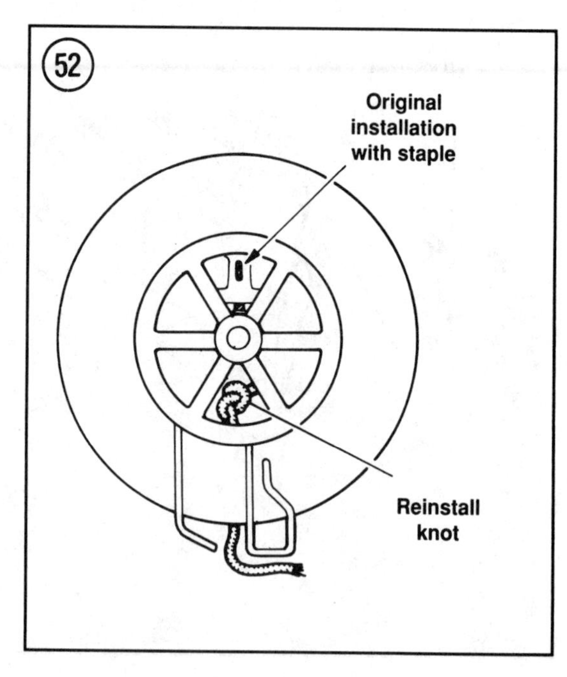

manual. Typical rope lengths are 65 in. (165 cm) and 98 in. (249 cm).

1. If the rope is unbroken, remove the rope handle and let the rope wind onto the rope pulley. On some models, the rope is retained in the rope handle by a staple (**Figure 51**). Pry out the staple to release the rope.

2. Drive or press out the pin (14, **Figure 49**) by placing the starter over a deep-well socket and forcing the pin into the socket.

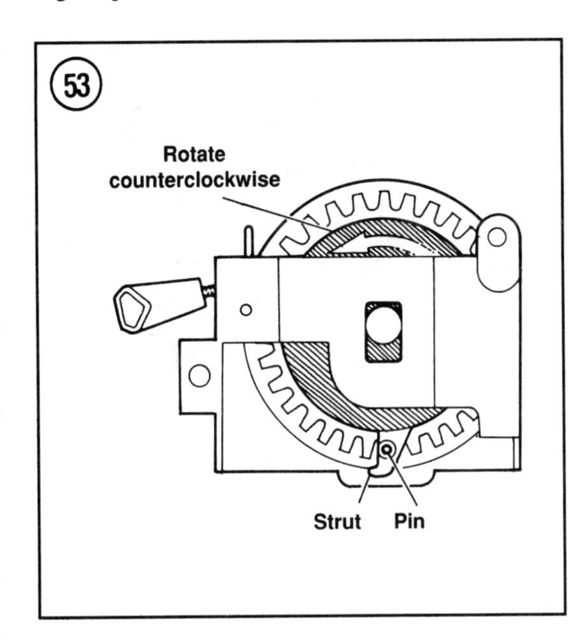

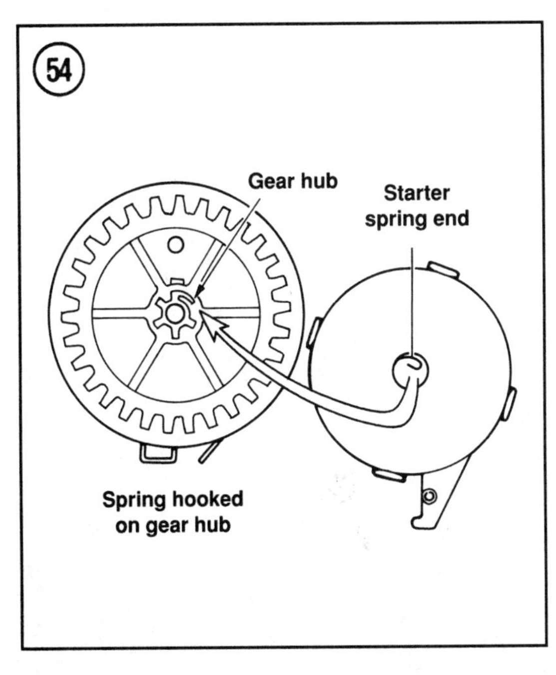

3. Turn the spring housing so the strut aligns with the legs on brake spring (**Figure 49**). Prevent housing rotation by inserting a pin or rod through the strut hole and into the gear teeth as shown in **Figure 53**.

4. Remove the pulley assembly from the starter bracket.

CAUTION

Do not allow the spring housing to separate from the rope pulley until spring tension has been relieved.

5. Hold the spring housing (**Figure 49**) against the rope pulley so the spring housing cannot rotate.

6. Withdraw the pin or rod in the spring housing strut (**Figure 53**) and allow the spring housing to rotate thereby relieving rewind spring tension.

7. If necessary, separate the spring housing from the rope pulley.

8. Do not attempt to remove the rewind spring from the spring housing. The spring and housing are available only as a unit assembly.

Assembly

1. If rope replacement is necessary, note in **Figure 52** that the inner end of the rope was originally retained by a staple, while the inner end on replacement ropes is inserted through a hole in the pulley and knotted. Pry out the staple, if so equipped, to release the rope. Route the new rope through the pulley hole and tie a knot in the rope end. Pull the knot into pulley cavity so rope end does not protrude.

NOTE

Do not lubricate any starter components.

2. When viewed from the gear side of the rope pulley, wind the rope onto the pulley in a clockwise direction.

3. Install brake spring on the rope pulley while being careful not to distort the spring legs.

4. If removed, place the spring housing on the rope pulley while being sure the inner spring end engages the anchor on the pulley. See **Figure 54**.

5. Rotate the spring housing four turns counterclockwise (**Figure 55**) then align the legs on brake spring with the hole in the strut. Prevent housing

rotation by inserting a pin or rod through the strut hole and into the gear teeth.

6. If so equipped, install clip, key and pawl. See **Figure 49**.

7. Install the rope pulley assembly in the starter bracket while inserting the brake spring legs into slots in the starter bracket (**Figure 56**).

8. Route the outer rope end past the rope guide and install the rope handle.

9. Withdraw the pin or rod in the strut hole. The strut will rotate until it contacts the starter bracket.

10. Press or drive in a new pin (14, **Figure 49**).

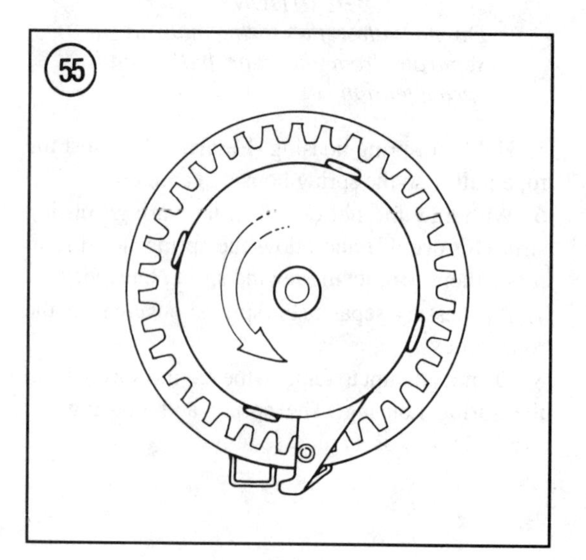

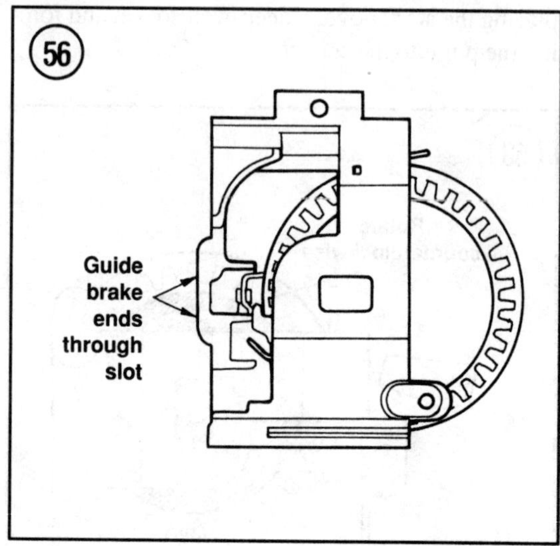

CHAPTER NINE

ELECTRICAL SYSTEM

The electrical system on Tecumseh engines uses an alternator adjacent to the flywheel to provide electrical energy. Current from the alternator may be used to power lights or directed to a battery and other electrical components.

ELECTRICAL PRECAUTIONS

Observe the following when operating the engine or servicing the electrical system

1. Be sure the battery cables are properly connected. The battery's negative terminal must be grounded to the engine or equipment. Improper connection of the battery cables can damage electrical components.
2. Do not disconnect battery cables while the engine is running.
3. Do not connect alternator leads together while the engine is running.
4. Disconnect battery cables when charging the battery to prevent damage to electrical components due to accidental connection to the battery charger.
5. The regulator-rectifier should be removed if arc welding is performed on the equipment.
6. Automotive type ignition switches should not be used as battery voltage may be routed to the ignition circuit, which may damage ignition components.

TESTING

Before testing any components, isolate the problem as much as possible. For instance, if the electric starter is inoperable, check the battery's condition, and if satisfactory, check for voltage at the starter to be sure the switch is operating properly. The electrical systems on Tecumseh engines are relatively simple. Common sense troubleshooting will reduce testing time and prevent the purchase of unnecessary parts.

Most of the following test procedures can be accomplished using a volt-ohmmeter and an ammeter. In some cases, a simple continuity tester or voltage tester will suffice, on the other hand, it may be necessary to have the component tested in a shop if expensive test equipment is not available.

WARNING
Some of the test procedures involve working with high voltages. All safety precautions related to working with or around harmful electrical circuits and devices must be followed.

BATTERY

A 12-volt, wet cell battery is used on most electrical systems requiring a battery. Some models may be equipped with a nickel-cadmium battery pack.

On models equipped with a wet cell battery, the battery's negative terminal must be grounded to the engine or equipment. The size of the battery depends on the application and should meet the specifications of the equipment manufacturer. Service and testing procedures applicable for wet cell batteries as found in automobiles and other vehicles also apply to batteries used with Tecumseh engines.

Service on nickel-cadmium electrical system components is limited to replacement of either the battery pack or 110-volt charger. The system can be tested by connecting two 4001 headlights to the outlet of a fully charged battery pack (14-16 hours of charging is required). If the lights do not burn for at least six minutes, then either the battery pack or charger is faulty. The charger must be tested using special equipment according to the manufacturer's procedure.

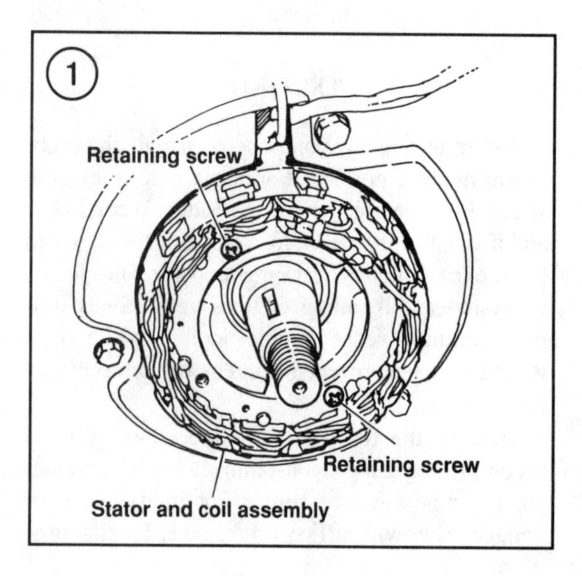

CHARGING/LIGHTING SYSTEM

The engine may be equipped with an alternator to provide battery charging direct current, or provide electricity for accessories, or both. The alternator coils may be located under the flywheel (**Figure 1**), or on some models with an external ignition module, the alternator coils are attached to the legs of the ignition coil (**Figure 2** and **Figure 3**). Rectification is accomplished either with a rectifier panel, regulator-rectifier unit, external or internal, or by an inline diode contained in the harness.

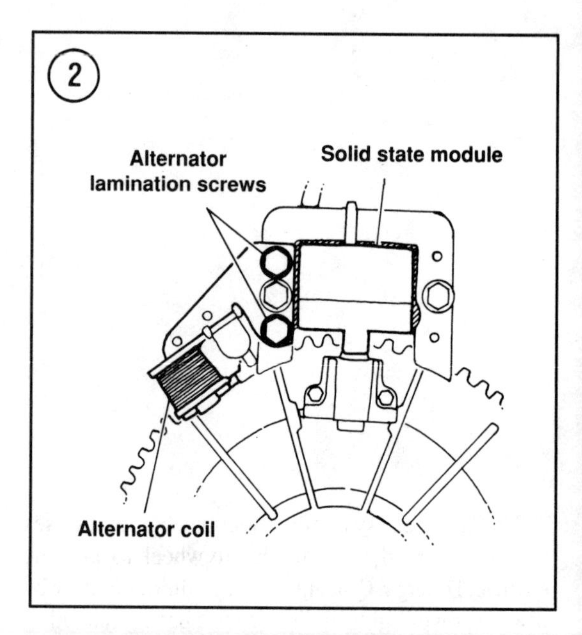

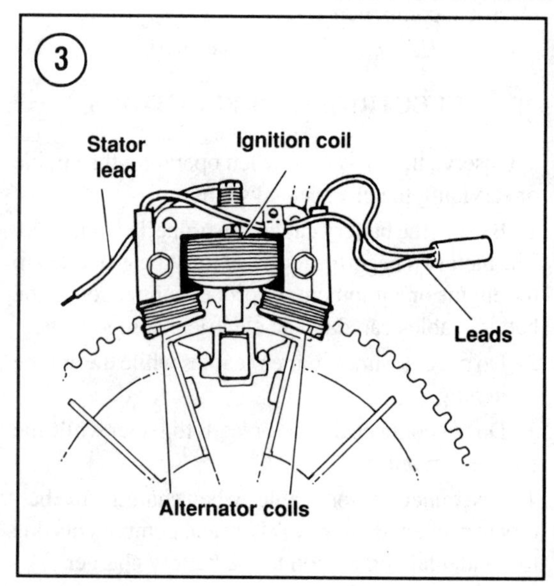

A lighting system consists of just the alternator. The alternating current produced by the alternator is used to power lights on the equipment.

Some systems are designed to use alternator output for both a charging system and a lighting system. The charging system provides direct current for battery charging as well as powering accessories. The lighting system powers the lights.

Refer to the following sections.

External Ignition Module Alternator

Engines with an external ignition may have an alternator coil attached to the ignition module (**Figure 4**). The alternator produces approximately 350 milliamperes for battery charging. To check alternator output, connect a DC voltmeter to the battery (battery must be in normal circuit) as shown in **Figure 5**. Run the engine. Voltage should be higher when the engine is running; if not, the alternator is defective.

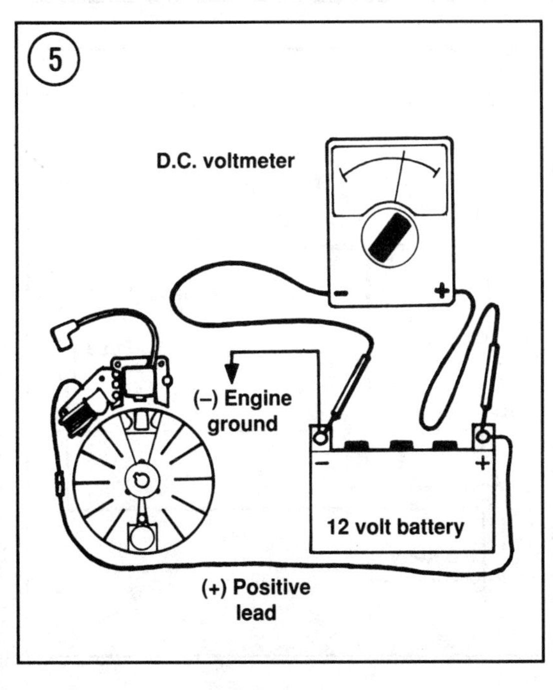

1 Amp Add-On Lighting System

The alternator is attached to the rewind starter and is used to power an AC lighting system. The alternator is driven by a shaft that attaches to the flywheel nut and extends through the rewind starter. Output is determined by engine speed. An exploded view of the system is shown in **Figure 6**.

To check alternator output, disconnect the alternator lead and connect a 4414, 18-watt bulb to the

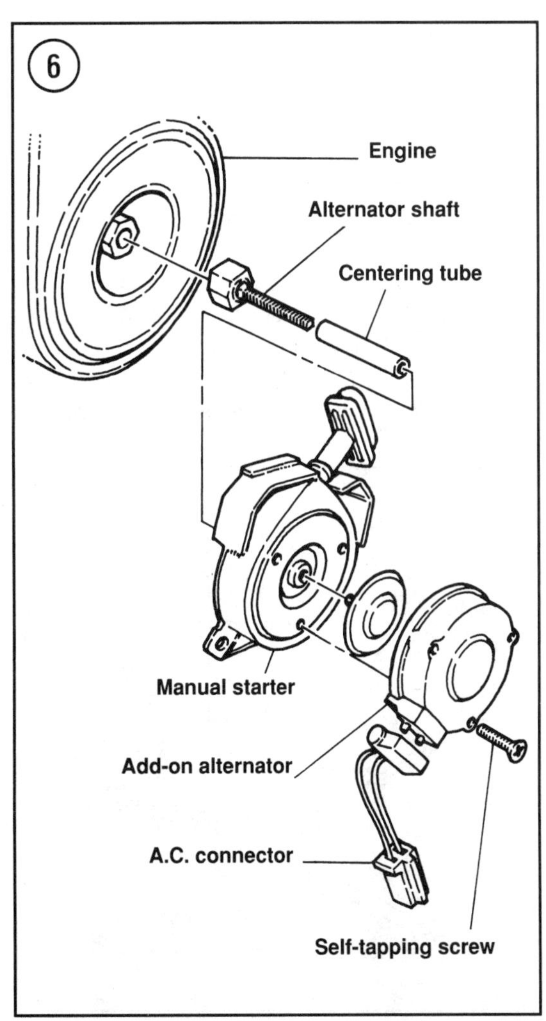

connector terminals as shown in **Figure 7**. Connect the test leads of an AC voltmeter to the bulb leads or connector terminals as shown in **Figure 7**. With the engine running at 3600 rpm, the voltmeter should indicate at least 12 volts. If the voltage reading is insufficient, then the alternator is faulty.

3 Amp Lighting System

This AC alternator is used to power 12-volt lights. Output is determined by engine speed. A typical wiring diagram is shown in **Figure 8**. Recommended bulbs are 4416 or 4420 for the headlights and 1157 for the tail light/stop light.

To check output, disconnect the wiring connector and connect the red lead of a voltmeter to the connector terminal for the stop light circuit (red wire). Run the engine at 3600 rpm. Voltmeter should indicate at least 11.5 volts, otherwise, the alternator stator must be replaced.

Inline Diode System

The inline diode system has a diode connected into the alternator wire leading from the engine. Two systems using an inline diode have been used. The system diagrammed in **Figure 9** provides direct current only, while the system in **Figure 10** provides both alternating and direct current.

The system in **Figure 9** produces approximately 3 amps direct current. The diode rectifies alternator alternating current into direct current and a 6 amp fuse provides overload protection. To check the system, disconnect the harness connector and, using a DC voltmeter, connect the tester positive lead to the red wire connector terminal and ground the negative tester lead to the engine. At 3600 rpm engine speed,

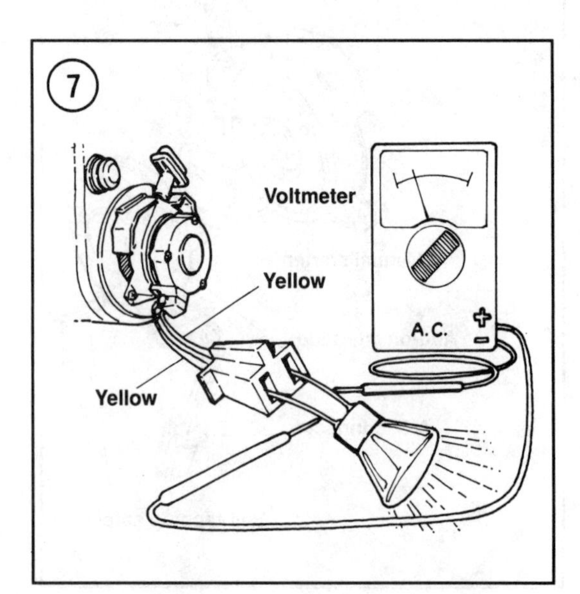

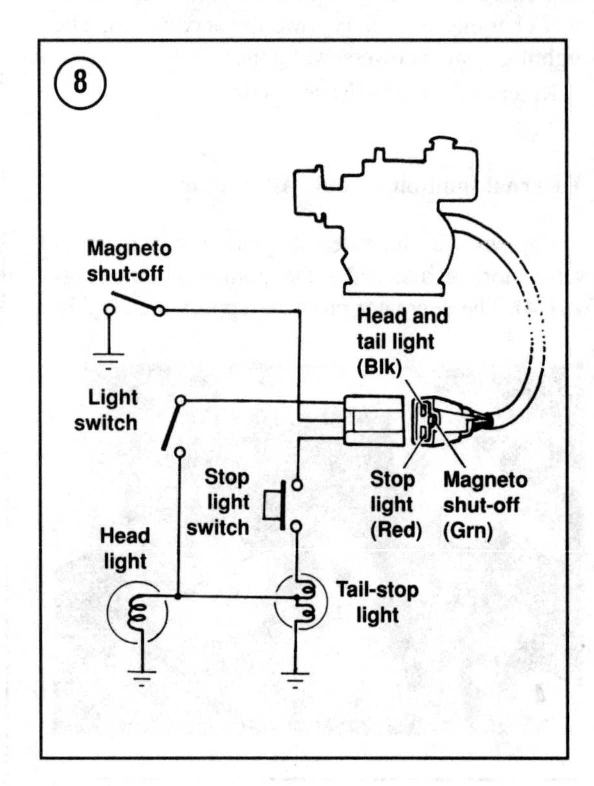

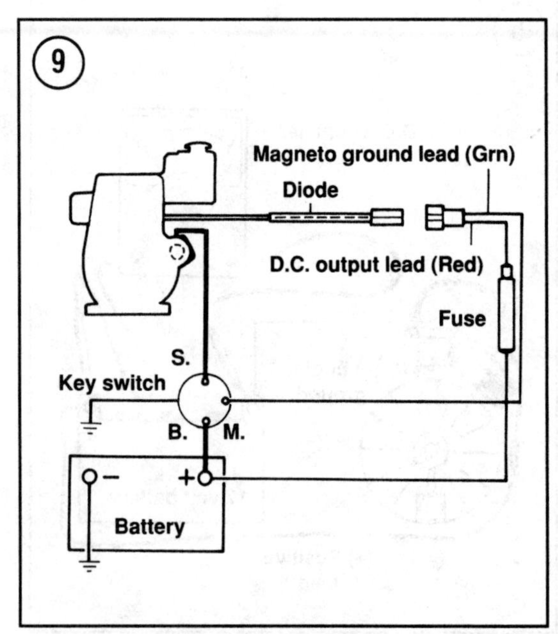

the voltmeter reading should be at least 11.5 volts. If engine speed is less, the voltmeter reading will be less. If the voltage reading is unsatisfactory, check the alternator coils by taking an AC voltage reading. Connect one tester lead between the diode and engine and ground the other tester lead to the engine. At an engine speed of 3600 rpm, the voltage reading should be 26 volts, otherwise the alternator is defective. If the engine cannot attain 3600 rpm, the voltage reading will be less.

The inline diode type system in **Figure 10** provides 3 amps direct current and 5 amps alternating current. This system has a two-wire pigtail consisting of a red wire and a black wire; the diode is inline with the red wire and covered by sheathing. To test the system, check the voltage at the pigtail connector. At an engine speed of 3600 rpm (less engine speed will produce less voltage), voltage at the red wire terminal should be 13 volts DC, while voltage at the black wire terminal should be 13 volts AC. If the voltage reading is unsatisfactory, check the alternator coils by taking an AC voltage reading. Pull back the wire sheathing and connect one tester lead to the red wire between the diode and engine and ground the other tester lead to the engine. At an engine speed of 3600 rpm, the voltage reading should be at least 29 volts, otherwise the alternator is defective. If the engine cannot attain 3600 rpm, the voltage reading will be less. If at least 29 volts is obtained, then the diode is defective.

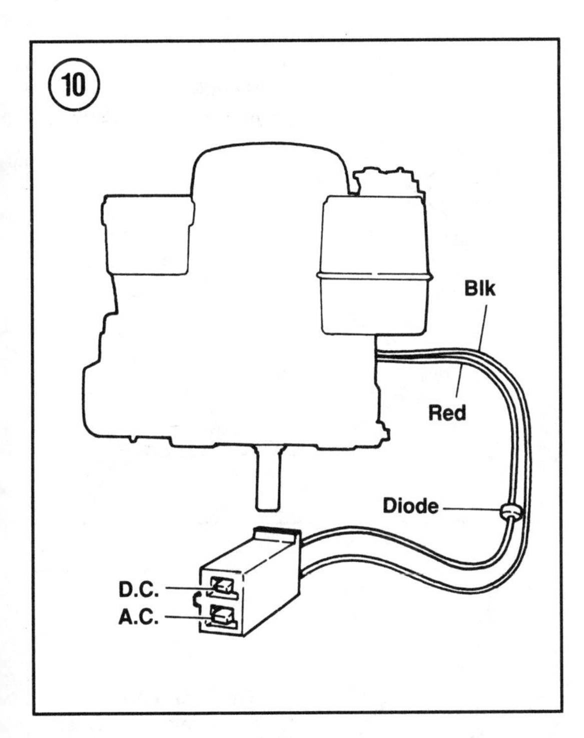

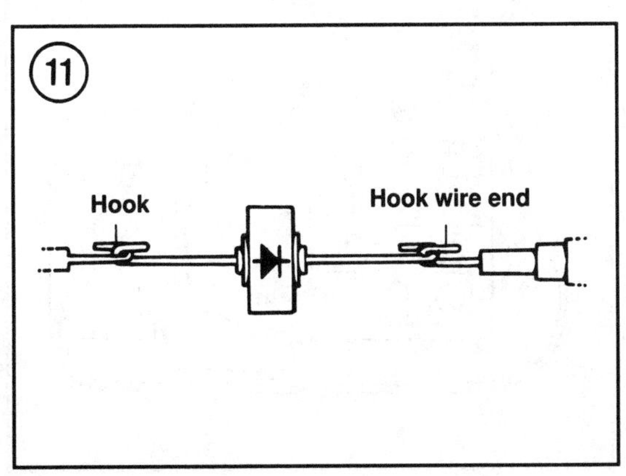

Diode replacement

1. Pull back the wire sheathing so the diode is accessible and cut the diode wires.
2. If using heat-shrink tubing, slide the tubing over the wires.
3. Bend the wire ends into hooks and install the new diode as shown in **Figure 11**. The diode must be installed so the arrowhead on the diode points toward the output end of the wire. Tightly squeeze the wire ends together.
4. Solder the wire ends together using rosin core solder.
5. If not using heat-shrink tubing, apply insulating tape.

Rectifier Panel System

The system shown in **Figure 12** has a maximum charging output of about 3 amperes at 3600 rpm. No current regulator is used on this low output system. The rectifier panel includes two diodes (rectifiers) and a 6 ampere fuse for overload protection.

To check output, first be sure the fuse is good. Disconnect the battery lead from the rectifier panel. Connect the positive (+) lead of a DC voltmeter to the BAT+ terminal on the rectifier panel (**Figure 13**) and the negative (–) voltmeter lead to engine ground. Run the engine at 3600 rpm. The voltmeter should indicate at least 18 volts. If the voltage reading is insufficient, check alternator output by connecting the leads of an AC voltmeter to the GEN terminals on the rectifier panel (**Figure 14**). Run the engine at

3600 rpm. The voltmeter should indicate at least 35 volts. If the voltage reading is insufficient, the alternator stator is faulty. If the AC voltage test was satisfactory, but the DC voltage test was unsatisfactory, check the rectifier panel diodes.

Check for a faulty rectifier diode as follows.

1. Remove the rectifier diodes (**Figure 12**) from the rectifier panel.
2. Connect the leads of an ohmmeter to each end of the diode.
3. Check for continuity.
4. Reverse the ohmmeter leads and again check for continuity. The ohmmeter should show a continuity reading (low ohms) for one test and infinity for other test.
5. Replace the diode if tests indicate the diode is faulty.

External Regulator-Rectifier System

The system shown in **Figure 15** may produce 7 or 10 amperes and uses a solid-state regulator-rectifier outside the flywheel that converts the generated alternating current to direct current for charging the battery. The regulator-rectifier also allows only the required amount of current flow for existing battery conditions. When the battery is fully charged, current output is decreased to prevent overcharging the battery.

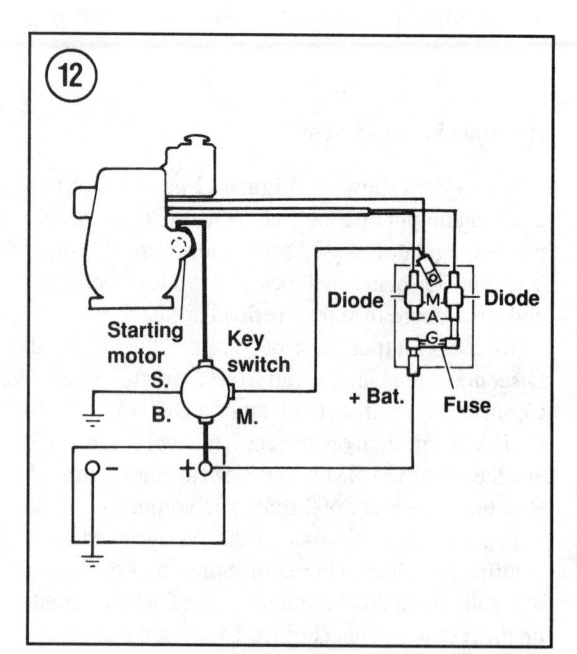

To test the 7 or 10 amp system, disconnect the B+ lead of the regulator-rectifier connector from the ammeter (or key switch on early models) and connect a DC voltmeter to the B+ lead as shown in **Figure 16**.

NOTE

On early models, the B+ wire is routed from the regulator-rectifier to an "R" terminal on the key switch and the ammeter is located between the battery and key switch.

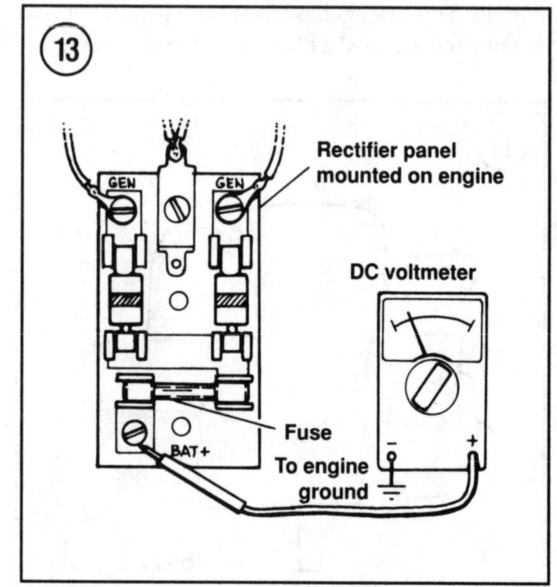

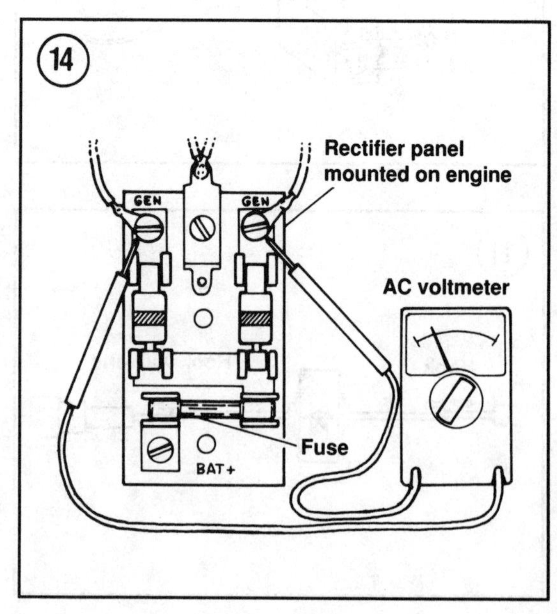

With the engine running at 3600 rpm, voltage should be at least 14 volts on 7 amp system and 20 volts on 10 amp system. If the reading is excessively high or low (less engine speed will produce less voltage), the regulator-rectifier unit may be defective. To check the alternator stator, disconnect the connector from the regulator-rectifier and connect an AC voltmeter to the AC (outer) connector terminals as shown in **Figure 17**. With the engine running at 3600 rpm, check AC voltage. The alternator stator is defective if voltage is less than 18 volts on 7 amp system or 24 volts on 10 amp system.

Internal Regulator-Rectifier System

The regulator-rectifier unit is either covered with epoxy or secured in an aluminum box with epoxy, and mounted under the blower housing. Units are not interchangeable. The system produces 7 amps at full throttle. A schematic is shown in **Figure 18**.

15

Solid state regulator/rectifier
A.C.
A.C.
Magneto shut-off
Charging coils
Starting motor
Key switch
Ammeter
Light, etc.
Switch

16

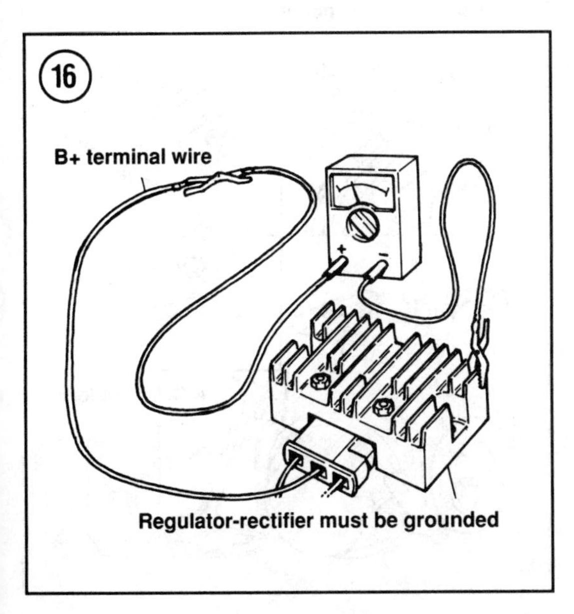

17

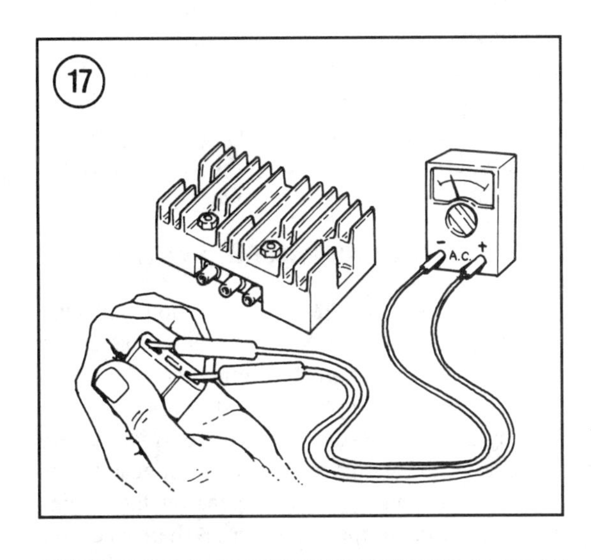

18

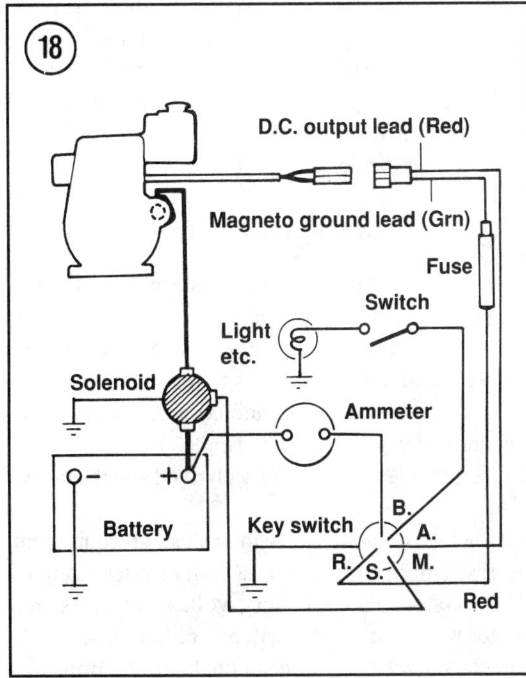

It is not possible to perform an open-circuit DC test. To check the alternator stator, remove the regulator-rectifier unit from the blower housing, but do not disconnect the wire connector from the regulator-rectifier. Reinstall the blower housing.

CAUTION
Do not run the engine without the blower housing installed.

Connect AC voltmeter leads to the "AC" terminals of regulator-rectifier unit as shown in **Figure 19**. At an engine speed of 3600 rpm (less engine speed will produce less voltage), the voltmeter should indicate at least 23 volts. If the voltage reading is insufficient, then the alternator is defective. If the voltage reading is satisfactory and a known to be good battery is not charged by the system, then the regulator-rectifier unit is defective.

ELECTRIC STARTER SYSTEM

This section covers 12-volt starter motors that use current provided by a wet-cell storage battery. A switch located either on the engine or on the equipment is used to energize the starter motor by connecting the wires leading from the positive battery terminal to the starter motor. A pinion gear on the starter motor engages a ring gear on the engine flywheel to rotate the engine crankshaft when the starter motor is energized. When the motor is energized, a helix moves the pinion gear towards the flywheel ring gear, then when the engine is running, the faster moving ring gear "unscrews" the pinion gear so it moves away from the ring gear, thereby disengaging the gears.

TROUBLESHOOTING

If the starter motor does not rotate the engine when the starter switch is actuated, or the motor turns slowly, use the following steps to isolate the problem before assuming the motor is faulty.

1. Check for loose, dirty and corroded wiring connections as well as faulty wiring.
2. Be sure the battery is fully charged and the proper size.
3. Check for faulty safety interlocks on equipment.
4. If so equipped, check for a faulty starter solenoid. The solenoid relays battery voltage to the starter motor when the starter switch is closed. Use a voltmeter to check for voltage at the battery terminal on the solenoid and at the starter terminal when the starter switch is actuated. Generally, if a solenoid "snaps" when energized, it is in good condition.
5. Check for a faulty starter switch. A jumper wire can be connected between the battery and starter motor to determine if the starter switch is faulty.
6. Operate the manual starter to determine if excessive force is required to start the engine (may be due to excessive load connected to engine, wrong oil viscosity or internal engine damage).

ELECTRIC STARTER MOTORS

While several electric starter motors have been used on Tecumseh engines, the motors are basically divided into those with a round frame or a square frame (**Figure 20**). Several variations of each type have been produced, but service is basically similar except as noted in the following paragraphs.

Tecumseh does not provide test specifications for electric starter motors, so service is limited to replacing components that are known or suspected faulty.

CAUTION
Before working on the starter motor, disconnect the negative battery terminal lead.

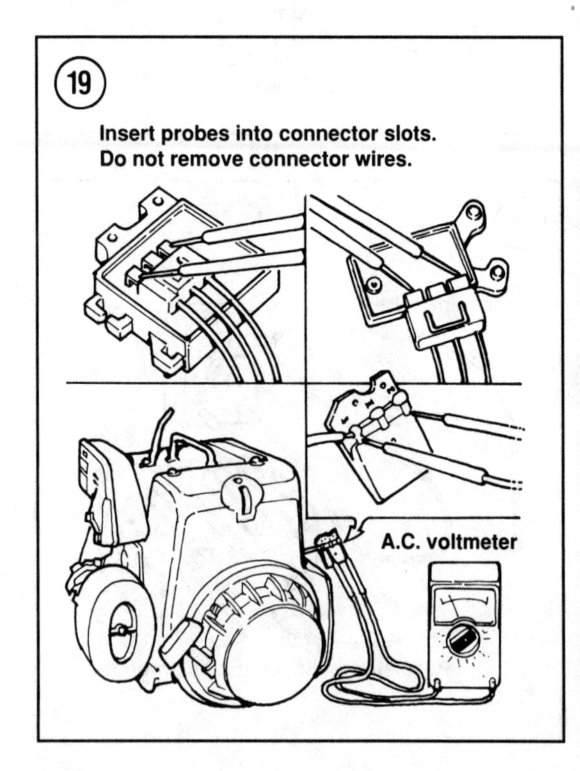

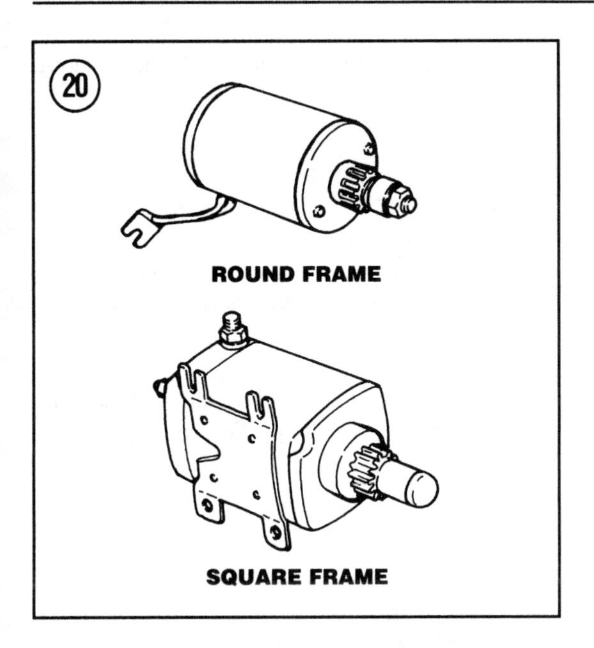

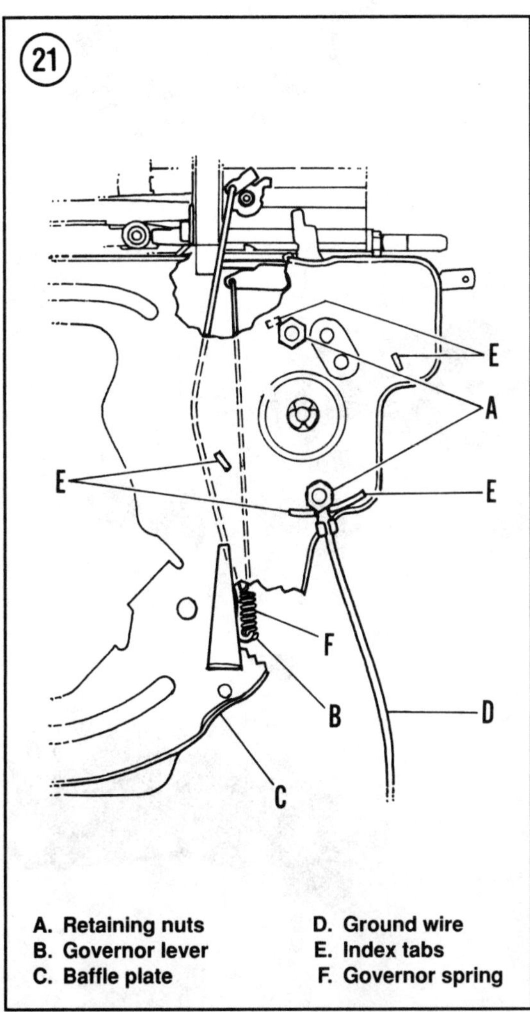

A. Retaining nuts
B. Governor lever
C. Baffle plate
D. Ground wire
E. Index tabs
F. Governor spring

CAUTION

The starter motor field magnets may be made of ceramic material. Do not clamp the starter housing in a vise or hit the housing as the field magnets may be damaged.

CAUTION

Do not run the starter motor continuously for more than 30 seconds. Allow the motor to cool for 10 minutes before further operation.

NOTE

Be sure the proper engine oil viscosity is used. In extremely cold temperatures, oil with the improper viscosity can prevent or inhibit engine starting due to excessive oil drag against moving parts.

Starter Motor Removal/Installation (Except Vector VLV50, VLXL50, VLV55 and VLXL55 Engines)

1. The starter motor can be removed on most engines after removing the blower housing.
2. Disconnect the negative lead from the battery.
3. Disconnect the starter wire.
4. Remove starter mounting screws and remove the starter motor.
5. Install the starter motor by reversing the removal Steps 1 through 4.

9

Starter Motor Removal/Installation (Vector VLV50, VLXL50, VLV55 and VLXL55 Engines)

Refer to **Figure 21** for the following procedure.

1. The fuel tank and blower housing must be removed for access to the starter motor retaining nuts.
2. Disconnect the negative lead from the battery.
3. Disconnect the motor power lead.
4. Unscrew the retaining nuts and remove the starter motor.
5. When installing the starter motor, note that the index tabs must fit into slots on the baffle plate and the governor spring passes between the tabs as shown in **Figure 21**. After installing the motor, attach the ground wire to the starter motor stud near the governor lever.

Starter Drive Operation

A pinion gear is located on the armature shaft of the starter motor. The pinion gear engages a ring gear on the engine flywheel to rotate the engine crankshaft when the starter motor is energized. When the motor is energized, a helix moves the pinion gear towards the flywheel ring gear. When the engine is running, the faster moving ring gear "unscrews" the pinion gear, thereby disengaging the gears.

On motors with a covered drive assembly (**Figure 22**), the pinion gear is drawn towards the motor when the motor is energized. On motors with an exposed drive assembly, the pinion gear may be driven away from the motor on motors with an inboard gear (**Figure 23**) or drawn towards the motor on motors with an outboard gear (**Figure 24**) when the motor is energized.

Starter Drive Overhaul

Refer to **Figure 22** for the following procedure.

1. Position the starter motor so it is upright with the drive end up.

2A. On motors with a covered drive, detach the retaining ring from the groove on the armature shaft. Unscrew the cap retaining nuts and remove the cap while sliding the drive assembly off the armature shaft.

NOTE
When the nuts are unscrewed, the armature, frame and end cap can separate. When handling the motor, hold the armature, frame and end cap so they cannot move.

2B. On motors with an exposed drive, detach the retaining ring (**Figure 24** or **Figure 25**) from the

STARTER DRIVE ASSEMBLY

1 2 3 4 5 6 7

1. Drive cap
2. Nut
3. Drive hub
4. Pinion gear
5. Spring
6. Spring cup
7. E-ring

(22)

(23)

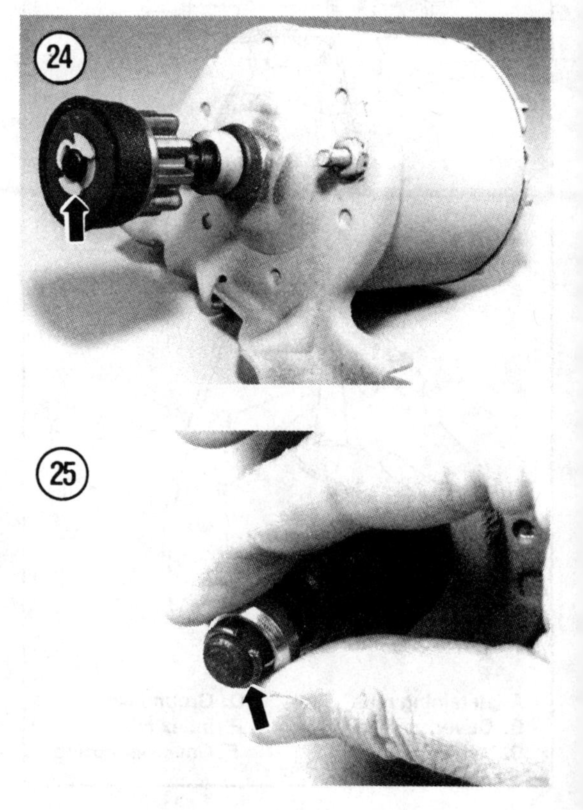

(24)

(25)

groove on the armature shaft. Slide the drive assembly off the armature shaft.

NOTE
*Refer to **Figure 26** for an exploded view of an inboard gear drive assembly or to **Figure 27** for an exploded view of an outboard gear drive assembly.*

3. Clean and inspect the drive assembly components. Broken, cracked or excessively worn parts should be replaced.

4. Apply a light coat of grease to the armature shaft helix. All other parts should be assembled dry.

STARTER DRIVE ASSEMBLY

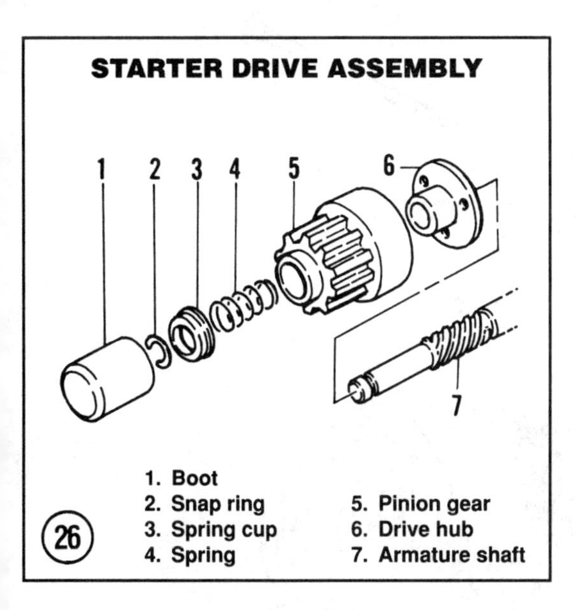

26

1. Boot
2. Snap ring
3. Spring cup
4. Spring
5. Pinion gear
6. Drive hub
7. Armature shaft

STARTER DRIVE ASSEMBLY

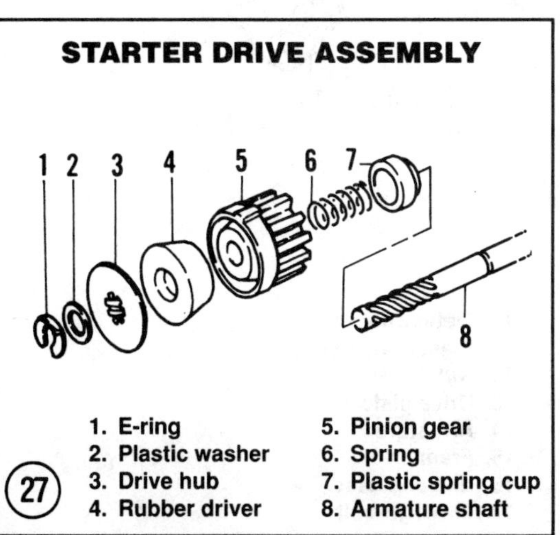

27

1. E-ring
2. Plastic washer
3. Drive hub
4. Rubber driver
5. Pinion gear
6. Spring
7. Plastic spring cup
8. Armature shaft

5. Assemble the drive components on the armature shaft in the order shown in **Figure 22**, **Figure 26** or **Figure 27**.

Starter Motor Disassembly

Refer to **Figure 28** or **Figure 29** for an exploded view of a typical round frame starter motor and to **Figure 30** for an exploded view of a typical square frame starter motor.

1. Remove the drive assembly as discussed in the previous section.

NOTE
Note the position of any spacers on the armature shaft.

2. Note at which end of the motor the through-bolts and nuts are located so that they can be installed in their original positions, then unscrew the nuts and bolts securing the frame to the drive plate and end caps. (On square frame motors, the retaining nuts and drive housing were removed when the drive assembly was removed.)
3. Remove the end cap, then withdraw the armature from the frame.
4. On square frame motors, unscrew the nuts on the wire terminals and remove the brush card.
5. Before removing the brushes, note the position of the brushes and wires.

Starter Motor Inspection

1. Clean the starter components, but do not use cleaning solvents that will damage the armature.
2. Minimum brush length is not specified. If the brush wire bottoms against the slot in the brush holder or the brush is less than half its original length, replace the brush. Be sure the brushes do not bind in the holders.

NOTE
A brush set is available for square frame starter motors, but not for round frame motors. The brushes are available only as a unit assembly with the end cap on round frame starter motors.

3. On square frame motors, the brush leads connected to the frame field coils must be cut so new brushes can be installed. The new brushes must be

connected and soldered to the field coil leads using rosin core solder.

4. Check the strength of the brush springs. The spring must force the brush against the commutator with sufficient pressure to ensure good contact.

5. On square frame motors, replace the brush holder card (**Figure 30**) if the card is warped or otherwise damaged.

6. Inspect the condition of the commutator (**Figure 31**). The mica between the commutator bars should be slightly undercut as shown in **Figure 32**. Undercut the mica or clean the slots between the commutator bars using a piece of hacksaw blade. After undercutting, remove burrs by sanding the commutator lightly with crocus cloth.

7. Inspect the commutator bars for discoloration. If a pair of bars is discolored, grounded armature coils are indicated.

8. Replace the armature if the commutator is damaged or excessively worn.

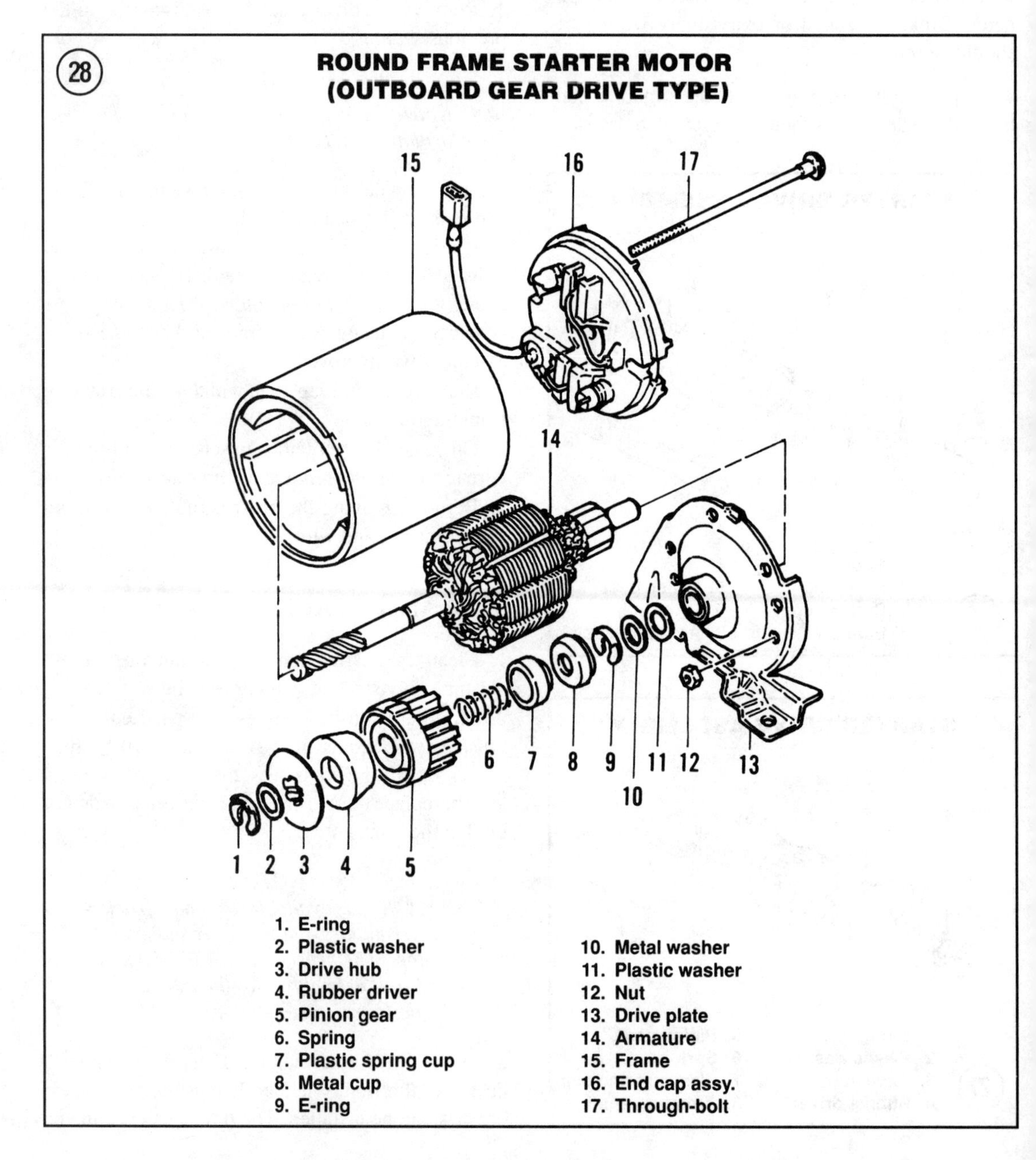

ROUND FRAME STARTER MOTOR (INBOARD GEAR DRIVE TYPE)

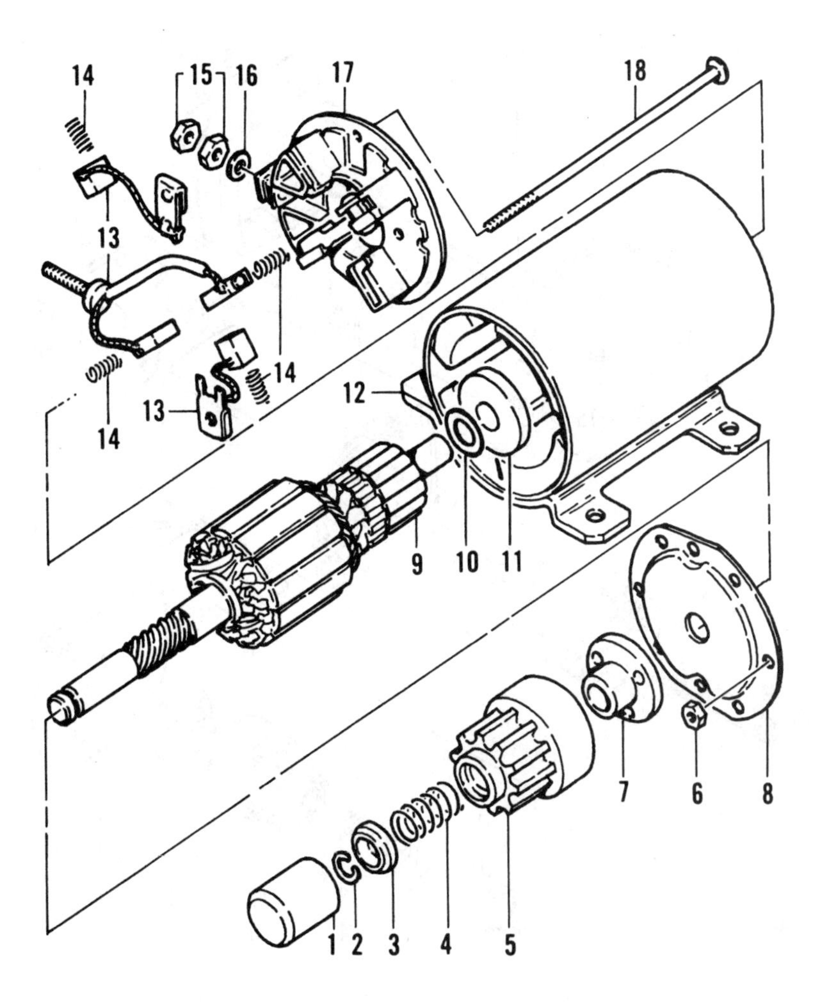

1. Boot
2. Snap ring
3. Spring cup
4. Spring
5. Pinion gear
6. Nut
7. Drive hub
8. Drive plate
9. Armature
10. Washer
11. Washer
12. Frame
13. Brushes
14. Brush springs
15. Nuts
16. Washer
17. End cap
18. Through-bolt

9

(30)

SQUARE FRAME STARTER MOTOR

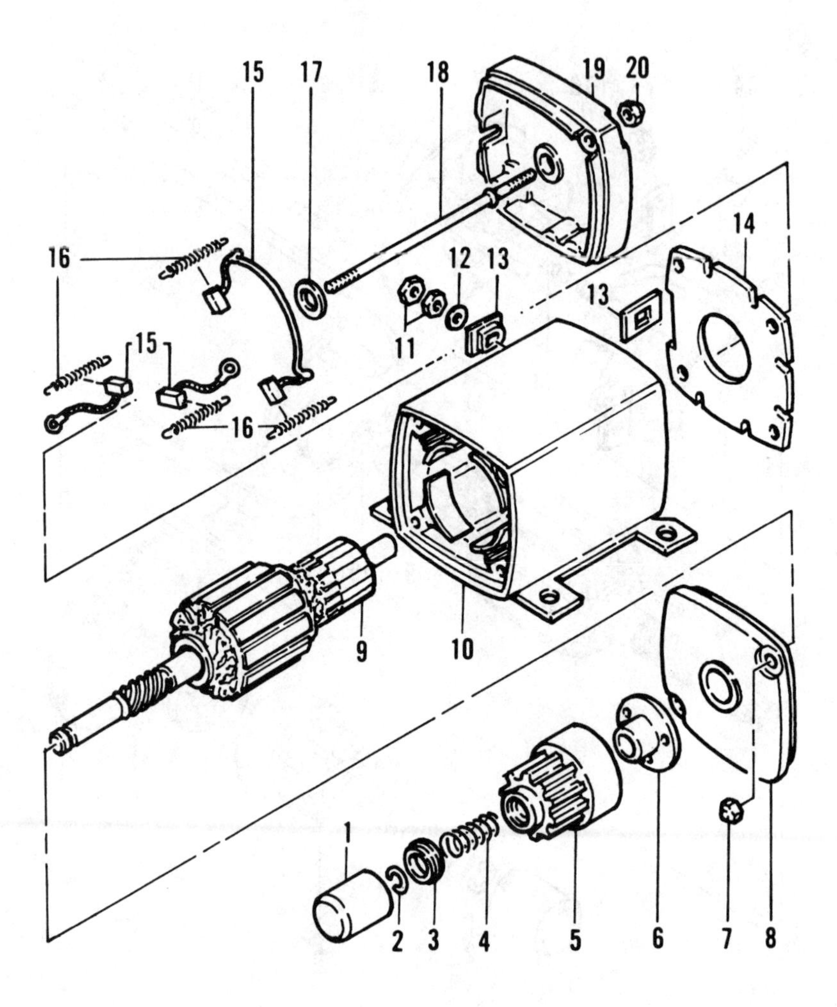

1. Boot
2. Snap ring
3. Spring cup
4. Spring
5. Pinion gear
6. Drive hub
7. Nut
8. Drive plate
9. Armature
10. Frame
11. Nuts
12. Washer
13. Grommet
14. Brush holder card
15. Brushes
16. Brush springs
17. Washer
18. Through-stud
19. End cap
20. Nut

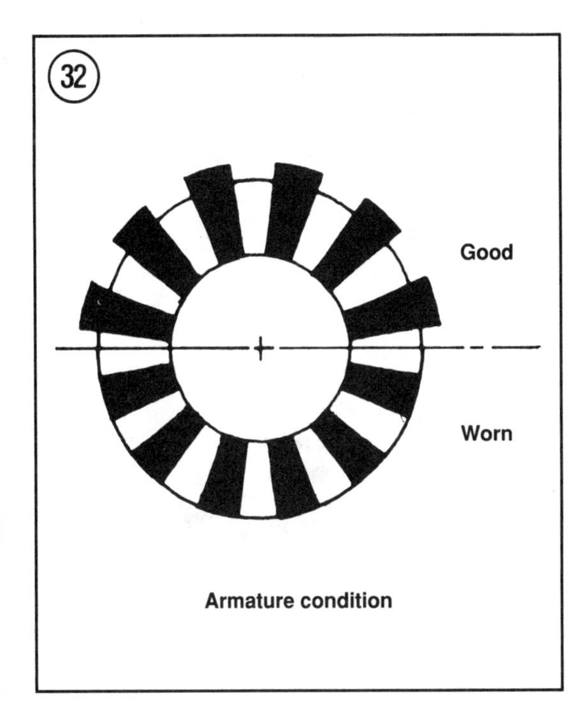

Armature condition

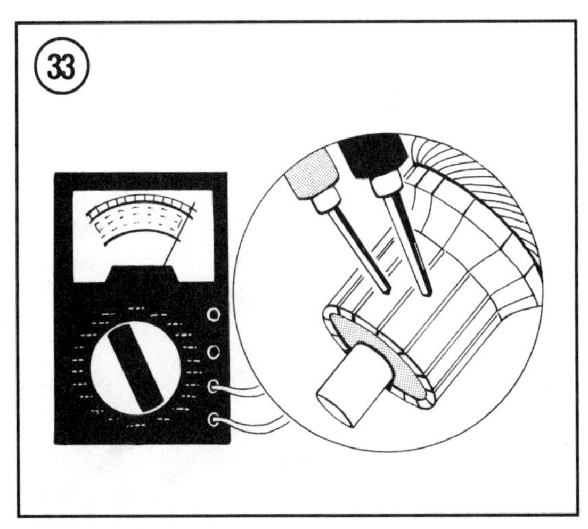

9. Use an ohmmeter and check for continuity between the commutator bars (**Figure 33**); there should be continuity between pairs of bars. If there is no continuity between pairs of bars, the armature is open. Replace the armature.

10. Connect an ohmmeter between any commutator bar and the armature shaft (**Figure 34**); there should be no continuity. If there is continuity, the armature is grounded. Replace the armature.

11. To check the field coils on square frame motors, connect an ohmmeter between the field coil connections shown in **Figure 35**. The ohmmeter should indicate continuity. Connect an ohmmeter between

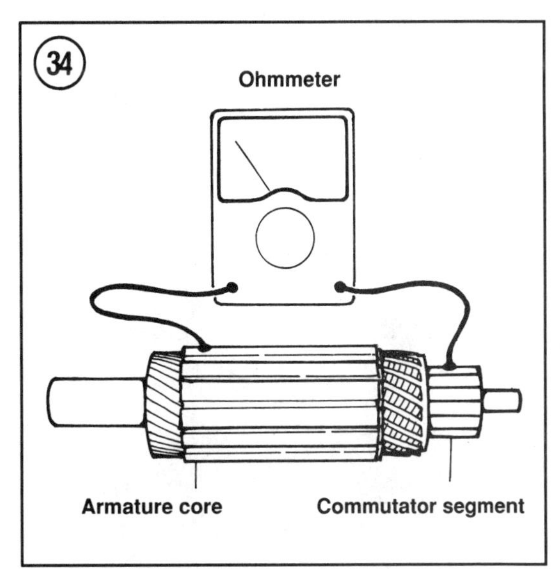

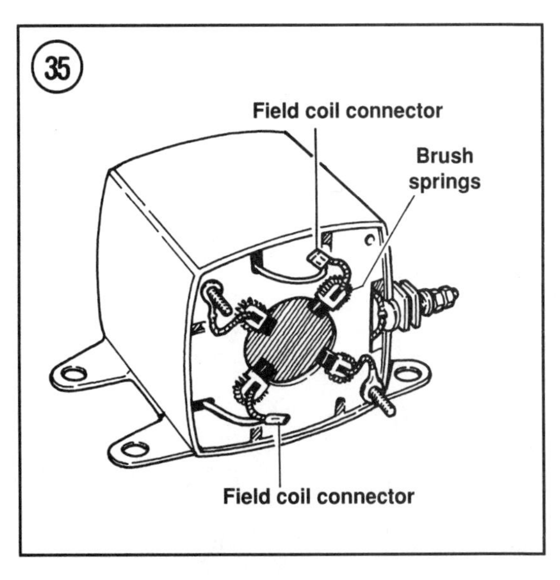

9

each field coil connection and the frame. The ohmmeter should indicate infinity.

NOTE
If more extensive starter motor testing is required, then the motor should be taken to a professional shop. However, this cost should be compared with the cost of a replacement starter motor.

12. Inspect the bushings in the drive plate (**Figure 36**) and end cap. If the bushings are damaged or excessively worn, then the drive plate and/or end cap must be replaced; individual bushings are not available.

Starter Motor Assembly

1. Position the armature in the field coil frame.
2. Install the spacers and thrust washers in their original positions.
3. On some motors, a brush holding tool may be helpful to retain the brushes and springs during assembly. If the end cap has two brushes, a piece of manual rewind starter spring can be modified to hold the brushes as shown in **Figure 37**.
4. Align the marks on the drive plate, housing and end cap. Note that some motors have notches on the housing and end cap that must be aligned during assembly.

NOTE
Use care when installing the end cap so the commutator and brushes are not damaged.

5. Install the through-bolts and nuts.
6. After assembling the motor portion of the starter, rotate the armature. The armature should rotate without binding.
7. Install the starter drive assembly.

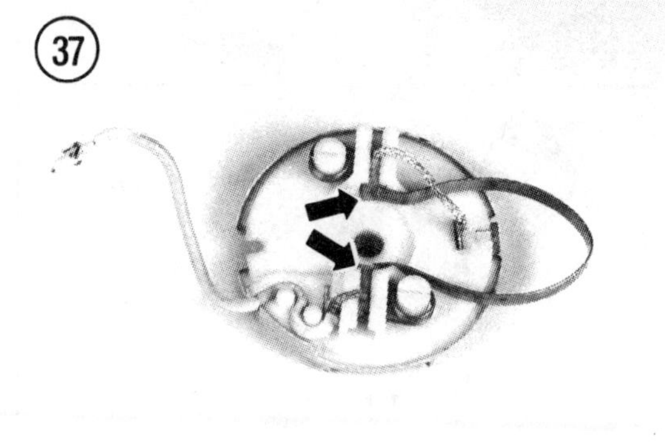

CHAPTER TEN

GENERAL INSPECTION AND REPAIR TECHNIQUES

FAILURE ANALYSIS

All mechanics hate to repeat a job. To a professional mechanic, repeating a job means lost money and a blemished reputation, either of which hurts business. To a do-it-yourselfer, repeating a job means frustration, money wasted on parts and possibly turning the work over to a professional shop.

The cause for most engine failures is usually easily identified and fixed. The trap that some mechanics fall into is overlooking or misidentifying the primary cause for engine failure or malfunction. It is often easy to merely replace parts rather than taking the time to inspect parts for clues that will identify the cause for failure. Not determining and correcting the primary cause will result in another repair job, often replacing the very parts that were just installed. Failure analysis is detective work that pays off in a long-lasting repair job.

Most small-engine failures are due to abuse and neglect as exemplified by the three leading causes of engine failure: abrasive grit, lack of lubrication and overheating. All of the foregoing are usually caused by improper maintenance. Small engines often operate in a harsh environment and receive minimal service. Knowing the engine's operating conditions as well as its service and repair history can provide clues to the cause for the engine failure and prevent future problems.

NOTE

Be sure any engine-driven equipment is operating properly. Damaged or improperly adjusted equipment that is driven by the engine (such as a loose lawn mower blade) can affect engine operation or cause engine damage.

CAUTION

Exercise caution when operating engine-driven equipment. Safety devices must be operable except as specified by the manufacturer for troubleshooting purposes.

Presuming that troubleshooting has determined that an internal engine problem exists, learn as much as possible before undertaking engine repair. If possible, operate the engine and note any symptoms such as engine smoke, excessive vibration and abnormal engine noise. Before removing the engine from the equipment, look for loose mounting fasteners. Look for clues during disassembly, such as missing or maladjusted parts. Inspect parts before and after cleaning. Before cleaning, look for and identify any grit or debris in the engine. After cleaning, look for discoloration and scoring, then measure the parts and compare dimensions with the manufacturer's specifications.

Some major problem areas and their effect on engine parts are discussed in the following sections.

Abrasive Grit

Particles that can cause engine wear are defined as abrasive grit. Grit can be made of dirt, sand, coal dust, cement dust or any other abrasive particle. Many small engines operate in environments in which abrasive grit is present, making the engine particularly vulnerable to the entrance of grit. Abrasive grit causes premature wear which results in symptoms of hard starting, loss of power, high oil consumption and oily exhaust smoke.

Grit generally enters the engine either through the intake system or through the oil fill hole. A dirty air cleaner element or poorly fitting air cleaner element is generally responsible for grit entering the intake system. Not cleaning around the oil fill hole before adding oil to the engine or a poorly fitting oil fill tube will allow grit to contaminate the engine oil. Grit can also enter the engine due to poorly fitting oil seals or gaskets.

Because grit can contaminate both the air and oil in an engine, all wear surfaces in an engine can be damaged. Look for signs of wear due to grit on the following components:

1. *Intake system*—Grit-blasting produces a dull finish on passages. The carburetor throttle shaft bearing area may be excessively worn.
2. *Valves*—A groove will form on the valve face (**Figure 1**) and the valve guide and valve stem will be worn (**Figure 2**). Excessive valve stem wear creates a ridge that can be detected by running a fingernail along the stem.
3. *Piston and piston rings*—Heavily scratched piston and piston rings (**Figure 3**). Worn piston rings will have excessive piston ring end gaps. The oil control ring face will be worn flat. Compare the worn oil control ring shown on the left in **Figure 4**; a new oil control ring is on the right. Dirty oil caused the wear.
4. *Cylinder bore*—The cross-hatch pattern will be worn away and a ridge will form at the top of the cylinder bore.
5. *Connecting rod*—The big-end bearing surface will have a dull finish.
6. *Crankshaft*—One or more journals will be scored. Note the crankpin journal shown in **Figure 5**.

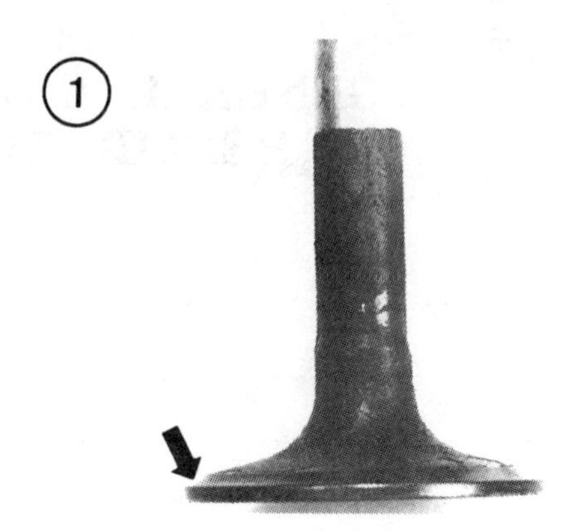

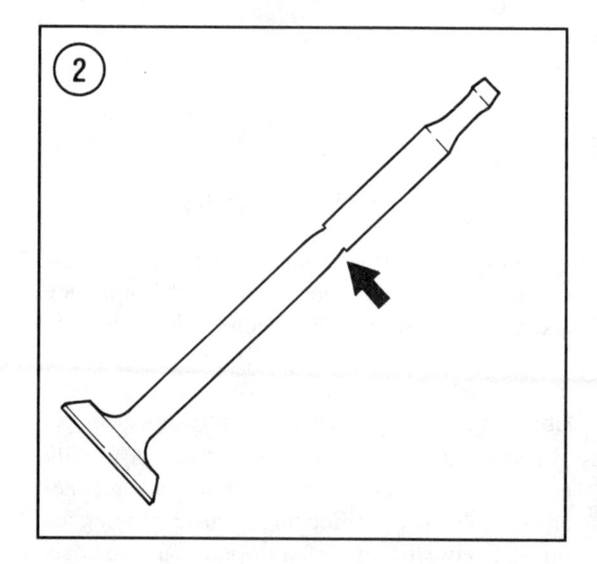

7. *Camshaft*—The journals and lobes will be scored, as well as the wear surfaces on the tappets.

Note that grit may be embedded in softer metals such as the aluminum of the connecting rod and crankcase bearing surfaces. If the rod or bushing is not replaced, then the new crankshaft will be damaged again.

Another cause of engine damage due to abrasive grit is a poor cleaning procedure during an engine overhaul. Particles produced during machining and not removed before assembling the engine can damage the engine when it runs. Parts can be contaminated if handled with greasy hands or if stored in a dirty environment.

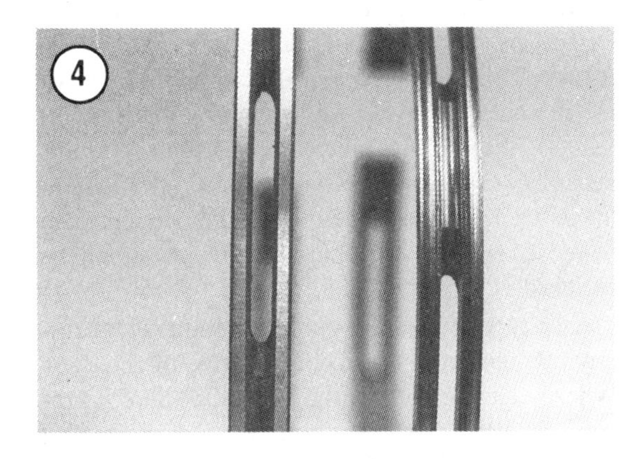

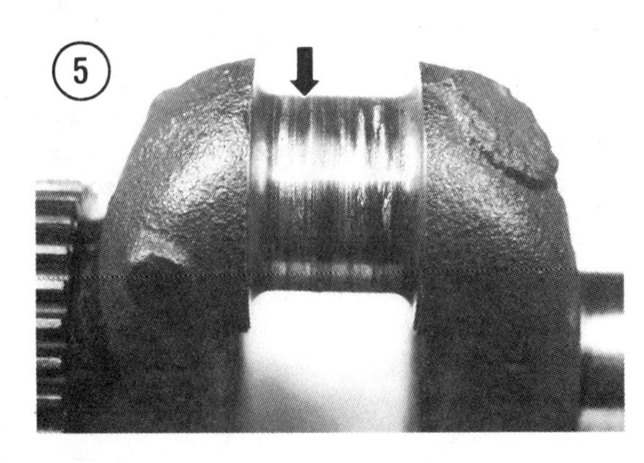

Insufficient Lubrication

Engine oil lubricates and cools engine parts. Insufficient oil in the crankcase or poor quality oil can lead to scoring or discoloration of engine parts. Insufficient lubrication can be a result of:

a. Low oil level.
b. Wrong oil viscosity.
c. Engine running at a high operating angle.
d. Fuel entering crankcase and diluting oil.
e. Overfilling (the splash lubrication device cannot operate effectively).

Insufficient lubrication is characterized by the discoloration and scoring of engine parts. The temperature of rotating parts that touch becomes excessive and aluminum parts often melt. The aluminum is transferred to the contacting part, such as metal transfer from the connecting rod to the crankshaft crankpin shown in **Figure 5**, or from the main bearing to the crankshaft journal shown in **Figure 6**. The bearing surface becomes scored and, in some cases, seizure occurs, which can result in broken parts, such as broken connecting rod. Burnt oil may also be found in localized areas where excessive heat is generated by moving parts.

NOTE
The area near the crankshaft crankpin may be blue due to a heat treating process used during manufacture and should not be incorrectly identified as a damaged crankpin.

Overheating

Overheating is a condition that is caused by either insufficient external cooling or excessive temperature in the cylinder. Insufficient external cooling is generally a result of debris in the cylinder fins, missing shrouds or poor ventilation around the engine. Excessive temperature in the cylinder is usually caused by a lean air:fuel carburetor mixture setting. If improperly adjusted, carburetors with adjustable mixture screws may provide a lean mixture that is harmful to engine operation. Nonadjustable carburetors may provide a lean mixture if dirty.

Some characteristics of an overheated engine are:
a. Loss of power.
b. High oil consumption.
c. Hot spots on the cylinder bore.
d. Cylinder head mounting surface discolored.
e. Exhaust valve seat loose.
f. Tar-like oil in the bottom of the engine.

Overspeeding

Engines are designed to operate reliably up to a specified upper engine speed. If the specified engine speed limit is exceeded, physical forces may exceed the design limits of the engine components and failure will result. The most common component failure is breakage of the connecting rod just below the rod small end.

Overspeeding is identified by a lack of other indications for engine failure, such as overheating or scoring. An obvious cause for overspeeding is an altered or malfunctioning governor system.

Breakage

An engine failure may be due to components that have failed due to structural faults. In these instances, attempt to determine the cause for the failure. If the engine had been previously repaired or serviced, did a mistake cause the part to fail. For instance, applying an incorrect bolt torque or cracking the flywheel during removal can cause parts failure. If the engine had not been serviced previously, a manufacturing flaw may have caused failure (check for the availability of a redesigned part).

When checking broken parts, be sure to check the condition of associated parts that may have been struck or otherwise damaged. Check all parts that may have contributed to the breakage. A chain of events may have occurred that caused the broken part, but the prime cause may not be apparent at first.

Another cause of breakage is vibration. Single-cylinder engines are particularly prone to vibrate, but the manufacturer accommodates the effects of vibration in the engine design. If, however, the engine is not securely mounted, excessive vibration can loosen and possibly break parts. Look for loose mounting bolts, elongated mounting holes and a polished mounting surface, any of which will indicate that the engine was not securely mounted during operation.

PRECISION MEASUREMENTS

This section covers precision measuring tools, how to use them and specific measuring applications. Note that a list of tools that are commonly required for engine service and overhaul is provided in Chapter One.

Precision Measuring Tools

The ability to measure engine components accurately is essential to accomplish successful engine overhauls. Although engine overhauls can be performed by eyeballing and an educated guess, the results are hit-or-miss, and usually a costly miss. Engines are built to close tolerances, and it is essential that the mechanic be equipped with the tools necessary to obtain consistent and accurate measurements to determine which engine components can be reused reliably and which components should be replaced.

Each type of measuring tool is designed to measure a dimension with a certain degree of accuracy. When selecting the measuring tool, be sure it provides the proper degree of accuracy. For instance, a ruler is inappropriate for measuring a crankshaft journal and a micrometer is overkill when measuring rewind starter rope diameter.

Precision measuring tools are available in varying degrees of quality and price, and in this case, the price usually reflects the quality. However, adequate measuring tools are available at a reasonable price for small engine work. A professional mechanic will usually choose high quality tools that will provide the best service life.

As with other tools, precision measuring tools will provide their best service life if cared for properly.

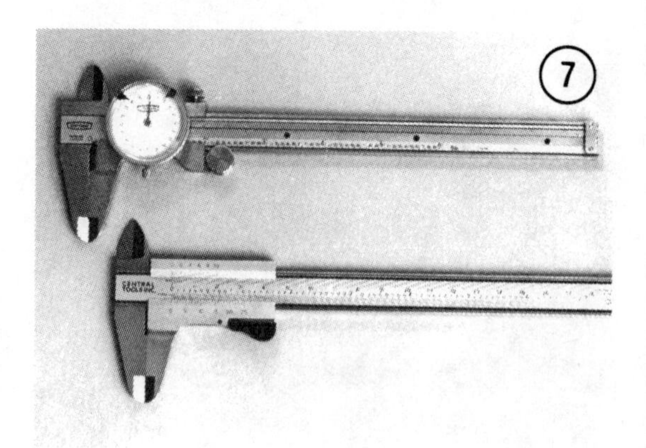

Particular attention must be paid to the care of precision measuring tools, in part due to the cost of replacement, but more importantly because the tool may read inaccurately. Dropping, hitting or improper use can damage the tool resulting in measurements that may be as small as one thousand of an inch off, but the inaccuracy may be sufficient to affect the reuse or replacement of expensive parts. A mechanic should keep in mind the condition of the measuring tool and verify any measurement by using another measuring tool if the measurement appears questionable. A standard gauge of a specific dimension is often used to check the accuracy of precision tools.

A key ingredient when measuring an object accurately is the person performing the measurement. Highly accurate measurements are only possible if the mechanic possesses a "feel" for using the measuring tool. The most sensitive portion of the hand is the fingertips. Heavy-handed use of measuring tools will produce less accurate measurements than if the tool is grasped gently by the fingertips so the point at which the tool contacts the object can be readily felt. The feel acquired when using a tool not only produces more accurate measurements, but there is less chance of damaging the tool's measuring surfaces.

Refer to the following sections for information on specific precision measuring tools.

Dial and Vernier Calipers

Precision measurements can be taken using good quality, direct reading calipers. Calipers are useful in that quick measurements are possible for a variety of situations. Inside and outside dimensions can be measured as well as determining depths. Most calipers can measure dimensions of 6 in. or greater.

Direct reading calipers are available either in dial or Vernier versions (**Figure** 7). Dial calipers are equipped with a calibrated dial that reads in thousands of an inch (or metric equivalent) which provides convenient reading. Vernier calipers have scales marked on the fixed and movable pieces that must be compared to determine the reading. For instance, to read the measurement shown in **Figure 8**, note that the scale on the fixed piece is graduated in increments of 0.025 in. and the movable scale is marked in increments representing thousandths. The first number established is determined by the location of the "0" line on the movable scale in relation to the first line to the left on the fixed scale. In this case the number is 0.150 in. To determine the next number, note which of the lines on the movable scale is aligned with a mark on the fixed scale (or is most nearly aligned). The number of the mark on the movable scale is the number of thousandths, in this case 17, that is added to the first number. In this example, adding 0.017 to 0.150 results in a measurement of 0.167 in. Note that in **Figure 8** the caliper is also calibrated to read metric dimensions; the ability to read both U.S. and metric dimensions is something to consider when purchasing a Vernier caliper.

Vernier and dial calipers are handy, versatile measuring tools, although they are not quite as accurate as a micrometer.

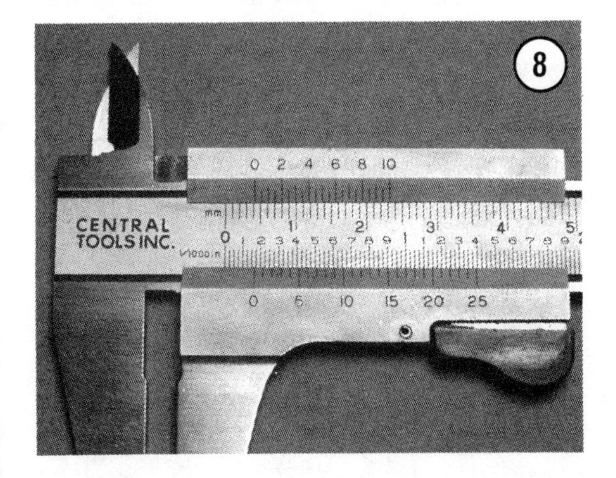

Micrometer

The micrometer is the measuring device most used when precision measurement is required. Micrometers are available in a variety of shapes and sizes to measure both inside and outside dimensions, although outside micrometers are most often used.

The actual measuring device of a micrometer is based on the rotation of a precision-machined screw thread that travels a specified distance (usually 0.025 in.) for each complete rotation. Dividing the distance the screw travels in each rotation into units (thousandths) provides a means to measure a dimension.

Figure 9 shows the parts of a typical micrometer. The part being measured is placed between the anvil and spindle, then the thimble is rotated so the contact surfaces of the anvil and spindle just contact the part.

Acquiring a feel when using the micrometer is essential to obtain an accurate measurement. Depending on the quality of the micrometer and the ability of the user, measurements of ten thousandths inch is possible, although measurements requiring that degree of accuracy are best performed with a Vernier micrometer.

To read a micrometer, note the nearest numbered line to the left of the thimble. The number denotes tenths. In the example shown in **Figure 10** the nearest tenths line is 4, so the measurement is at least 0.4 in.. Then note the number of lines between the numbered line (4) and the edge of the thimble, in this example there are three lines. Each line equals 0.025 in., so 3×0.025 in. equals 0.075 in., which added to 0.4 equals 0.475 in. Now note which horizontal line on the thimble is most nearly aligned with the reference line on the sleeve. In this example the 18 line is aligned, so 0.018 in. is added to the previously read 0.475 in. to equal the measured dimension of 0.493 in.

Care should be exercised in using and storing a micrometer. Rough use can affect the accuracy of a micrometer. The micrometer's accuracy should be checked periodically or any time a reading is suspect. Micrometers can be zeroed by adjusting the position of the anvil or sleeve.

Dial Indicator

A dial indicator or gauge is a precision measuring tool that is generally used to measure the travel or out-of-roundness of a part. Measuring crankshaft end play is a typical use for a dial indicator (**Figure 11**). A consideration when purchasing a dial indicator is whether apparatus is included for securing the dial indicator during measuring.

NOTE
When positioning the dial indicator for a reading, the dial indicator must be securely mounted to prevent any deflection in the set-up apparatus from creating a false reading.

Dial indicators are available to measure a variety of ranges and in various graduations. A special type of dial indicator is available to measure cylinder bore taper and out-of-roundness, but a cylinder bore dial gauge is expensive and generally used only by a professional shop.

Telescoping Gauges

A telescoping gauge (sometimes called a snap gauge) is used to measure hole diameters. The telescoping gauge (**Figure 12**) does not have a scale gauge for direct reading. Thus an outside micrometer is required in conjunction with the telescoping gauge to determine bore dimensions.

Particular attention must be taken when using a telescoping gauge to make sure the true diameter is measured and the gauge is not off center or cocked.

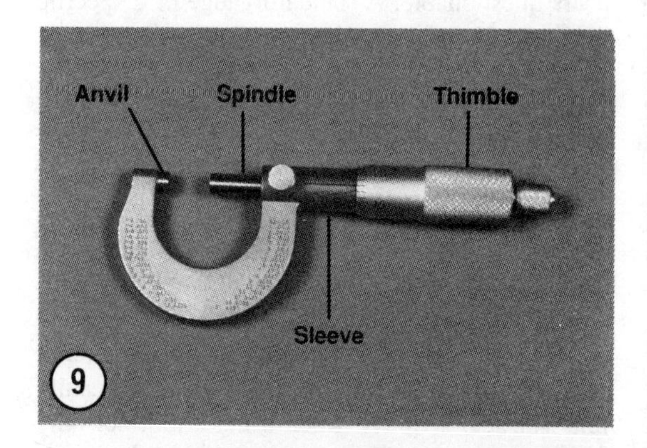

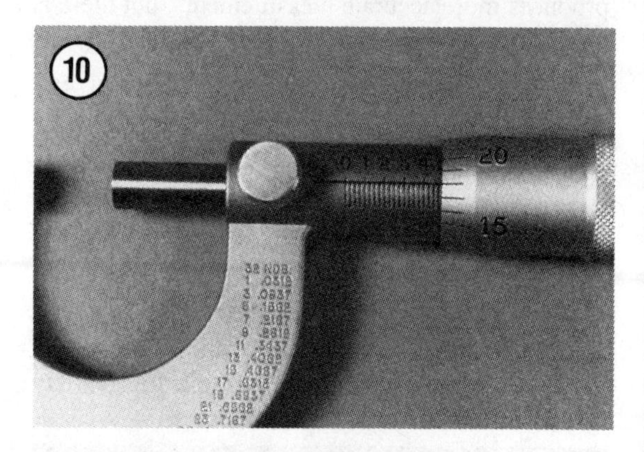

Several measurements should be taken to be sure the true measurement is obtained.

Small Hole Gauges

Small hole gauges are used to measure hole diameters, such as valve guides, or the widths of grooves or slots (**Figure 13**). An outside micrometer must be used together with the small hole gauge to determine the measured dimension.

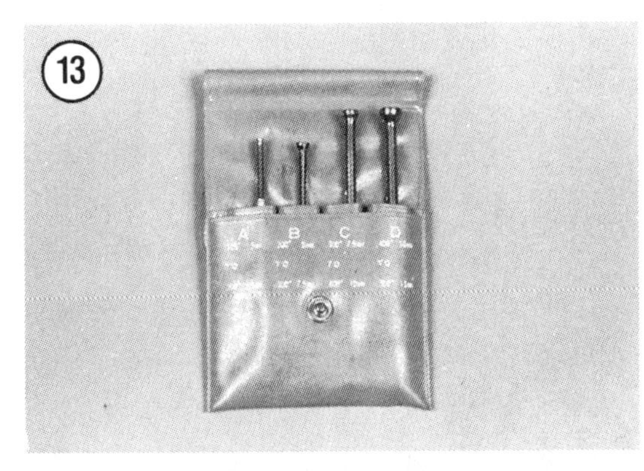

Go-No Go Gauges

These types of gauges are used to quickly and accurately determine if a given hole is within the specifications specified by the manufacturer. For instance, the manufacturer may state that if the gauge will fit into a particular hole, then the hole is too large and a bushing must be installed to return the hole to the desired specification. Go-no go gauges are generally available only from the engine or equipment manufacturer.

REPAIR TECHNIQUES

Removing Frozen Nuts and Screws

Removing nuts and screws that are frozen due to rust can be frustrating and time consuming. But the problem can be lessened if excessive force is not used when initially trying to turn the nut or screw. A broken screw or damaged threads usually result if more than normal force is placed on the wrench.

Before turning a rusty nut or screw, apply a penetrating oil such as Liquid Wrench or WD-40 (available at hardware or auto supply stores). Apply it liberally and let it penetrate for 10-15 minutes. Tap the fastener several times with a small hammer; do not hit it hard enough to cause damage. Reapply the penetrating oil if necessary.

Particular care should be taken when removing a frozen nut on a stud because removal may break the stud. Observe the end of the stud when attempting to turn the nut to be sure the nut is turning and the stud is not. If penetrating oil will not loosen the nut, a nut splitter tool (**Figure 14**) can be used to separate the nut from the stud.

For frozen screws, apply penetrating oil as previously described, then insert a screwdriver in the slot and rap the top of the screwdriver with a hammer. This loosens the rust so the screw can be removed in the normal way. If the screw head is disfigured, grip the head with locking pliers and twist the screw out.

Repairing Damaged Threads

Fastener threads can be damaged for a variety of reasons, and in some cases, restoring the threads is required.

NOTE
Be sure to identify the thread size and type (U.S. or metric) before attempting to restore the threads. The only positive method of identification is to use a screw pitch gauge or known fastener.

Often the threads can be cleaned by running a tap (for internal threads) or die (for external threads) through the threads. See **Figure 15**. To clean or repair spark plug threads, a spark plug tap or thread chaser can be used.

If an internal thread is damaged, it may be necessary to install a thread repair insert (**Figure 16**). Thread repair inserts are available in a wide variety of U.S. and metric thread sizes at auto supply stores and some hardware stores. Note that a specific size drill bit must be used, which must be purchased separately. To install a typical thread repair insert, proceed as follows:

1. Drill out the old threads using the drill bit recommended for the thread size being repaired (**Figure 17**). Be sure the hole is straight and the centerline of the hole is not moved while drilling.

2. Cut new threads in the hole using the tap provided in the kit (**Figure 18**). This is a special tap that cuts threads to fit the outer threads on the thread repair insert.

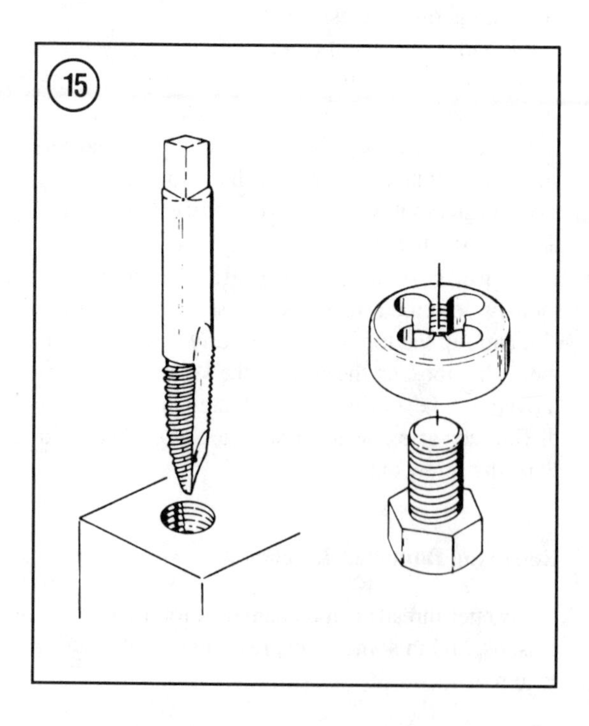

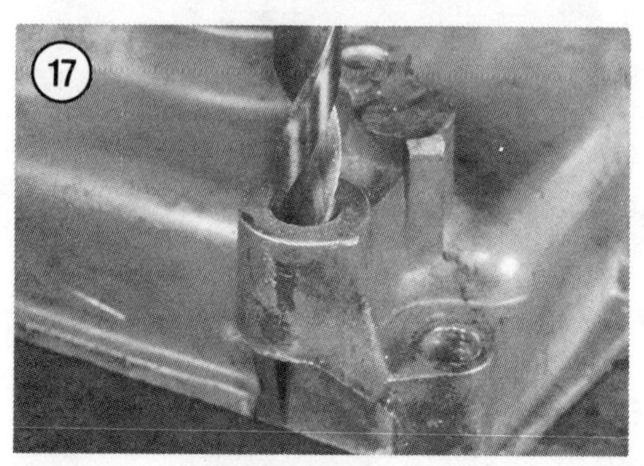

3. Turn the thread repair insert into the hole (**Figure 19**) using the special tool until the top of the insert is a quarter to one-half turn below the surface (**Figure 20**).

4. Snap off the insert tang by pushing down on the tang; don't attempt to twist off the tang.

Removing Broken Screws or Bolts

When the head breaks off of a screw or bolt, several methods are possible for removing the remaining portion.

If a large portion of the screw or bolt projects out, try gripping it with locking pliers. If the projecting portion is too small, file it to fit a wrench or cut a slot in it to fit a screwdriver as shown in **Figure 21**.

If the head breaks off flush, use a screw extractor using the following procedure while referring to **Figure 22**.

1. Center punch the exact center of the remaining portion of the screw or bolt.
2. Drill a small hole in the screw.
3. Tap the screw extractor into the hole.
4. Back the screw out with a wrench on the extractor.

If the preceding procedures are unsuccessful, contact a professional shop. A technique known as electric discharge milling (EDM) is available at some machine shops to remove broken screws and studs, but be sure to get an estimate first.

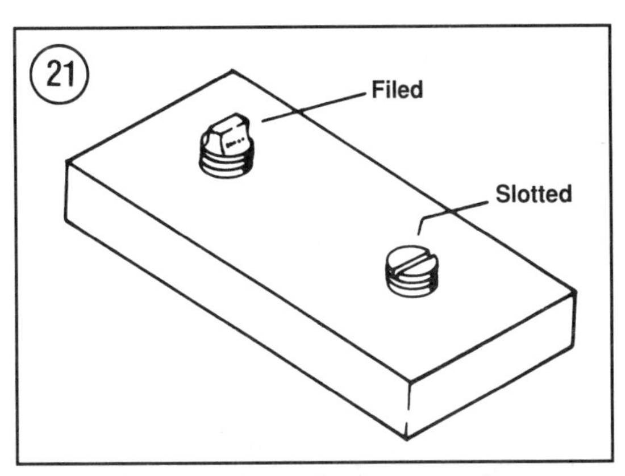

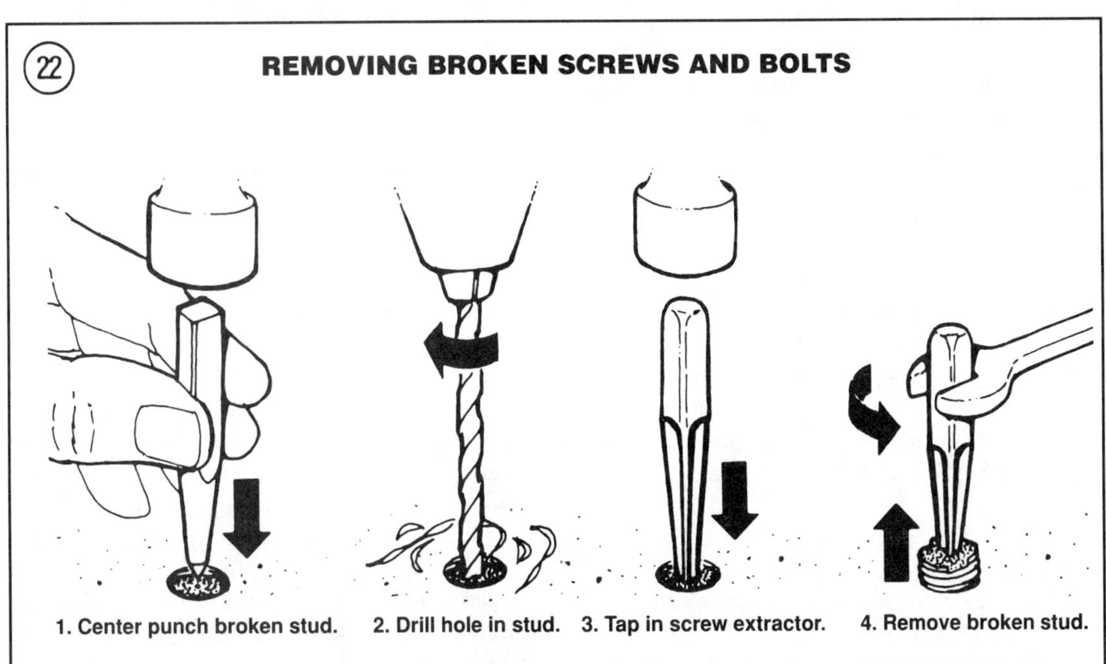

CHAPTER ELEVEN

ENGINE OVERHAUL

The following sections provide disassembly, inspection and reassembly information for the major engine components of a basic engine.

Due to the many variations of Tecumseh engines, it is not possible to address all possibilities when outlining disassembly and reassembly steps. The mechanic must take a thoughtful approach and address any additional steps necessary to service a particular engine.

In some cases, a service procedure may be performed with the engine mounted on the equipment, such as cylinder head removal. In most cases, however, the engine must be removed from the equipment before undertaking the procedure. Before removing the engine from the equipment, note the location of any wiring, cables or brackets.

CAUTION

Be sure to contact the equipment manufacturer for instructions regarding the removal and installation of engine-driven components. Improper service can damage the engine and equipment. Be sure all safety devices operate correctly.

Table 1 provides engine specifications. **Table 2** lists special tightening torques. **Table 3** and **Table 4** list general tightening torques for inch series and metric series bolts and nuts. **Tables 1-4** are located at the end of this chapter.

Tecumseh tools, or suitable equivalent tools, may be recommended for some procedures outlined herein. The Tecumseh tools can be obtained through a Tecumseh dealer or distributor. See Chapter Twelve. Good quality tools are also available from small engine parts suppliers. Be careful when substituting a "home-made" tool for a recommended tool. Although time and money may be saved if an existing or fabricated tool can be made to work, consider the possibilities if the tool does not work properly. If the substitute tool damages the engine, the cost to repair the damage may exceed the cost of the recommended tool.

NOTE
Some procedures may require that work be performed by a professional shop. Be sure to get an estimate, then compare it with the cost of a new or rebuilt engine or short block (a basic engine assembly).

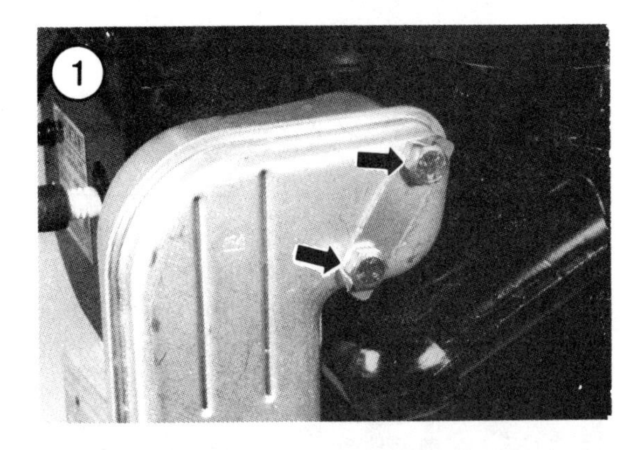

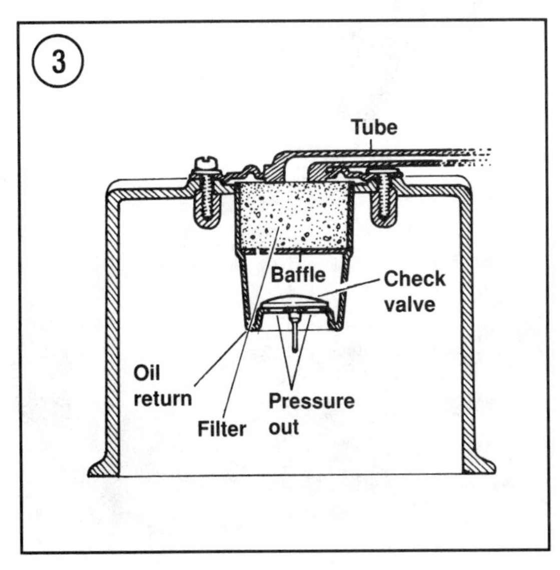

MUFFLER

In most cases, the muffler is secured to the engine by screws or it is screwed directly into the engine. In some applications, the muffler is connected to the engine by a pipe.

Mufflers that are secured by screws can be removed after bending back the locking tabs (**Figure 1**) and unscrewing the retaining screws. It may be necessary to "work" the screws back and forth and use penetrating oil to successfully remove screws that are corroded.

If the muffler is screwed into the engine, grasp the muffler with a suitable tool and unscrew the muffler (**Figure 2**). On some engines, a lock ring is used to lock the muffler in place. The lock ring must be loosened before unscrewing the muffler. If the muffler body collapses due to rust, it may be necessary to cut away the muffler body so the threaded pipe can be grasped.

NOTE
Tecumseh offers an adapter plate that can be installed on engines using a screw-on type muffler if the original threads in the engine are damaged. The adapter plate is retained by screws that fit in the two screw holes near the exhaust port.

If the muffler is significantly damaged due to rust, it should be replaced.

CRANKCASE BREATHER

The crankcase breather (**Figure 3**) vents gases in the crankcase to the carburetor, air cleaner or atmosphere, which provides a negative pressure in the crankcase, if the breather is operating properly. Oil may be forced past the piston rings, oil seals and gaskets if the breather malfunctions.

Four types of crankcase breather have been used: an integral breather, a top-mounted breather, a side-mounted breather, and the breather used on Vector VLV50, VLXL50, VLV55 and VLXL55 engines.

Side-Mounted Crankcase Breather

This type of breather is located on the side of the engine (**Figure 4**) and covers the valve chamber. A disc valve in the breather regulates pressure in the crankcase. Unscrew the retaining screws to remove the breather.

On some engines the breather can be separated for access to the filter element (**Figure 5**), while on other engines, the breather is a unit assembly. On the unit type breather, a removable filter element resides inside the breather (**Figure 6**). A barb inside the housing holds the element in place. Insert a smooth blade between the barb and filter element to remove the element. Clean the filter element in a suitable solvent.

Install either type breather so the drain hole (**Figure 7**) is down. Some units have two gaskets located between the breather and engine. Install both gaskets if so equipped.

Top-Mounted Crankcase Breather

Some vertical crankshaft engines are equipped with a crankcase breather that is mounted on top of the crankcase (**Figure 8**) under the flywheel.

Remove the flywheel as outlined in this chapter for access to the crankcase breather. Unscrew the breather cover and lift out the breather assembly. Remove and clean the filter element (**Figure 9**) in a suitable solvent. Inspect the check valve (**Figure 10**) for damage. Removal may damage the valve. Lubricate the stem of a new check valve (**Figure 11**) to ease installation. Install the baffle plate (**Figure 12**) above the check valve.

Integral Crankcase Breather

The integral type breather is found on some ECV series engines. This type breather is mounted on the top of the crankcase and functions using passages in the crankcase (**Figure 13**). Gases are vented out the

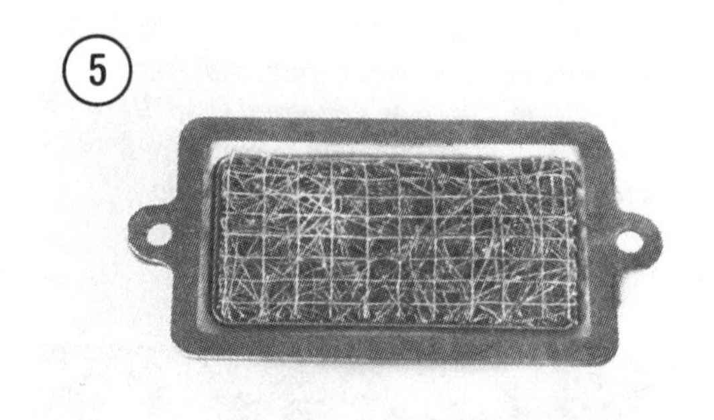

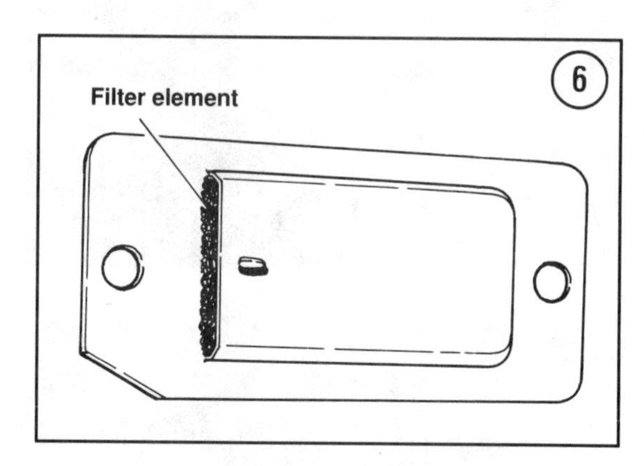

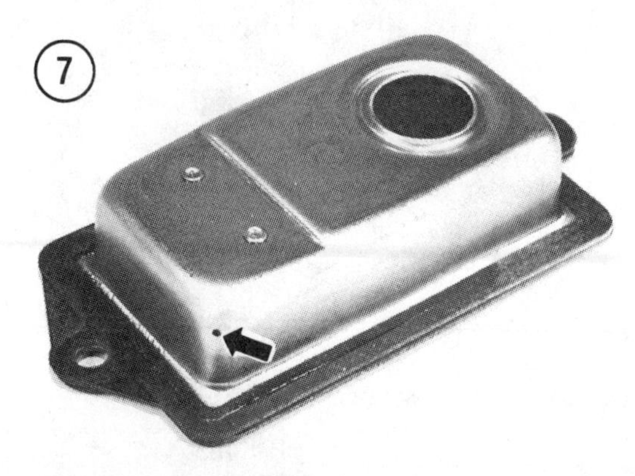

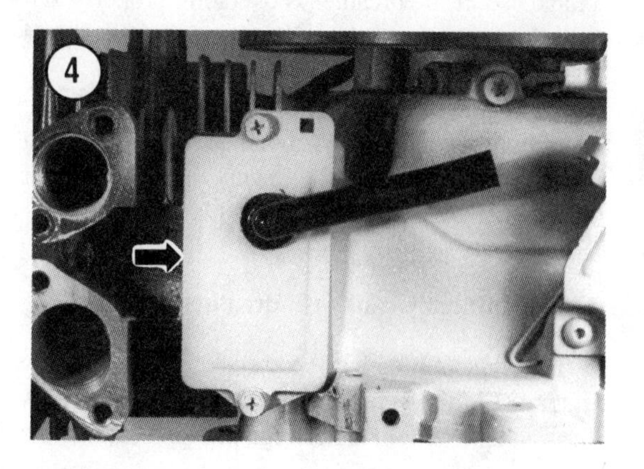

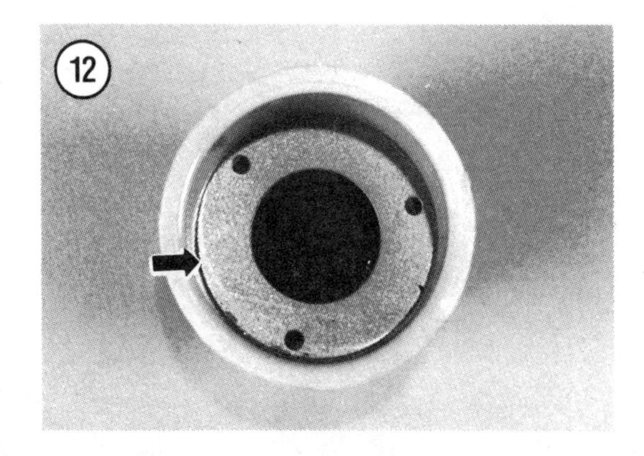

back of the crankcase, sometimes behind the identification plate.

Remove the flywheel as outlined in this chapter for access to the crankcase breather. Check for blocked passages and a damaged valve. A replacement parts set is available.

Vector VLV50, VLXL50, VLV55 and VLXL55 Crankcase Breather

The crankcase breather on Vector VLV50, VLXL50, VLV55 and VLXL55 engines is located under the flywheel beneath the cover shown in **Figure 14**. A disc type valve (A, **Figure 15**) maintains a vacuum in the crankcase. Crankcase gases are routed via a breather tube to the air cleaner. A filter (B) traps contaminants, while any oil is returned to the crankcase through a return passage (C) that exits through a hole in the cylinder bore.

Remove the flywheel as outlined in this chapter for access to the crankcase breather cover. Remove

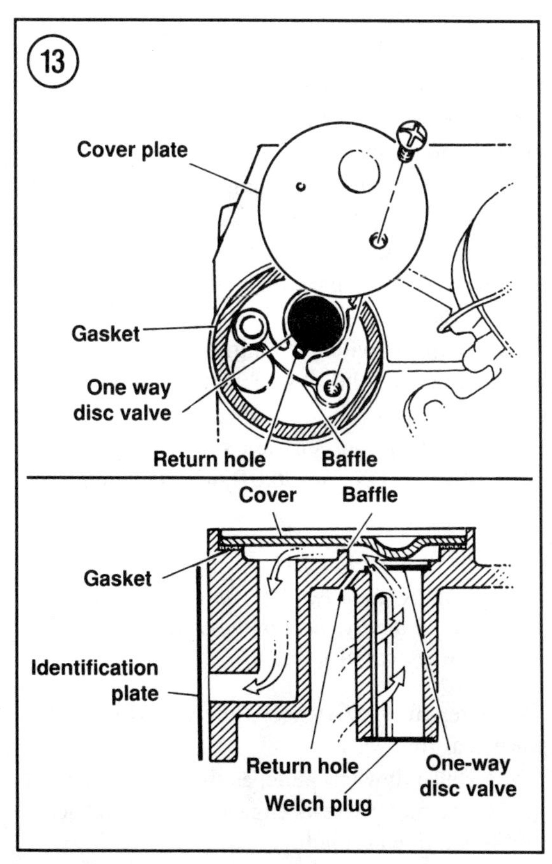

and clean the filter element (B, **Figure 15**) in a suitable solvent.

Do not remove the check valve (A) unless faulty as removal may break off the valve stem, which will fall into the crankcase. The edge of the valve should fit snugly against the valve seat. The valve seat is a press fit in the engine cover and should not be removed unless faulty. Inspect the valve for cracks and other damage. Check for a blocked return passage.

Lubricate the stem of the new check valve (**Figure 11**) to ease installation.

NOTE
*Oil leaking around the cover may be due to the filter element (B, **Figure 15**) being trapped under the outer edge of the cover or a plugged lubrication hole (D).*

FLYWHEEL

Removal

The flywheel is secured to the crankshaft by a retaining nut (**Figure 16**). On models with a rewind starter, the nut also holds the starter cup. A separate fan is attached to the flywheel on some engines.

1. Disconnect the spark plug wire and properly ground it to the engine.
2. Remove the blower housing and any other components so the flywheel is accessible.
3. If equipped with a flywheel brake, refer to Chapter Seven and disengage the brake so the flywheel turns freely.
4. If equipped with an ignition coil or module mounted outside the flywheel, remove the coil or module.
5. Use a suitable tool, such as a strap wrench, to hold the flywheel and unscrew the flywheel retaining nut (**Figure 17**).

NOTE
*If the flywheel has fins, **do not** attempt to hold the flywheel by inserting a tool between the fins.*

6A. To remove the flywheel using a flywheel puller, proceed as follows:

a. Install a flywheel puller as shown in **Figure 18** so the puller screws engage the holes adjacent to the flywheel hub.

14

15

16

17

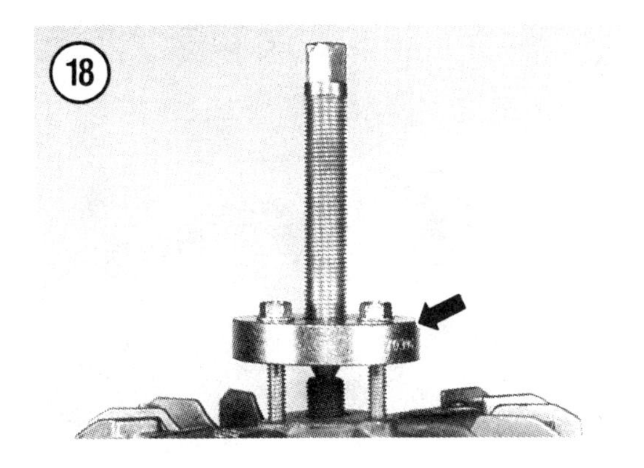

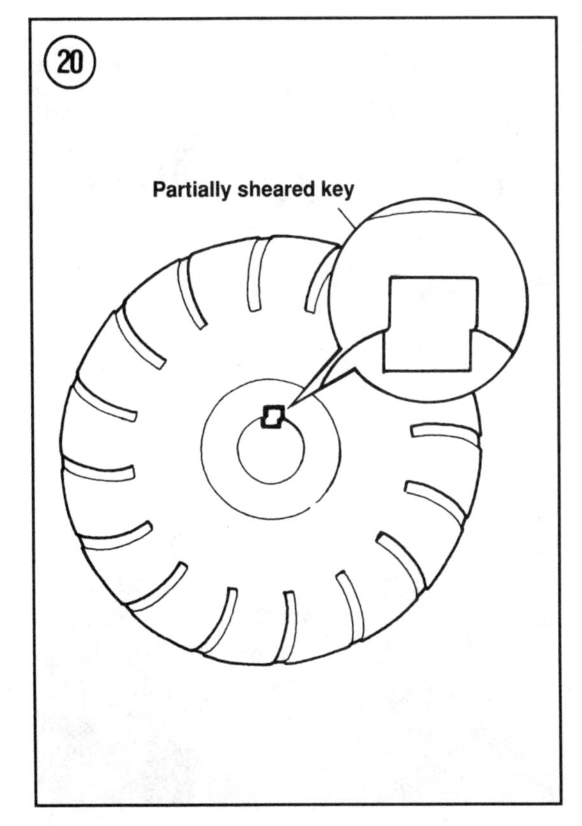

NOTE
The puller holes in the flywheel may not be threaded. The flywheel puller screws must be self-tapping or threads must be cut into the holes using a 1/4 × 20 tap.

b. Rotate the puller center screw until the flywheel "pops" free.

c. Remove the flywheel and the flywheel key. If additional engine work is planned, it is a good practice to wrap the crankshaft threads with tape to prevent damage to the threads.

6B. To remove the flywheel using a knock-off tool, proceed as follows:

a. Screw the tool onto the crankshaft until the tool bottoms against the flywheel, then turn the tool back two turns.

b. Insert a screwdriver or other tool under the flywheel and pull up on the screwdriver so pressure is applied to the underside of the flywheel.

c. While pulling up on the screwdriver, strike the knock-off tool sharply (**Figure 19**). The flywheel will "pop" free.

WARNING
Do not strike the flywheel. The flywheel must be replaced if the flywheel is cracked or any fins are damaged.

Inspection

1. Inspect the flywheel and the crankshaft. The tapered portion of the flywheel and crankshaft must be clean and smooth with no damage due to movement between the flywheel and crankshaft.

2. Check the fit of the flywheel on the crankshaft. There should be no looseness or wobbling.

3. Replace the flywheel if any cracks are evident or any fins are broken.

4. Be sure the keyway in the crankshaft and flywheel is not damaged or worn.

5. Inspect the flywheel key. If any indications of shearing (**Figure 20**) are apparent, then the key must be replaced. If key replacement is required, be sure the proper key is installed. The key must match the type of ignition system used. See Chapter Seven.

Installation

1A. Install a stepped flywheel key so the stepped end is towards the engine (**Figure 21**).

1B. Install a tapered flywheel key so the big end is towards the engine as shown in **Figure 22**.

2. If a spacer is used, install the spacer so the protrusion fits in the keyway and is towards the end of the crankshaft (**Figure 23**).

3. Install the flywheel.

4. On engines with a plastic fan that grips a cast iron flywheel, install the fan by first heating the fan in boiling water, then place the fan on the flywheel so the indexing bosses are properly positioned. On Vector VLV50, VLXL50, VLV55 and VLXL55 engines, install the fan so the Tecumseh emblem is adjacent to the flywheel magnet as shown in **Figure 24**.

21

22

5. On models so equipped, be sure the debris screen is positioned on the starter cup (**Figure 25**).

6. The washer under the retaining nut may be a flat washer or a Belleville washer, which has a rounded cross-section. Install the Belleville washer so the cupped side is towards the flywheel.

7. Tighten the flywheel retaining nut to the torque listed in **Table 2**.

8. If equipped with an ignition system located outside the flywheel, install the coil or module. Adjust the ignition coil armature air gap as outlined in Chapter Five.

23

CYLINDER HEAD

Removal

1. Remove the blower housing as well as any brackets that will interfere with cylinder head removal. On some engines, components attached to the cylinder head with a bracket must also be removed.

2. Unscrew the cylinder head retaining screws (**Figure 26**).

NOTE

Some engines are equipped with cylinder head screws with different lengths. Make a diagram and mark the length and location of the cylinder head screws. Also note the location of any washers.

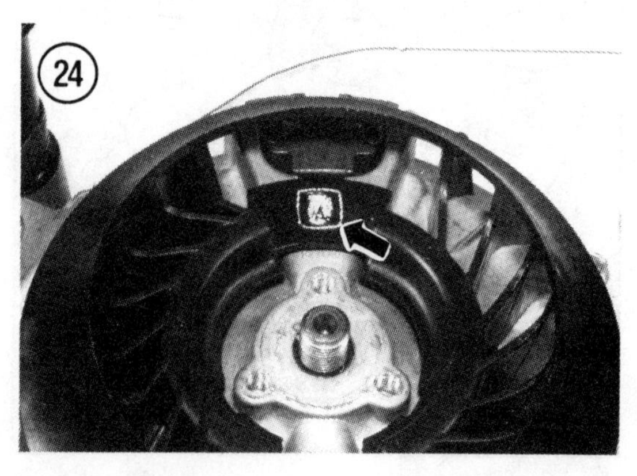

24

25

26

27

28

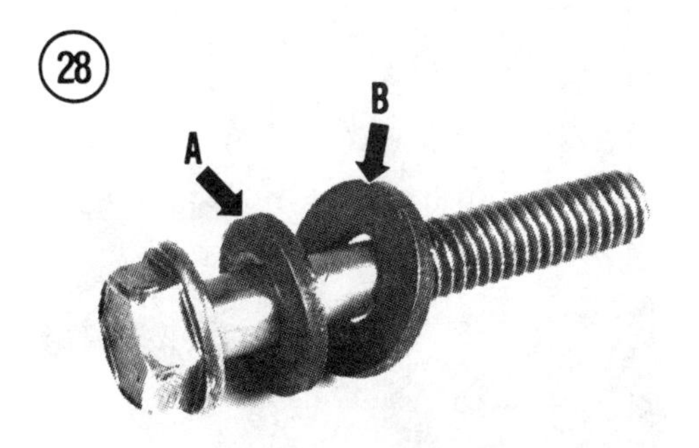

3. Remove the cylinder head. It may be necessary to tap the cylinder head to break it free from the gasket.
4. Remove the cylinder head gasket (**Figure 27**).

Inspection/Cleaning

1. Look for black carbon tracks on the cylinder head and the top of the engine which indicate that the gasket was leaking. Evidence of leakage past the gasket may indicate that the cylinder head is warped.
2. Use a wooden or plastic scraper to remove carbon deposits from the cylinder head.
3. Rotate the crankshaft so the piston is at the top of the cylinder and both valves are closed, then remove carbon from the piston, valves and cylinder block. Spray a suitable solvent on hardened deposits to soften them. Be careful not to damage engine surfaces.
4. Two methods may be used to check for a warped cylinder head and correct it. A large file can be drawn across the cylinder head surface to indicate the high spots. Another method is to place the head on a flat plate, such as glass, that has dabs of valve grinding compound on it, then move the head in a figure-eight pattern on the plate. Either method can be used satisfactorily. Using a file will remove metal more quickly, but care should be used to remove no more metal than is necessary to produce a level surface.

Installation

1. Install a new head gasket. Do not apply any type of sealer to the head gasket.
2. Apply graphite grease to the cylinder head screws before installation.
3. Some cylinder head screws are equipped with flat washers, while some screws are equipped with a Belleville washer (A, **Figure 28**) as well as a flat washer (B). The Belleville washer must be installed so the concave side is towards the screw threads.
   a. On H50, H60, H70 and V50 series engines, install cylinder head screws equipped with a Belleville washer in the locations shown in **Figure 29**.
   b. On HM100 and TVM220 series engines, install cylinder head screws equipped with a Belleville washer in the locations shown in **Figure 30**.

4. The cylinder head screws must be tightened incrementally in a sequence to prevent distortion of the cylinder head.
   a. Tighten the cylinder head screws on engines with eight cylinder head screws in the sequence shown in **Figure 31**.
   b. Tighten the cylinder head screws on engines with nine cylinder head screws in the sequence shown in **Figure 32**.
   c. Tighten the cylinder head screws on Vector VLV50, VLXL50, VLV55 and VLXL55 engines in the sequence shown in **Figure 33**.
5. Initially tighten the cylinder head screws finger-tight. Using a torque wrench, tighten the screws in 50 in.-lb. (5.6 N•m) increments until the final torque reading is obtained. The final torque specification is 180-220 in.-lb. (20.3-24.9 N•m) on Vector VLV50, VLXL50, VLV55 and VLXL55 engines and 160-200 in.-lb. (18.1-22.6 N•m) on all other engines.

VALVE SYSTEM

The valves are located in the cylinder block portion of the engine. Each valve rides in a non-replaceable bushing and sits in a non-replaceable valve seat. If valve service is required, refer to the following sections.

Removal

1. Remove the cylinder head as previously outlined.
2. Remove components as required for access to the valve cover (**Figure 34**). On some engines, the crankcase breather is located behind the valve cover.
3. Rotate the flywheel so the piston is at the top of the cylinder and both valves are closed.

4. Using a valve spring compressor, compress the valve spring by positioning the compressor against the valve spring retainer (**Figure 35**).

NOTE
*Several types of valve spring compressors (**Figure 36**) are available. Be sure*

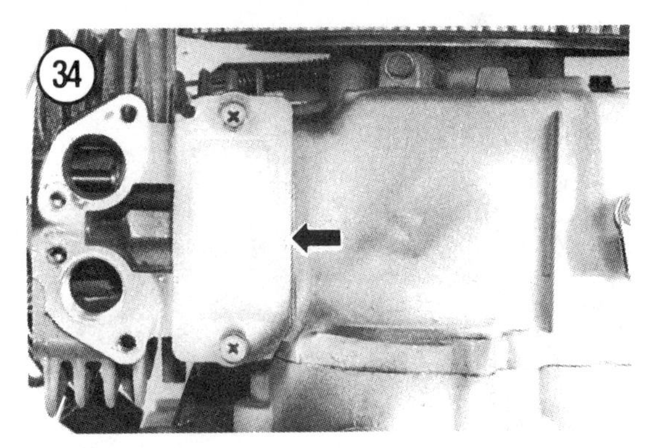

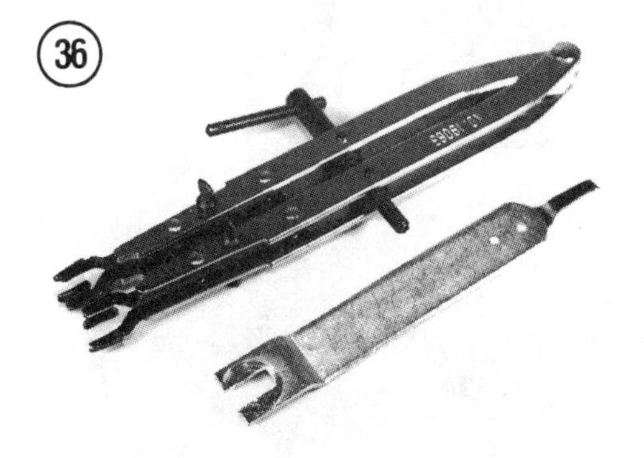

the compressor will work on a small engine.

WARNING

Safety eyewear should be worn when using a valve spring compressor. A compressed spring can travel several feet if it works free of the compressor.

5. Remove the valve, valve spring retainer and valve spring. The valve spring retainer is slotted and fits in a groove on the valve (**Figure 37**). Position the larger section of the retainer slot over the valve stem, then pull out the valve.

NOTE

If the intake and exhaust valves are similar in size, mark them to prevent misidentification.

6. On some large engines, a valve stem seal (**Figure 38**) is located on the intake valve.
7. On early engines, a spring cup (**Figure 39**) is located between the bottom of the spring and engine.

Inspection

The exhaust valve on some engines is identified by "EX" or "X" marked on the valve head (**Figure 40**), while the intake valve on some engines is identified by "I" marked on the valve head.

1. Inspect each valve for damage and excessive wear.
2. Inspect the valve before cleaning. Gummy deposits on the intake valve (**Figure 41**) may indicate that the engine has run on gasoline that was stored for an extended period. Hard deposits on the intake valve are due to burnt oil, while hard deposits on the exhaust valve (**Figure 42**) are due to combustion byproducts and burnt oil.
3. Moisture, which can accumulate in the engine during storage, can cause corroded or pitted valves which should be replaced.
4. Check the valve face for an irregular seating pattern. The seating ring around the face should be concentric with the valve head and equal in thickness all around the valve. If the seating pattern is irregular, then the valve may be bent or the valve face or seat is damaged.
5. Remove deposits either with a wire brush or soak the valve in parts cleaner.
6. Run a fingernail down the valve stem and check for a ledge that would indicate that the valve stem is worn. Replace the valve if the valve stem is worn.
7. The valve stem must be perpendicular to the valve head. To check for a bent valve, carefully install the valve stem in a drill chuck and rotate the drill. Be sure the valve stem is centered in the drill chuck.

8. Measure the valve head margin (**Figure 43**). The valve must be replaced if the margin is less than 1/32 in. (0.8 mm).
9. If the valve and valve seat (**Figure 44**) are worn, but serviceable, they can be restored by machining. Take the valves and engine to a professional shop that is equipped to perform valve machine work.
10. The valve face angle (**Figure 43**) should be 45° and the valve seat angle (**Figure 45**) should be 46°.
11. Valve seat width (**Figure 45**) should be 0.042-0.052 in. (1.07-1.32 mm) for models H40, H50,

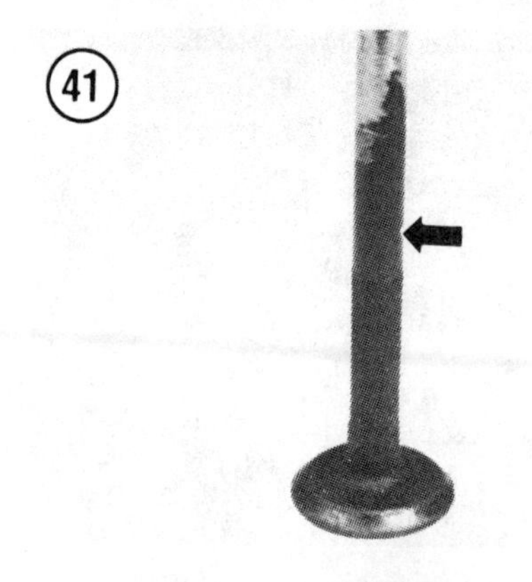

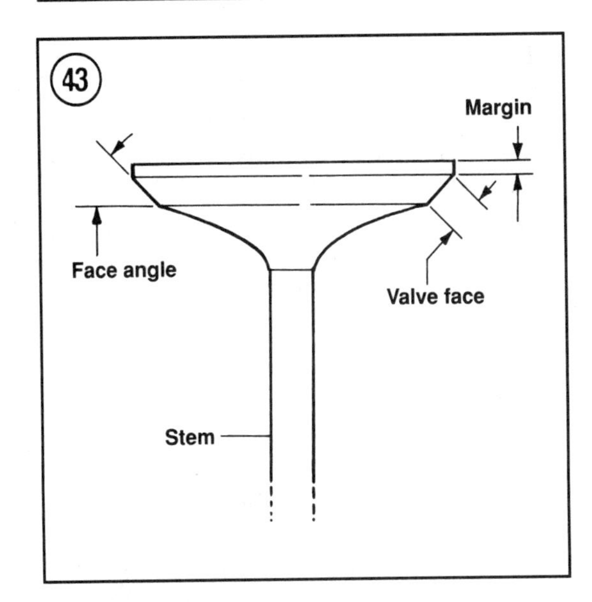

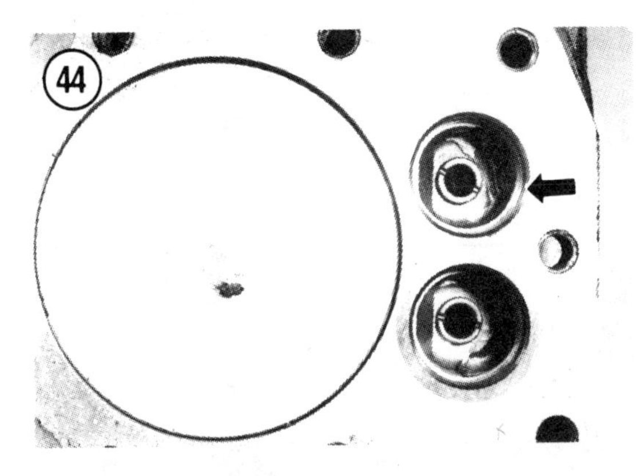

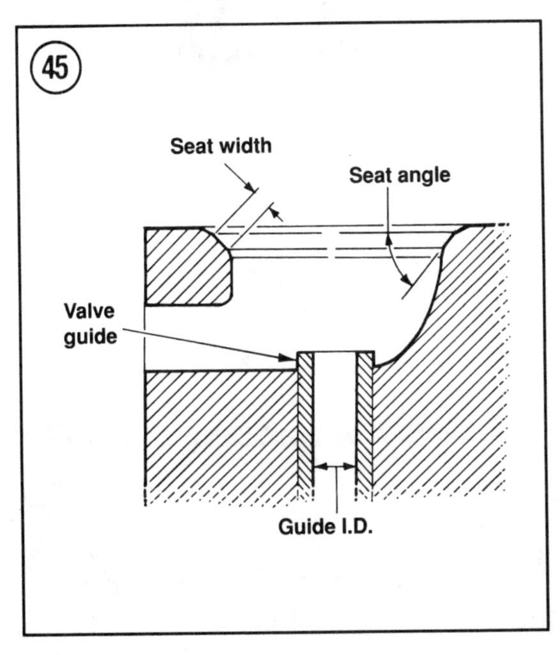

H60, H70, HM70, H80, HM80, HM100, TVM125, TVM140, TVM170, TVM195, TVXL195, TVM220, TVXL220, V40, V50, V60, V70, VM70, V80, VM80 and VM100; and 0.035-0.045 in. (0.89-1.14 mm) for all other models.

12. The valve seats are not replaceable. If the valve seat is damaged to the extent that machining will not restore the seat, then the engine block must be replaced.

13. The valve guides are not replaceable. To check for excessive clearance between the valve stem and valve guide, insert the valve into the guide and move the valve head from side to side. If there is excessive wobble, then the valve guide inside diameter should be enlarged with a reamer (**Figure 46**) to accept a valve with a larger valve stem. Tecumseh offers valves that have valve stems that are 1/32 in. (0.8 mm) oversize. Use Tecumseh reamer 670283 if the original valve stem diameter is 1/4 in. (6.35 mm) or use reamer 670284 if the original valve stem diameter is 5/16 in. (7.94 mm). The specified oversize valve guide diameter is 0.2807-0.2817 in. (7.130-7.155 mm) for oversize valves with a 9/32 in. (7.14 mm) stem. The specified oversize valve guide diameter is 0.3432-0.3442 in. (8.717-8.743 mm) for oversize valves with a 11/32 in. (8.73 mm) stem. Proceed as follows to ream a valve guide to accept an oversize valve.

NOTE
If the engine is not disassembled, prevent machining debris from entering the engine.

a. Fit the reamer in a suitable hand-holder.
b. Insert the reamer in the valve guide (**Figure 47**).

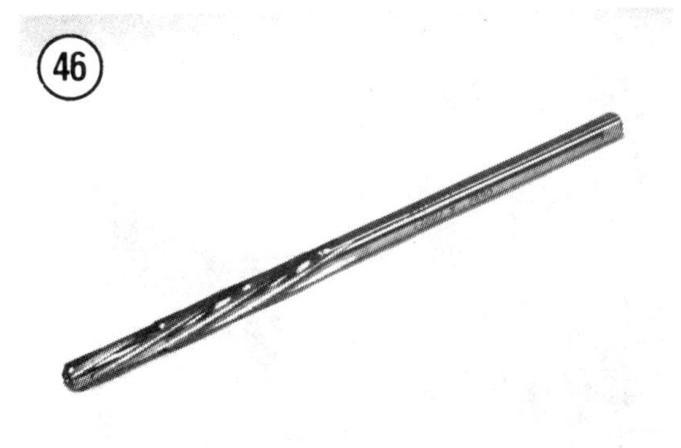

c. Carefully rotate the reamer, but do not use excessive force. Rotate the reamer until it travels completely through the valve guide.
d. Withdraw the reamer while turning clockwise.
e. Thoroughly clean the valve guide and engine.
f. Lubricate the valve guide, install the oversize valve and check for fit. If the valve will not rotate or binds in the guide, carefully run the reamer through the guide again.

NOTE
When using oversize valves, the hole in the valve retainer and spring cup, if used, must be enlarged with an appropriate size drill.

14. After a valve guide is reamed, the valve seat and guide may not be centered. Check the contact pattern between the valve and valve seat by lapping the valve. If the pattern is not concentric, then the valve seat must be machined.
15. Check the valve spring. Discard the spring if it is pitted, cracked or broken. The spring should be straight as shown in **Figure 48**.
16. Inspect the valve seal (**Figure 49**) on engines so equipped. Replace the seal if it is cracked or damaged or if it does not fit the valve stem snugly.

Valve Lapping

Lapping the valve against the valve seat is a simple operation which can be used to check the contact pattern between the valve and valve seat. Valve lapping is also a means of restoring the seating surfaces without machining if the amount of wear or distortion is not too great.

Lapping requires the use of a good lapping compound and a lapping tool (**Figure 50**). Lapping compound is available in either coarse or fine grade. Coarse compound is used first, followed by fine compound.

1. Apply small dabs of compound to the valve face (**Figure 51**). Too much can fall into the valve guide and cause damage.

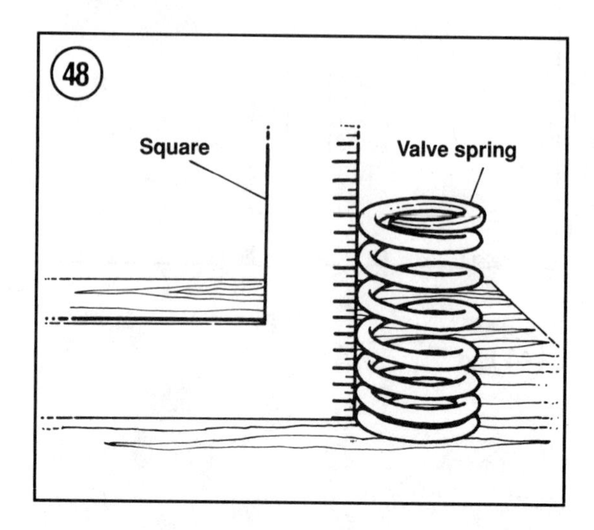

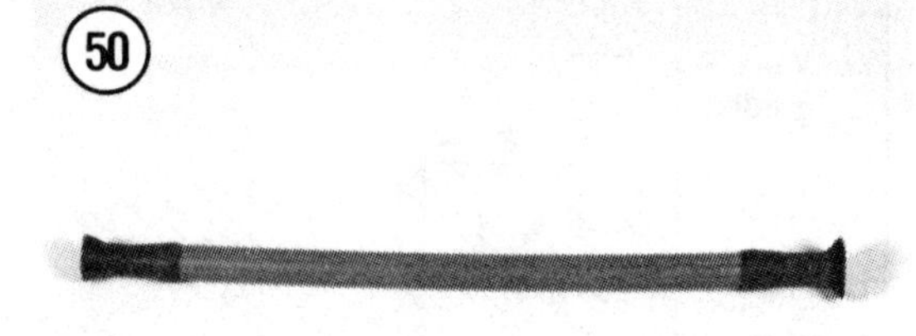

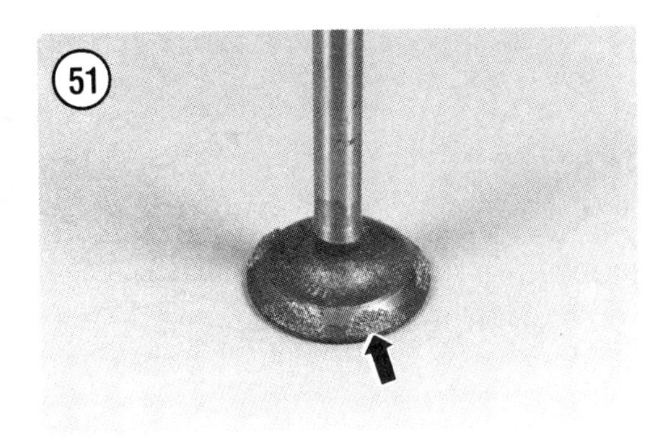

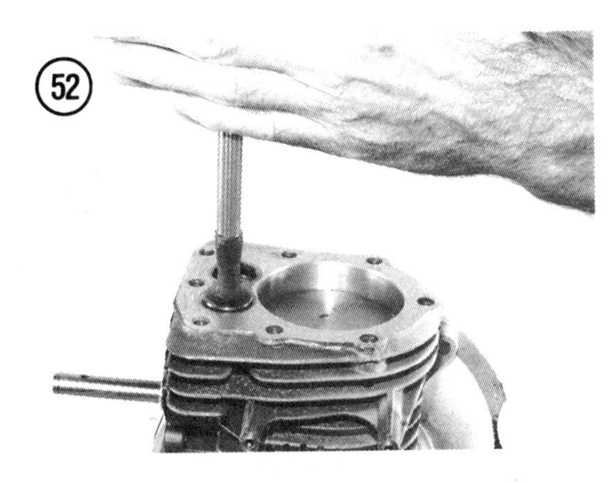

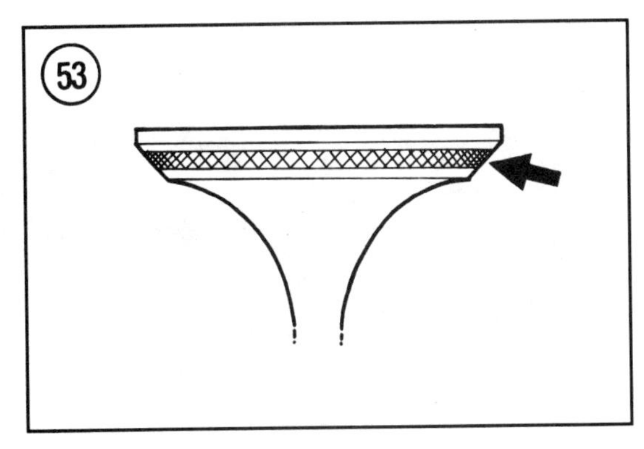

2. Insert the valve into the valve guide.
3. Moisten the end of the lapping tool suction cup and place it on the valve head.
4. Rotate the lapping tool back and forth between your hands several times (**Figure 52**).

NOTE
When the "grating" sound lessens, reapply lapping compound.

WARNING
Do not use a drill to rotate the valve. The lapping compound may be forced out so quickly that metal-to-metal contact will occur, thereby damaging the valve face and/or valve seat.

5. Relocate the valve head approximately 45°, then repeat the lapping motion.
6. Switch to fine compound after several lapping sequences.
7. Clean the compound from the valve and seat frequently to check progress of lapping. Lap only enough to achieve a precise seating ring around the valve head. The pattern on the valve face should be an even width as shown in **Figure 53**. The valve seat in the cylinder block should be smooth and even.
8. After lapping has been completed, thoroughly clean all lapping compound from the valve and engine. Any lapping compound residue will cause rapid wear if left in the engine.

Installation

1. On engines so equipped, install the intake valve seal so the raised side is towards the valve spring retainer as shown in **Figure 54**.
2. Lubricate the valve stems before installation.
3. Some springs are progressively wound, with the coils at one end more closely wound than the opposite end. The close-wound coils must be towards the engine and the wide-wound coils must be towards the valve spring retainer. See **Figure 55**.
4. Assemble the valve spring and retainer in the compressor as shown in **Figure 56** (note the position of the large hole of the retainer slot).
5. Position the spring and retainer in the engine, insert the valve so the valve stem passes through the large portion of the retainer slot. Then, push the retainer so the valve stem engages the narrow portion of the retainer slot.

6. Measure the valve tappet gap before installing the cylinder head (all valve components must be installed for an accurate measurement).
7. Install the crankcase breather and valve chamber cover.
8. Install the cylinder head.

Valve Tappet Gap

The valve tappet gap should be checked whenever the valves are serviced. All valve components must be installed for an accurate measurement.

1. Rotate the flywheel so the piston is at the top and both valves are closed.
2. Measure the gap between the end of the valve stem and the tappet end for each valve using a feeler gauge (**Figure 57**) with the engine cold.
3. The valve tappet clearance should be 0.010 in. (0.25 mm) for both valves on models H50 (ignition under flywheel), H60, H70, HM70, H80, HM80, HM100, TVM125 (ignition under flywheel), TVM140, TVM170, TVM195, TVM220, V50, V60, V70, VM70, V80 and VM100. The valve tappet clearance for all other models should be 0.008 in. (0.20 mm) for both valves.
4. If the tappet gap is less than specified, the valve must be removed so the valve stem end can be ground or filed to increase the gap. Be especially careful when grinding the stem as too much metal can be quickly removed. The end of the valve must be square with the stem after grinding. Dress the end of the valve stem to remove any burrs.
5. If the tappet gap is greater than specified, then the valve seat must be ground to lower the valve or renew the valve and/or tappet to decrease the clearance.

OIL SEALS (SERVICE WITH CRANKSHAFT INSTALLED)

Oil seals at the flywheel end of the crankshaft (**Figure 58**) and at the output end (**Figure 59**) prevent oil in the engine from leaking out. Wear or damage can reduce their effectiveness. Oil leakage can become a problem, particularly if the oil contaminates the ignition breaker points or excessive amounts of oil are lost.

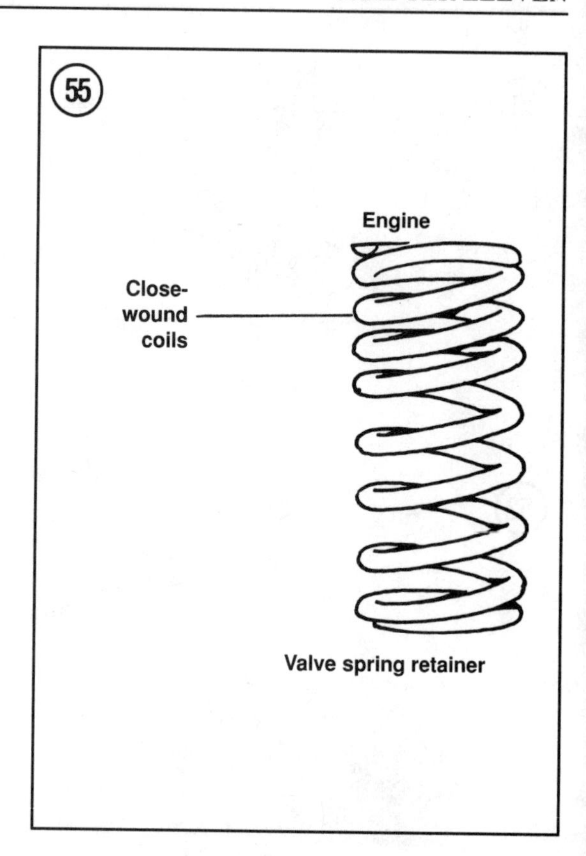

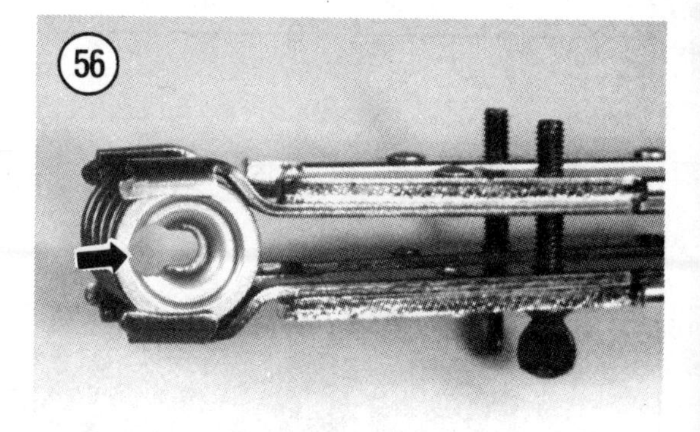

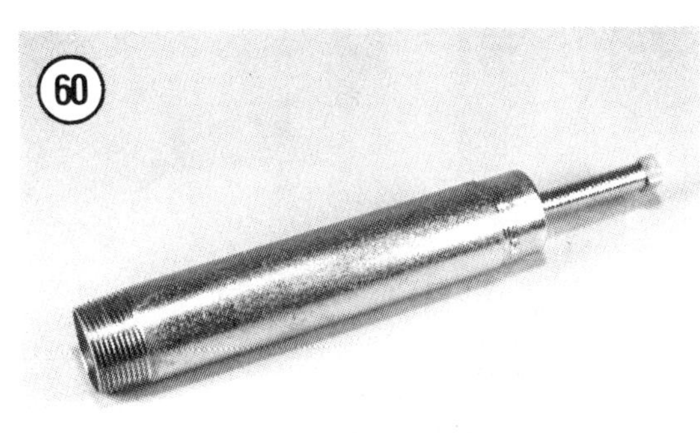

Removal

In some instances, it is possible to remove and install an oil seal without removing the crankshaft. Depending on the location of the engine and which oil seal is leaking, it may be necessary to remove the engine from the equipment.

If the oil seal at the flywheel end of the crankshaft is leaking, the flywheel must be removed (see *Flywheel* section). If the oil seal at the output end of the crankshaft is leaking, any parts attached to the crankshaft that deny access to the oil seal must be removed.

NOTE
On vertical crankshaft engines, drain the engine oil before replacing the lower crankshaft oil seal.

1. Before removing an oil seal, determine the depth of the existing seal so the new seal can be installed in the same position.
2. Tecumseh offers a seal remover tool (**Figure 60**). The tool screws into the seal (**Figure 61**), then the seal is extracted by turning the jack screw in the end of the tool.
3. Other methods of seal removal are possible, such as inserting a thin-blade screwdriver or pick under the seal and prying it out.

CAUTION
Care must be exercised when extracting a seal to prevent damage to the crankshaft or metal behind the seal.

Installation

1. Clean the seal seating area so the new seal will seat properly.
2. Check the oil seal bore in the engine and dress out any burrs created during removal. Remove only metal that is raised and will interfere with installation of the new seal. If a deep gouge is present, fill the depression with a suitable epoxy so it is level with the surrounding metal.
3. Inspect the crankshaft for a wear groove caused by the seal lip rubbing against the shaft.

NOTE
If a wear groove is present on the crankshaft, it may be possible to reposition the new seal slightly so the seal lip will

bear against an unworn portion of the shaft.

4. Apply nonhardening sealer to the periphery of the oil seal prior to installation.
5. Cover keyway and threads on the crankshaft with thin tape (**Figure 62**) or a seal protector sleeve so the oil seal lip will not be cut when passing the oil seal down the crankshaft.
6. Apply oil or grease to the lip of the seal.
7. Position the seal so the seal lip is slanted toward the inside of the engine.
8. Use a suitable tool with the same diameter as the oil seal to force the oil seal into the engine (**Figure 63**). A deep-well socket may work if the end of the crankshaft is short. Suitable sizes of tubing or pipe may be purchased at a hardware store and used as seal installing tools. Be sure the end of the tool is square and not sharp.

INTERNAL ENGINE COMPONENTS

The following discussion covers the following engine components: piston, piston rings, connecting rod, camshaft, tappets, crankshaft, bearings and governor. The overhaul procedure addresses a complete disassembly of internal engine components. Although some components can be serviced without a complete disassembly, damage to one internal engine component often affects some or all of the other parts inside the engine. Failing to inspect all components may result in another repair that could have been avoided.

NOTE
The following procedures apply to a basic Tecumseh engine. It may be necessary to modify a specific procedure to accommodate variations in engine components or special engine applications.

NOTE
Some specifications apply to engines with "external ignition." Engines with external ignition have all ignition components located outside the flywheel.

ENGINE DISASSEMBLY

Use the following procedure to disassemble the engine.

1. Drain the engine oil and remove the engine from the equipment.
2. Remove the electric starter motor, if so equipped.
3. Remove the muffler.
4. Remove the air cleaner, fuel tank and carburetor.
5. Remove the rewind starter and/or blower housing.
6. Remove the flywheel and key, and if so equipped, the plastic spacer on the crankshaft (**Figure 64**).
7. If equipped with a breaker point ignition system, check that the ignition timing marks on the stator

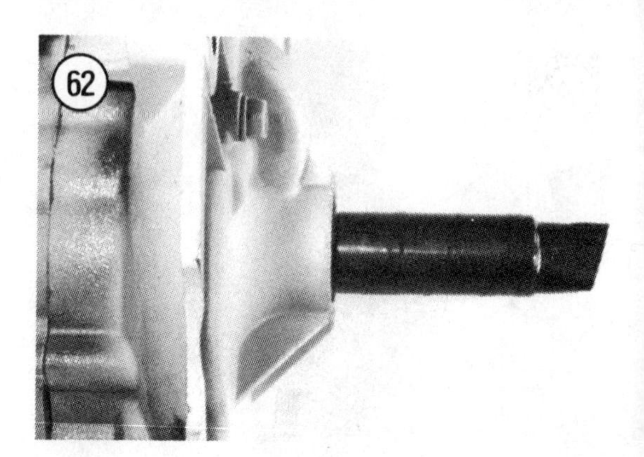

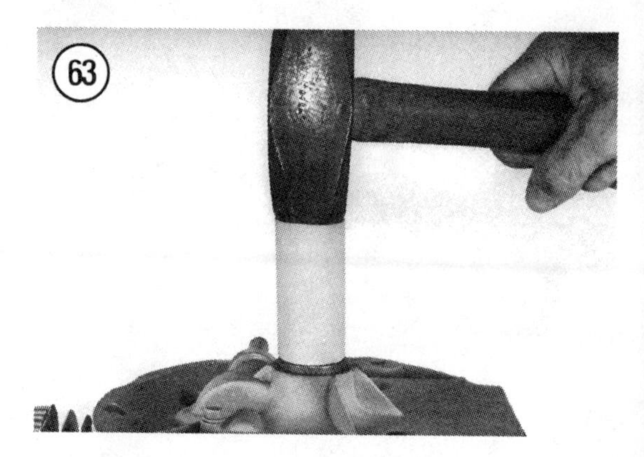

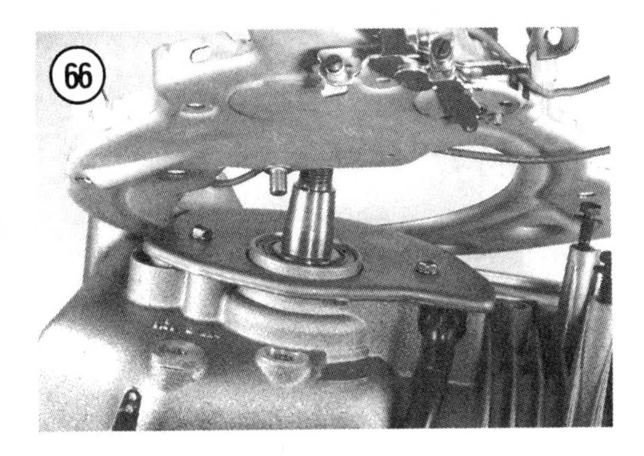

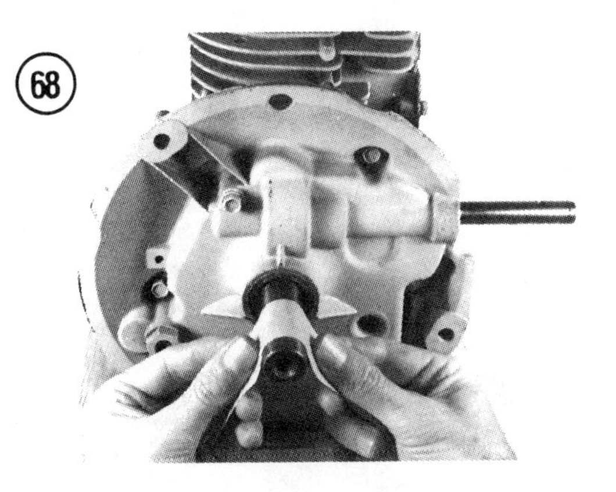

plate and mounting post (**Figure 65**) are aligned. If there are no marks, and the ignition timing is correct, make marks for reference during assembly.

8. Remove the ignition system components.

NOTE

*On Vector VLV50, VLXL50, VLV55 and VLXL55 engines, the blower plate can be removed (**Figure 66**) after unscrewing the three retaining screws and detaching the governor spring. Disassembly of the flywheel brake mechanism is not required.*

9. Remove the cylinder head and head gasket.
10. Remove the crankcase breather.
11. Remove the valve retainers and remove the intake and exhaust valves and springs.
12. Clean any carbon from the top of the cylinder bore.
13. Run a fingernail over the top portion of the cylinder and determine if a ridge exists. Use a ridge reamer (**Figure 67**) and remove the ridge at the top of the cylinder to prevent damage to the piston or rings when removing.
14. If not previously removed, remove any parts attached to the crankshaft.
15. Remove any rust or burrs from the crankshaft. This will prevent damage to the main bearing. A strip of emery cloth can be used as shown in **Figure 68** to remove rust. Burrs may be found adjacent to the keyway. It may be necessary to file down the end of the crankshaft to the original crankshaft diameter.

NOTE

*Refer to **Table 2** in Chapter Three for engine model number and displacement specifications.*

16A. On horizontal crankshaft engines with a displacement of 12.04 cu. in. (197 cc) or less (except Model H40), proceed as follows.

a. Remove the oil seal at the output end of the crankshaft to check for the presence of a snap ring on the crankshaft. Refer to the *Oil Seals* section for the seal removal procedure.
b. On engines with a ball bearing at the output end of the crankshaft, a snap ring is located on the crankshaft (**Figure 69**). Remove the snap ring. The ball bearing will be removed with the crankcase cover.

16B. On Model H40 and horizontal crankshaft engines with a displacement greater than 12.04 cu. in. (197 cc), proceed as follows.

a. Check for the presence of bearing lock screws on the crankcase cover (**Figure 70**). If present, the lock screws secure the ball bearing in the crankcase cover.
b. Loosen the locknuts on the screws, then rotate each screw counterclockwise so the flat on the screw faces away from the crankshaft. The bearing can now slide out of the crankcase cover with the crankshaft.

17. Unscrew the oil pan retaining screws on vertical crankshaft engines (**Figure 71**—typical engine shown) or the crankcase cover retaining screws on horizontal crankshaft engines (**Figure 72**—typical engine shown).

NOTE
Mark or make note of the location of any screws that have different lengths.

18. Remove the oil pan on vertical crankshaft engines or the crankcase cover on horizontal crankshaft engines by tapping with a soft-faced hammer. If the engine is equipped with an auxiliary drive shaft (A, **Figure 71**), turn the crankshaft while removing the oil pan so the internal gears turn and disengage.

CAUTION
Do not pry between the mating surfaces. Do not use excessive force against the oil pan or crankcase cover. Applying excessive force may damage the main bearing, oil pan or crankcase cover. If binding occurs, determine if an obstruction on the crankshaft, such as a burr, is causing the binding. If no obstruction is present, then the crankshaft may be bent. If the crankshaft is bent, then a replacement short block should be considered, depending on the condition of other internal engine components.

NOTE
Tools are available to straighten a bent crankshaft, however, Tecumseh does not recommend straightening crankshafts, but states that a bent crankshaft must be discarded.

69

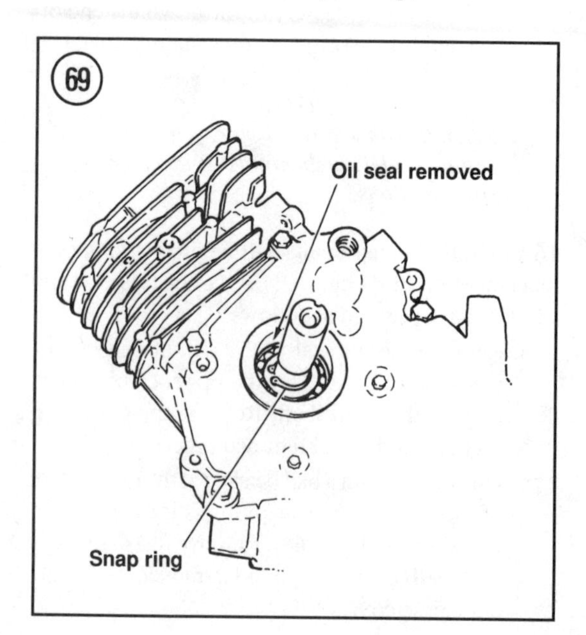

70

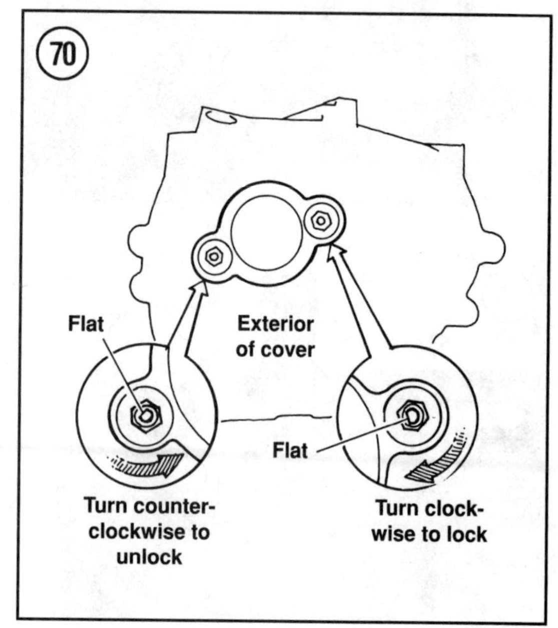

71

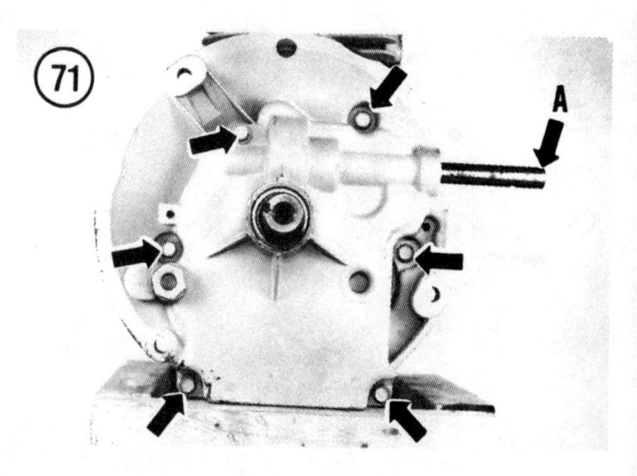

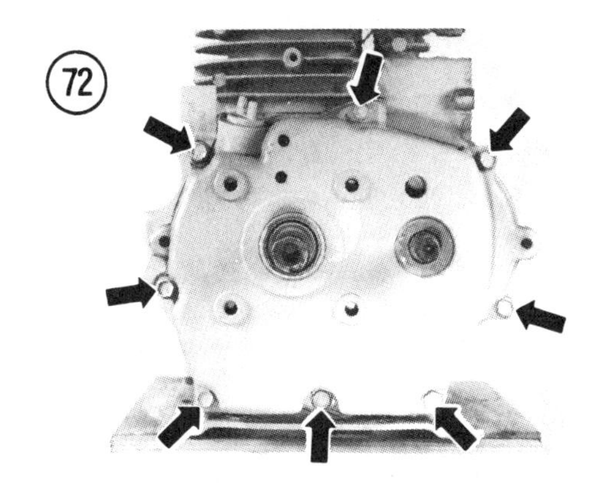

19. On engines with a vertical crankshaft, lift the oil pump (**Figure 73**) off the camshaft.

20. On engines equipped with a balancer, remove the balancer drive gear (A, **Figure 74**) and balancer shaft (B).

21. Position the engine so the tappets will not fall out.

22A. On HM70, HMXL70, HM80, HM100, TVM170, TVM195, TVXL195, TVM220, TVXL220, VM70, VM80 and VM100 engines, the camshaft must be properly positioned so that the compression release on the camshaft will not contact the exhaust valve tappet. Rotate the crankshaft counterclockwise so that the beveled tooth on the crankshaft gear and the mark adjacent to the hole in the camshaft gear are aligned, then continue turning the crankshaft so that the crankshaft beveled tooth and camshaft timing mark are out of alignment by three teeth (**Figure 75**).

22B. On all engines except those listed in Step 22A, rotate the crankshaft so that the timing marks are aligned (**Figure 76**). Note that the timing mark on the crankshaft gear may be either a round punch mark or a straight line mark.

NOTE
If there is no timing mark on the crankshaft gear, align the keyway in the crankshaft with the timing mark on the camshaft gear.

NOTE
*If the crankshaft is equipped with a ball bearing or the crankshaft gear is pressed on (no keyway), align the crankshaft gear tooth that is beveled or has a punch mark (**Figure 77**) with the mark on the camshaft.*

23. Remove the camshaft.

24. Mark then remove the tappets (**Figure 78**). The tappets must be marked so they can be reinstalled in their original positions.

25. Some engines are equipped with screws that retain the rod cap to the connecting rod, while other engines are equipped with bolts. Unscrew the connecting rod cap retaining screws or nuts (**Figure 79**) and remove the rod cap and oil dipper, if so equipped. Some engines may have lock tabs that must be bent back before the screw or nut can be unscrewed.

NOTE
Care must be exercised not to damage the bearing surface of the connecting rod and rod cap.

26. Push the connecting rod and piston out the top of the engine (**Figure 80**).

27. Remove the crankshaft. Do not lose any washers mounted on the crankshaft; they must be reinstalled in their original locations.

28. The major components of the engine should now be removed from the engine crankcase. Refer to the following sections for further disassembly and inspection procedures.

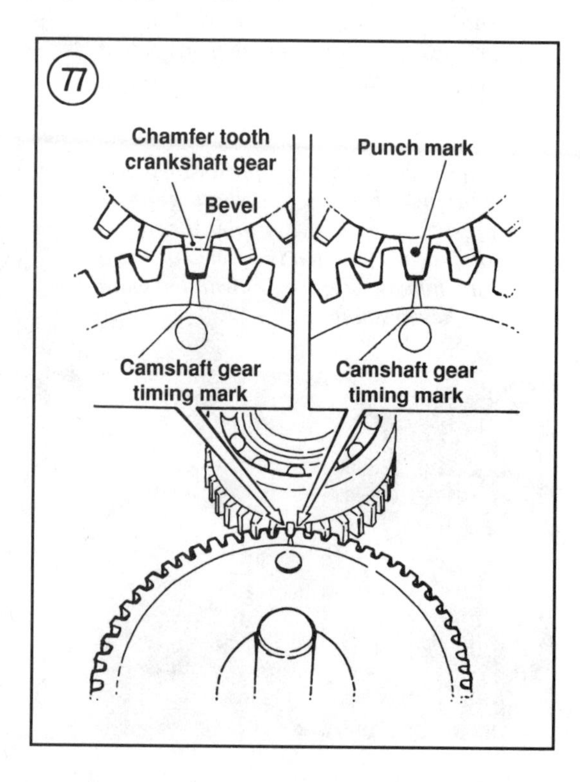

PISTON RINGS

Replacement

A suitable piston ring expander tool (**Figure 81**) should be used to remove or install the piston rings. Although the rings can be removed or installed by hand, there is less chance of piston ring breakage or gouging the piston ring grooves when the ring expander tool is used.

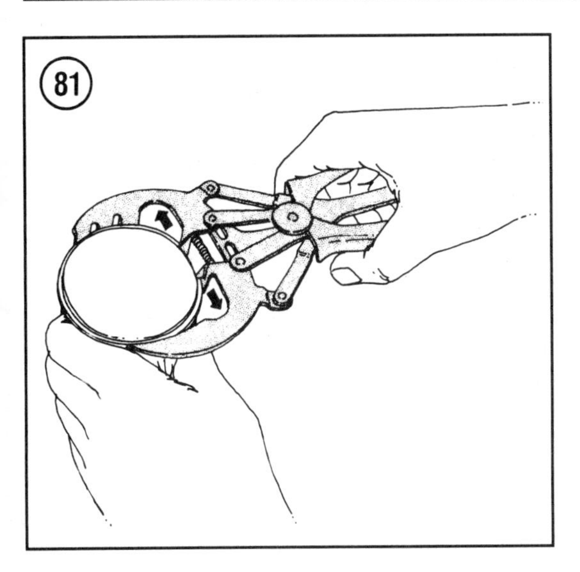

If new piston rings are being installed, the cylinder bore surface should be reconditioned with a hone or deglazing tool to restore the crosshatch pattern. Refer to the *Cylinder* section.

1. Remove the rings from the piston. Remove the top compression ring (closest to the piston crown) first, then the second compression ring and finally the oil control ring. Move the rings toward the piston crown for removal.

NOTE

Mark the rings so they can be returned to their original positions; note which side is towards the piston crown. Reverse the procedure when installing the rings.

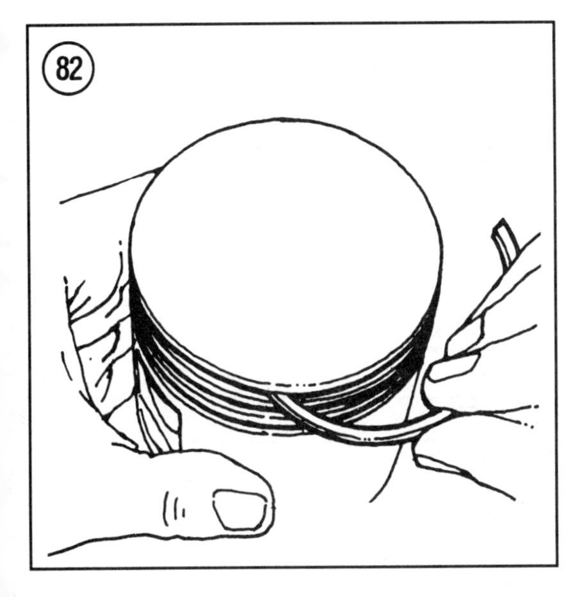

2. Clean the piston ring grooves. One method for cleaning piston ring grooves is to pull the end of a broken piston ring through the groove as shown in **Figure 82**. Care must be used not to gouge or otherwise damage the groove. Piston ring groove cleaning tools are also available.

3. If replacement rings are required, be sure the replacement piston ring size matches the piston size. Oversize pistons are marked on the piston crown (**Figure 83**) to indicate if the piston is oversize. A marking of ".010" would indicate that the piston is 0.010 in. oversize; a piston ring set that is 0.010 in. oversize must be installed.

4. Check the end gap of the two compression piston rings. Place the piston ring in the cylinder bore, then push the piston ring down into the bore with a piston (**Figure 84**) so the ring is square in the bore. Position the piston ring 1 in. (25 mm) down in the bore from the top of the cylinder and measure the piston ring end gap as shown in **Figure 85**. If the piston ring end

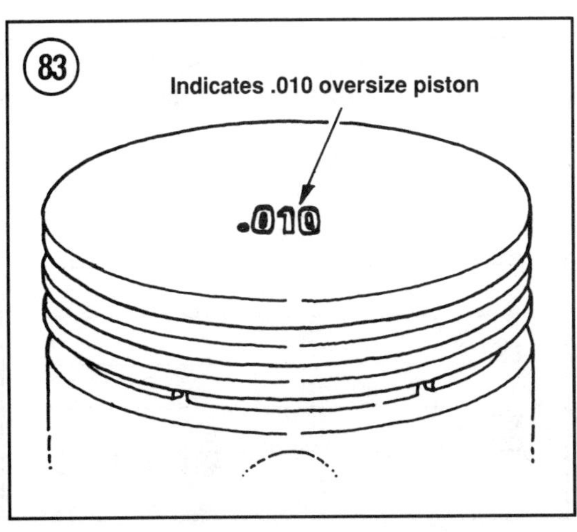

gap is excessive, the ring must be replaced. Refer to **Table 1** for piston ring end gap specification.

NOTE
*The end gap of new piston rings should also be measured. If the gap is greater than the specification in **Table 1**, the cylinder bore may be excessively worn. Measure the cylinder bore as outlined in the **Cylinder** section.*

4. Install the replacement rings by reversing the removal procedure in Step 1. Refer to **Figures 86-88** for the proper position of the rings. Note that on all top compression rings the beveled side of the ring must be towards the top side of the piston (towards the piston crown). On all engines except HM100, TVM220 and TVXL220, if the second compression ring has a beveled side, then the beveled side must be towards the top side of the piston (towards the piston crown). On HM100, TVM220 and TVXL220 engines, if the second compression ring has a beveled side, then the beveled side must be towards the bottom of the piston (towards the piston skirt). If there is a notch on the outside of the piston ring, install the ring with the notch towards the piston skirt.

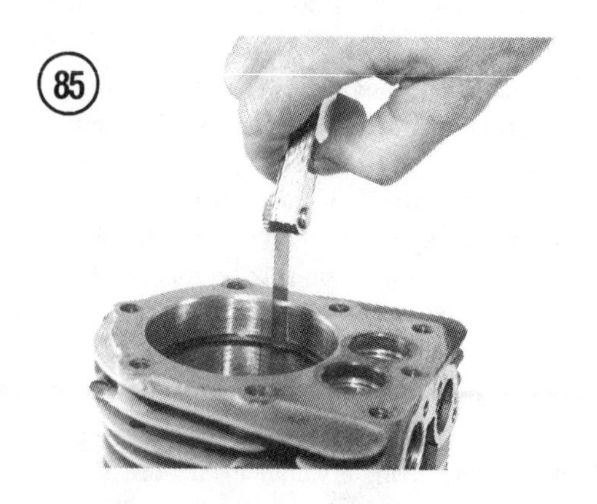

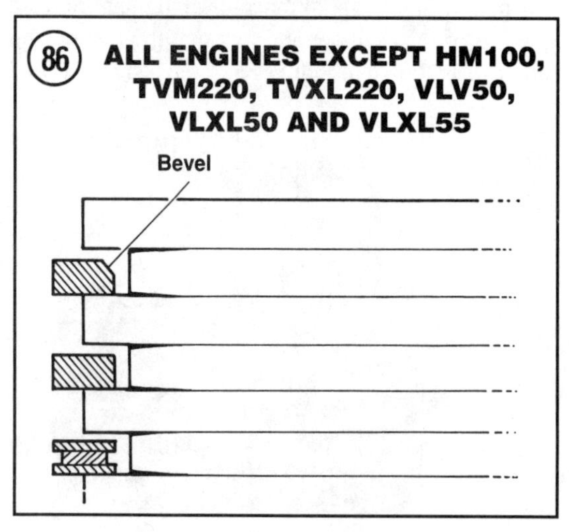

5. Measure the ring side clearance using a feeler gauge as shown in **Figure 89**. Refer to **Table 1** for the specified ring side clearance (if the engine has an external ignition, check **Table 1** for a different specification than engines not so equipped). If the piston ring side clearance is excessive, then the piston must be replaced.

PISTON AND PISTON PIN

Disassembly

If it is necessary to separate the connecting rod from the piston, use the following procedure.

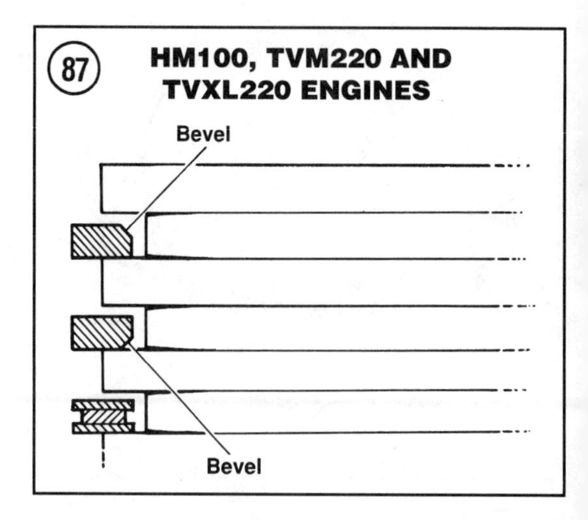

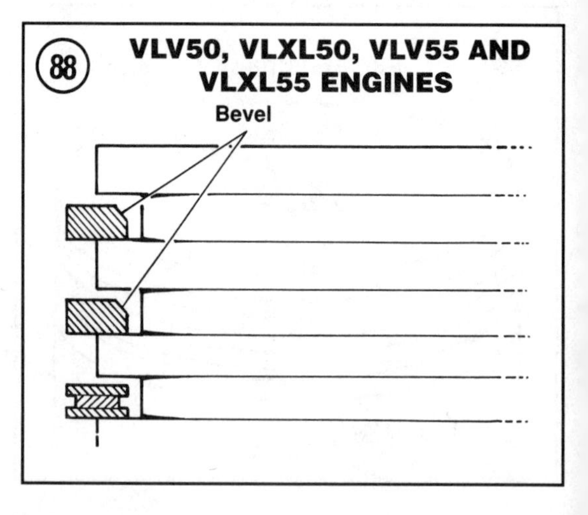

1. Extract the piston pin retaining clips (**Figure 90**). Place your thumb over the hole to help prevent the rings from flying out during removal.

WARNING
Because the piston pin retaining rings are compressed in the piston pin ring groove, safety glasses must be worn during their removal and installation.

2. Push or tap out the piston pin.

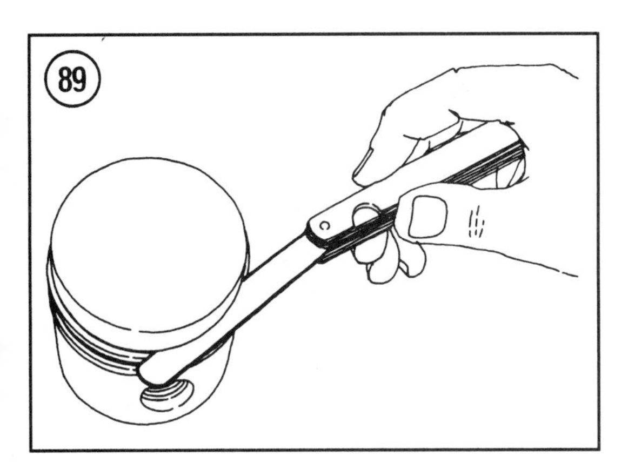

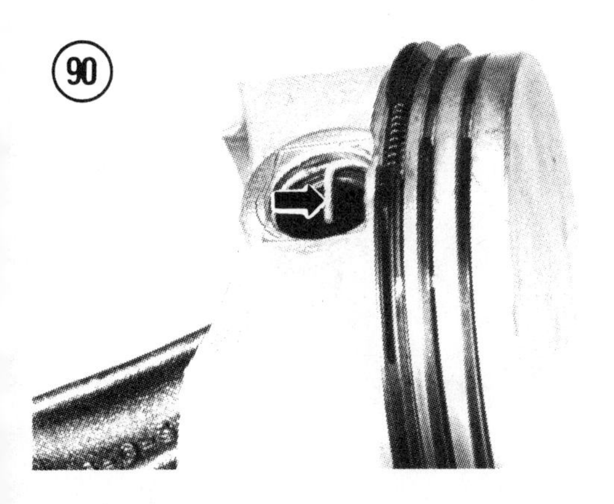

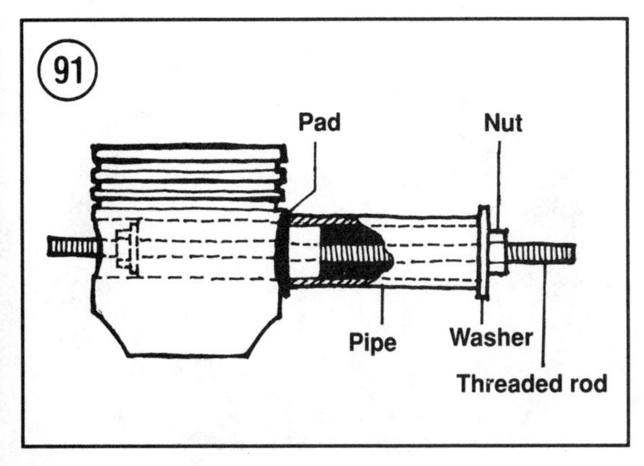

NOTE
*If the piston pin is stuck in the piston, a puller tool like that shown in **Figure 91** can be fabricated to extract the pin. Another method commonly used to remove a tight fitting piston pin is to heat the top of the piston to about 120°-140° F (50°-60° C) using a small propane torch or hair dryer. The pin should then push easily out of the piston.*

WARNING
Wear gloves when handling metal parts that have been heated to prevent burning your hands.

Inspection

1. Clean the top of the piston using a soft wire brush. Hard deposits can be loosened by soaking the piston in carburetor cleaner.
2. Replace the piston if it is cracked, scored, scuffed or scratched. Be sure to inspect the underside of the piston around the piston pin bosses for cracks, as well as the piston ring grooves.
3. The piston crown must be smooth except for machining or casting marks.
4. Inspect the piston pin and the piston pin hole in the piston for scoring and other damage. The pin should be a snug fit in the piston. The piston pin is available only as an assembly with the piston.

Piston Clearance

To determine the piston clearance in the cylinder, measure the cylinder bore as outlined in the *Cylinder* section, then measure the outside diameter of the piston. For the purpose of determining piston clearance, the piston diameter must be measured at the bottom of the piston skirt and 90° from the piston pin hole (**Figure 92**).

Subtract the piston diameter measurement from the cylinder bore diameter measurement to determine the piston clearance. Refer to **Table 1** for the specified piston clearance (if the engine has an external ignition, check **Table 1** for a different specification than engines not so equipped).

If the piston clearance exceeds the specified dimension, then the cylinder bore or the piston is worn. Check the piston diameter against the specified standard piston sizes listed in **Table 1**.

Note that the piston may be oversize. Oversize pistons are marked on the piston crown (**Figure 83**) to indicate if the piston is oversize. A marking of ".010" would indicate that the piston is 0.010 in. oversize. Add the oversize dimension to the standard size listed in **Table 1** to determine the specified piston diameter.

NOTE
If an oversize piston must be installed, be sure the proper size is available before machining or purchasing components.

Assembly

Before mating the piston and connecting rod, note the match marks (casting projections) on the rod and rod cap. On Vector VLV50, VLXL50, VLV55 and VLXL55 engines, the marks are straight casting indentations on the rod and cap (**Figure 93**). On all other engines, the match marks are in the form of the casting projections shown in **Figure 94** and **Figure 95**. Also note that on engines with five or more horsepower there may be an arrow on the piston crown (**Figure 96**) or adjacent to the piston pin hole (**Figure 97**). During assembly the arrow on the piston must point in the correct direction in relation to the match marks on the connecting rod.

1. Install one of the piston pin retaining clips in the piston pin bore.

WARNING
Safety glasses must be worn when installing the piston pin retaining clips in Steps 1 and 4.

2. Position the connecting rod small end in the piston. On engines with less than five horsepower, and Vector VLV50, VLXL50, VLV55 and VLXL55 engines, the piston may be installed on the connect-

92

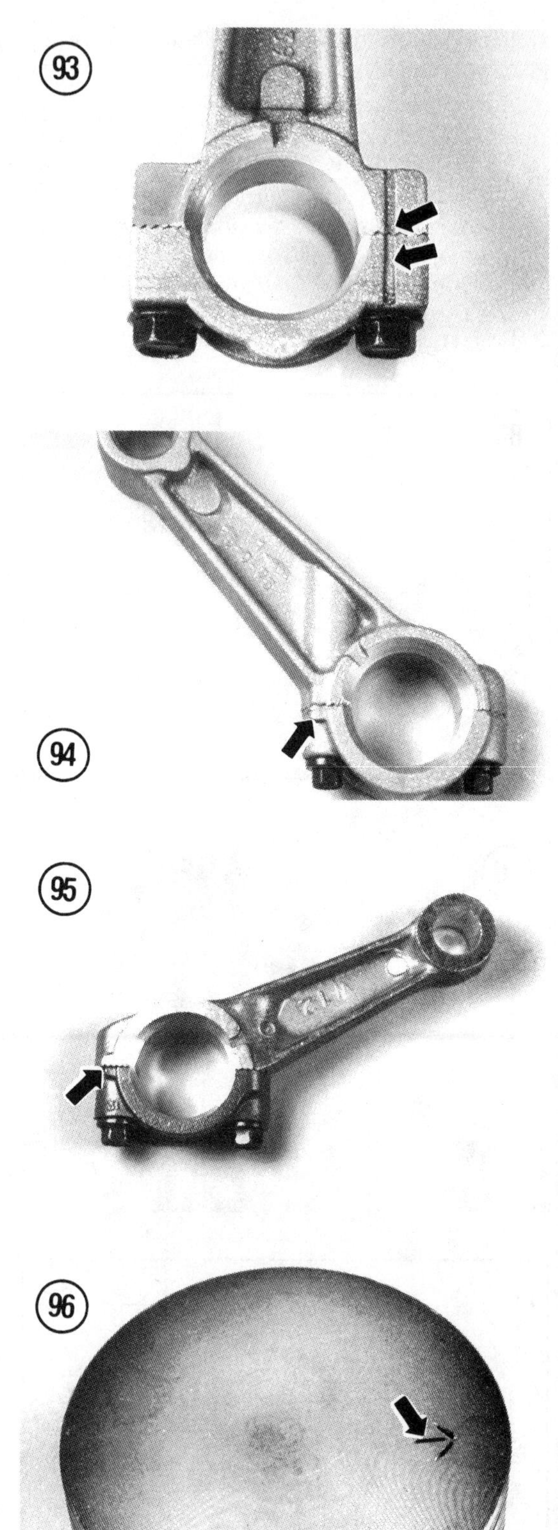
93 94 95 96

ing rod in either direction. On engines with five or more horsepower that have an arrow on the piston, refer to **Figures 97-99** for proper assembly.

NOTE
*On larger engines without an arrow on the piston, observe the casting number inside the piston skirt and install the rod so its long side is towards the casting number as shown in **Figure 100**.*

PISTON AND CONNECTING ROD ASSEMBLY (MOST LARGER ENGINES)

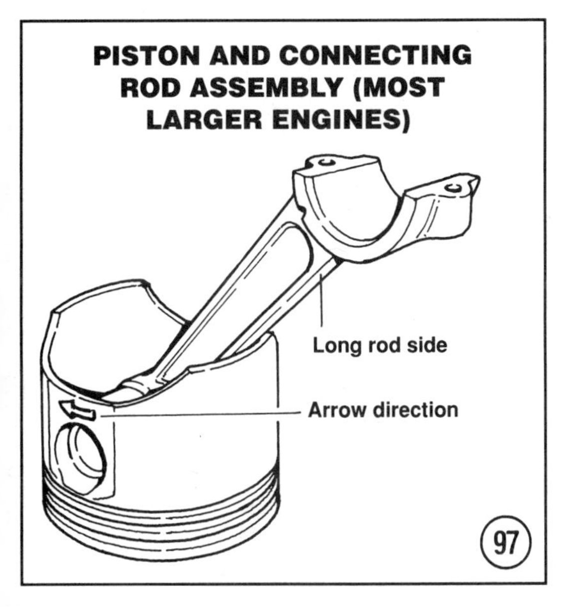

PISTON AND CONNECTING ROD ASSEMBLY (5 HORSEPOWER ENGINES, EXCEPT VLV50, VLXL50, VLV55, VLXL55)

Arrow
Match marks
98

3. Lubricate the piston pin with engine oil, then insert the piston pin in the piston and connecting rod.
4. Install the remaining piston pin clip.
5. Be sure the piston pin retaining clips are securely positioned in the piston grooves.

CONNECTING ROD

Disassembly/Assembly

Separate and assemble the connecting rod and the piston as described previously under the *Piston and Piston Pin Disassembly/Assembly* sections.

Inspection

1. On all engines, the connecting rod rides directly on the crankshaft crankpin. Inspect the bearing surfaces for signs of scuffing and scoring. If any damage is observed, also inspect the surface of the mating part, i.e., the crankpin or piston pin.

PISTON AND CONNECTING ROD ASSEMBLY (TVM220 AND TVXL220 ENGINES)

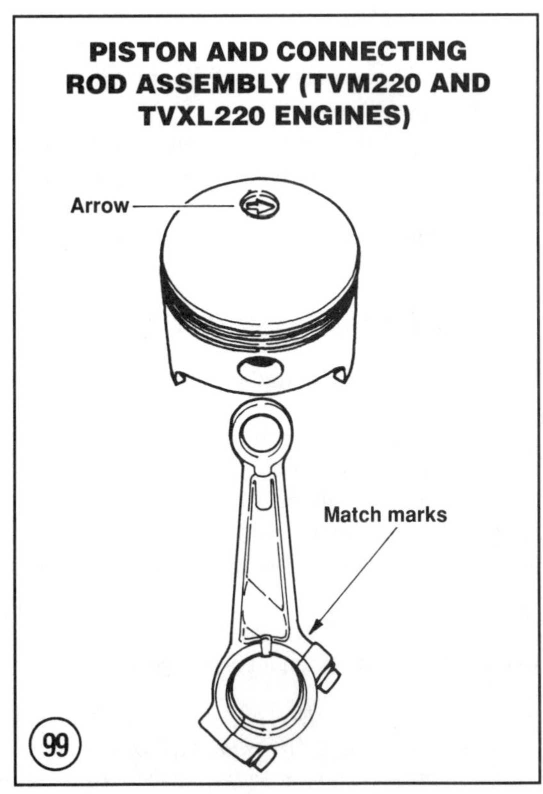

NOTE

If the rod bearing surface is worn due to abrasive particles (the surface texture is dull and rough), it should be replaced even if it is not worn beyond the specified wear limit. Grit may be embedded in the aluminum which will continue to cause wear on mating surfaces.

2. Replace the connecting rod if there is any indication that it is cracked, bent, twisted or otherwise damaged.
3. Measure the inner diameter of the connecting rod big end (**Figure 101**) after installing the rod cap.

NOTE

*Be sure the match marks are on the same side (see **Figures 93-95**) and the mating surface grooves on the rod and cap (**Figure 102**) are aligned. Tighten the connecting rod screws to the torque specified in **Table 2** after determining the type of fastener used. Older rods are equipped with standard type screws or bolts to secure the rod cap to the rod. Later rods are equipped with Durlock screws that can be identified by serrations on the underside of the screw head (**Figure 103**).*

4. Replace the connecting rod if the big end diameter is not as specified in **Table 1**.
5. Insert the piston pin in the connecting rod and check for excessive play. If there is perceptible wobble, and the piston pin is new or unworn, then the connecting rod should be replaced.

CAMSHAFT AND TAPPETS

Removal/Inspection/Installation

The camshaft rides directly in the aluminum of the crankcase, oil pan or crankcase cover. The camshaft gear is integral with the shaft.

1. Remove the camshaft and tappets as described under the *Engine Disassembly* section in this chapter.
2. The camshaft and gear should be inspected for wear on the journals, cam lobes and gear teeth.

NOTE

Some camshafts are equipped with a compression release ridge on the base

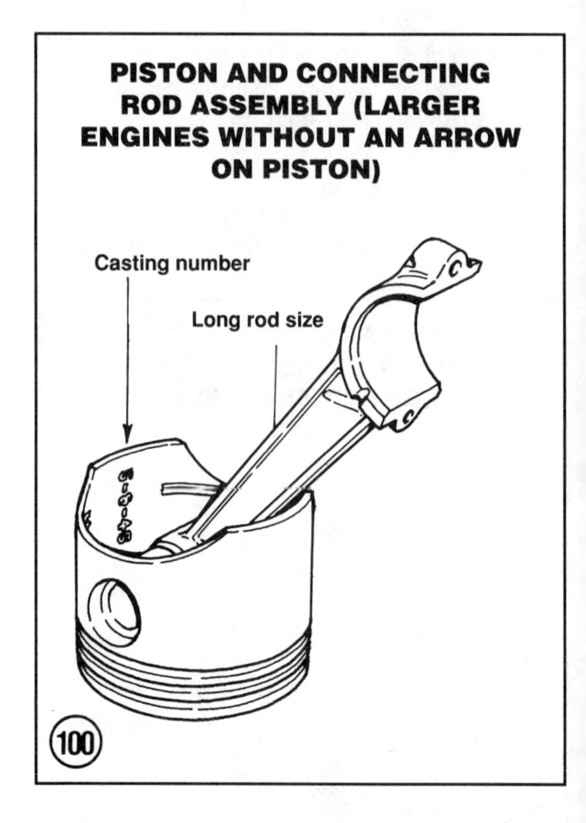

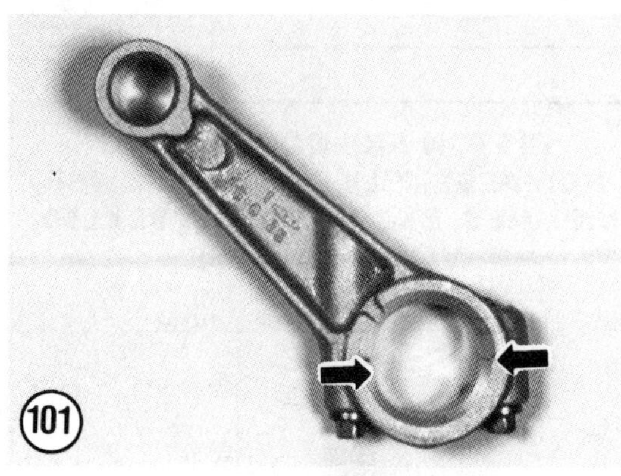

103

104

105

106

*portion of the exhaust cam. See the **Compression Release** section for a detailed description.*

3. Measure the diameter of the bearing journals (**Figure 104**). Replace the camshaft if the journal diameter is not as specified in **Table 1**.
4. Inspect the camshaft bearing surfaces in the crankcase and oil pan (on vertical crankshaft engines) or crankcase cover (on horizontal crankshaft engines). The surface must be smooth with no sign of abrasion.
5. Insert the camshaft into the bearing bore. The camshaft journal should fit snugly in the bearing bore without excessive side play.
6. The tappets should be inspected for excessive wear on the contact surface with the camshaft lobe. Replace the tappet if there are signs of galling or other damage.
7. Check for excessive side play with the tappet installed in the crankcase tappet bore.
8. Check the camshaft lifter lobes for damage and excessive wear. The camshaft should be replaced if there are signs of galling or other damage.

NOTE
If the camshaft is replaced, it is good practice to also replace the tappets.

9. Refer to the *Engine Assembly* section in this chapter for procedure to install the camshaft and tappets in the engine.

Compression Release

Some camshafts may be equipped with a mechanically operated compression release near the exhaust cam. With the engine stopped or at cranking speed, a spring-loaded pin or arm extends above the exhaust cam. The pin or arm holds the exhaust valve open slightly during the compression stroke. This compression release greatly reduces the power needed for cranking.

When the engine starts, the cam weight moves outward, overcoming the spring pressure and withdrawing the pin or arm to provide normal exhaust valve operation.

Three configurations have been used: gear-mounted (**Figure 105**), end-mounted (**Figure 106**) and Vector VLV50, VLXL50, VLV55 and VLXL55 engines (**Figure 107**). The weight and actuating pin or arm are indicated in **Figures 105-107**.

The compression release mechanism should move freely without binding. No individual components are available; the compression release is available only as a unit assembly with the camshaft.

Some engines are equipped with a camshaft that has a slight ridge machined on the base portion of the exhaust cam (**Figure 108**). The ridge raises the exhaust valve slightly. This has little effect when the engine is running, but assists engine starting by releasing compression pressure when the crankshaft is turning slowly.

OIL PUMP

Engines with a vertical crankshaft are equipped with a plunger type oil pump (**Figure 109**). The ball end of the pump plunger fits in a recess in the oil pan (**Figure 110**). The large end of the pump surrounds an eccentric on the camshaft so that when the camshaft rotates, the eccentric forces the pump plunger to reciprocate. Oil from the pump is routed through the center of the camshaft to the upper camshaft bearing. A passage in the crankcase routes oil from the upper camshaft bearing to the upper main bearing. Some engines are equipped with a crankshaft that is drilled so oil is routed from the upper crankshaft main bearing journal to the crankpin to lubricate the connecting rod.

Removal/Inspection/Installation

The oil pump is accessible after removal of the oil pan. Lift the pump (**Figure 109**) off the camshaft gear.

Individual pump components are not available. The pump must be serviced as a unit assembly.

When installing the pump, note that one side of the loop end of the pump is chamfered (**Figure 111**). The chamfered side must be towards the camshaft gear.

GOVERNOR

The engine is equipped with a flyweight-type mechanical governor. The governor is mounted on a shaft that is pressed into the crankcase cover or oil pan (**Figure 112**). The governor gear is driven by the camshaft gear on all engines except Vector VLV50, VLXL50, VLV55 and VLXL55 engine. On Vector VLV50, VLXL50, VLV55 and VLXL55 engines, the governor gear is driven by an idler gear (C, **Figure 113**) that also meshes with the camshaft gear.

107

108

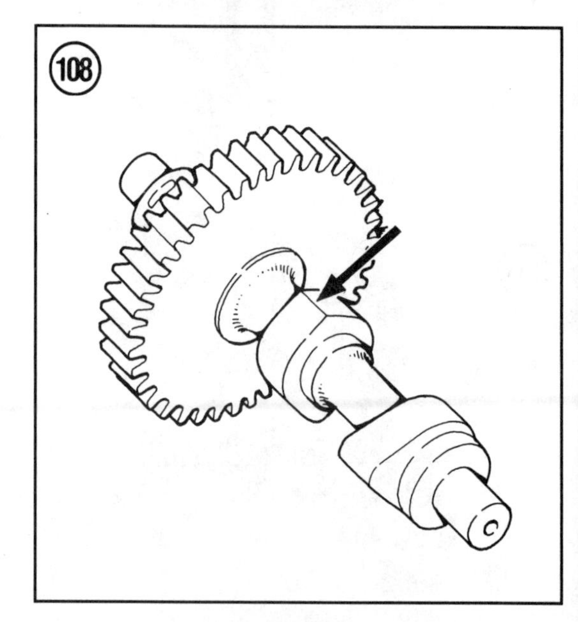

109

Removal

1. The governor is accessible after removing the crankcase cover on engines with a horizontal crankshaft or the oil pan on engines with a vertical crankshaft.

2. On most engines with a vertical crankshaft and on some low horsepower horizontal crankshaft engines, the governor spool and the governor gear and flyweight assembly are retained on the shaft by snap rings (**Figure 112** and **Figure 113**).

   a. See **Figure 112** and **Figure 113**. Remove the upper snap ring (A) and withdraw the spool (B).

   b. Remove the lower snap ring (A, **Figure 114**) and remove the gear and flyweight assembly (B).

3. A washer (**Figure 115**) is located under the flyweight assembly. On some engines, a spacer (A, **Figure 116**) is located under the washer (B).

4. On most horizontal crankshaft engines, the governor is retained by a bracket (**Figure 117**).

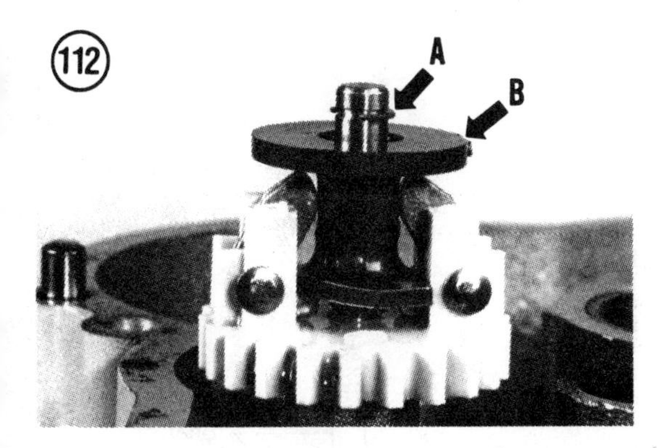

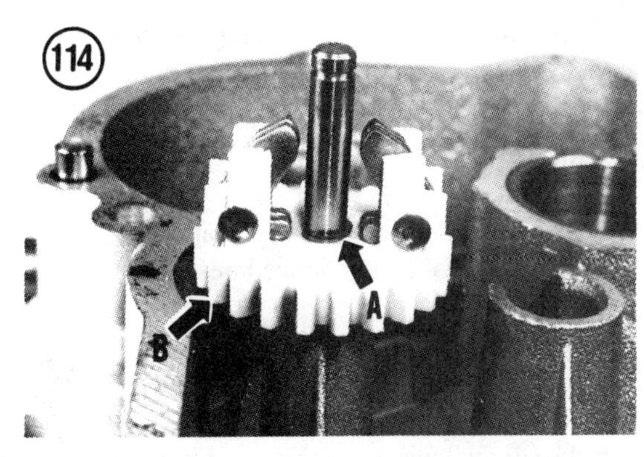

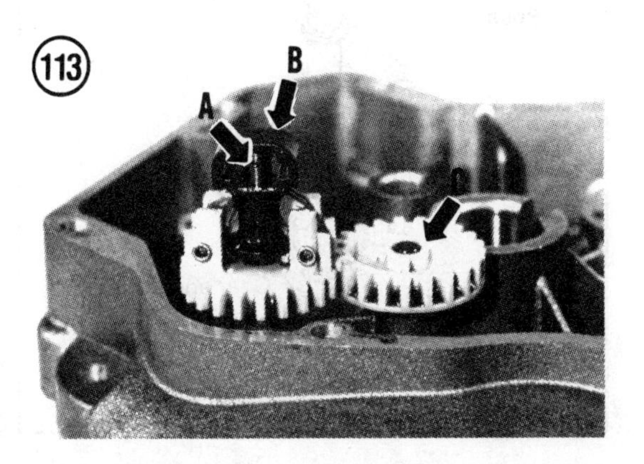

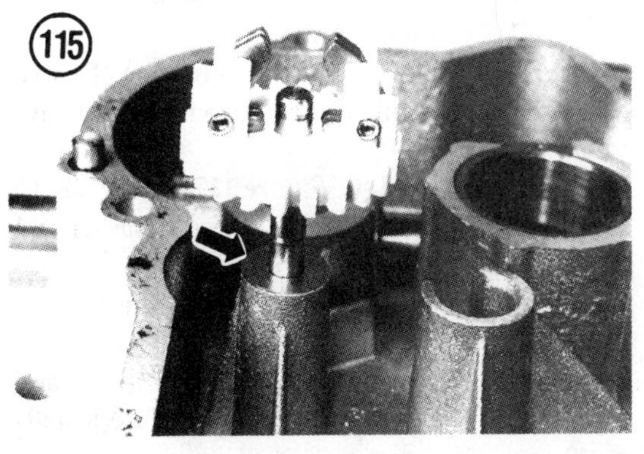

a. Detach the retaining bracket (A, **Figure 117**) and remove the governor gear and flyweight assembly (B).
b. Remove the washer and snap ring, located under the governor gear, and withdraw the spool (C, **Figure 117**) from the shaft.

5. On later small frame engines, including Vector VLV50, VLXL50, VLV55 and VLXL55 engines, no snap rings are used on the governor shaft. The governor is held in place by a boss on the governor shaft (**Figure 118**). The governor shaft must be removed as described under *Governor Shaft Replacement* to remove the governor gear and flyweight assembly.

NOTE
*Replacement governor shafts for small frame engines use the retainerless design (**Figure 118**). See **Governor Shaft Replacement** section.*

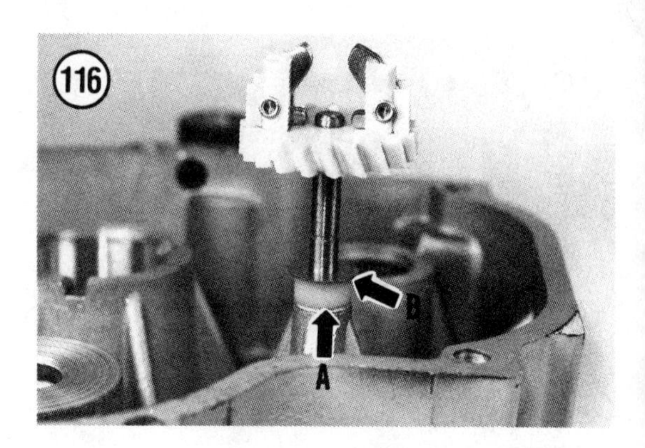

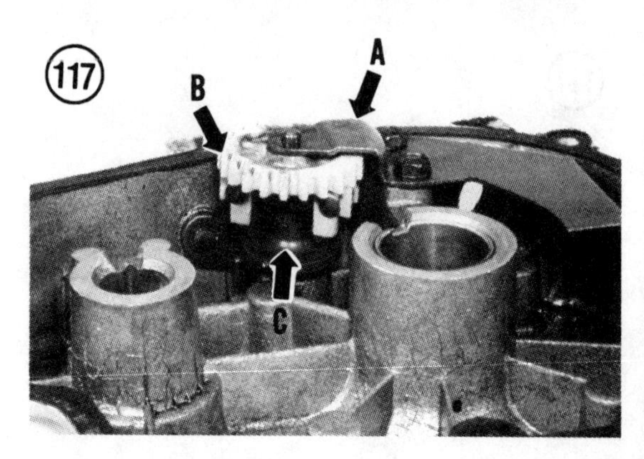

Inspection

1. Inspect the flyweight assembly for broken components, worn gear teeth and excessive clearance on the stub shaft. The flyweights and gear are available only as a unit assembly.
2. If the governor shaft requires replacement, refer to the *Governor Shaft Replacement* section.

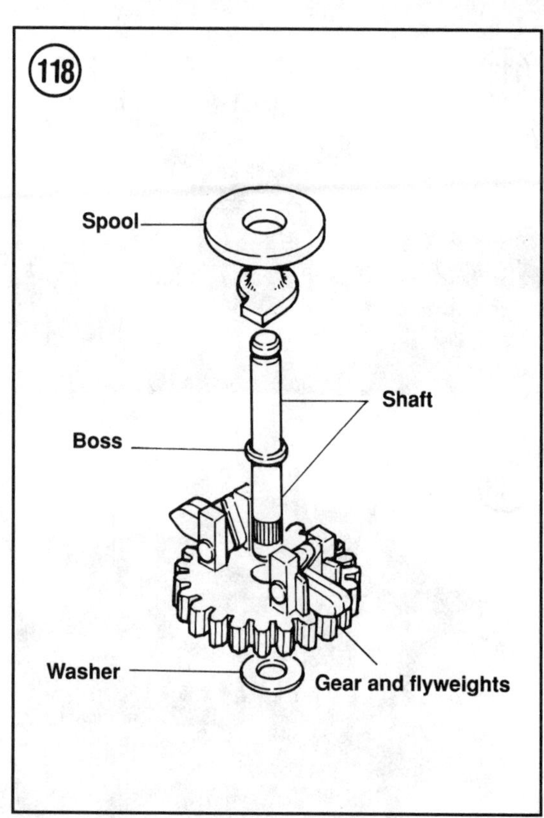

Installation

1. If the governor is retained by snap rings on the shaft, use the following procedure to install the governor.
a. Assemble the spacer (if used), washer and flyweight assembly on the shaft (**Figure 115** and **Figure 116**).
b. Install the lower snap ring (A, **Figure 114**).
c. Install the spool and the upper snap ring (**Figure 112** or **Figure 113**).

2. If the governor is retained by a bracket (**Figure 117**), use the following procedure to install the governor.
a. Install the spool (C, **Figure 117**) on the shaft.
b. Install the lower snap ring, thrust washer and governor gear (B, **Figure 117**).
c. Install the retainer bracket (A, **Figure 117**).

3. If the governor is retained by a boss on the governor shaft (**Figure 118**), position the washer and governor gear assembly on the shaft before install-

ing the shaft. Then, press the shaft into the boss until the governor gear has 0.010-0.020 in. (0.25-0.50 mm) end play.

Governor Shaft Replacement

The governor shaft is pressed into the crankcase cover or oil pan and may be replaced if the mounting boss is not damaged or the hole is not enlarged.

Removal/Installation

1. Clamp the shaft in a vise and, using a soft mallet, drive the crankcase cover or oil pan off the shaft.

CAUTION
Do not attempt to twist the governor shaft out of the boss. Twisting will enlarge the boss hole and the shaft will not fit securely in the hole.

2. The governor shaft must be pressed into the crankcase cover or oil pan. On all engines, apply Loctite 271 (red) to the shaft end that fits in the crankcase cover or oil pan before installation.

3A. If the shaft is the retainerless design (**Figure 118**), position the washer and governor gear assembly on the shaft before installing the shaft, then press in the shaft until the governor gear has 0.010-0.020 in. (0.25-0.50 mm) axial play.

3B. If snap rings are used on the governor shaft, press the shaft into the crankcase cover or oil pan until the height above the boss (**Figure 119**) is as specified in **Table 1**.

119

120

CRANKSHAFT AND MAIN BEARINGS

Removal/Inspection/Installation

1. Refer to the *Engine Disassembly* section in this chapter for procedure to remove the crankshaft from the engine.

NOTE
*The crankshaft gear is either pressed on the crankshaft or it is removable and located on the shaft by a key. If the gear is removable, note that the gear must be installed on the crankshaft so the side with the timing mark is visible (**Figure 120**). If the gear is pressed on, then the gear and crankshaft must be serviced as a unit assembly; the gear is not available separately.*

NOTE
*On Vector VLV50, VLXL50, VLV55 and VLXL55 engines, the crankshaft gear timing mark should be on the fifth tooth from the crankpin centerline (see **Figure 121**), otherwise the gear has moved. In some instances the gear can be relocated satisfactorily, but if not, the crankshaft must be replaced. Tecumseh does not specify timing mark position for other engines with a pressed-on timing gear.*

2. If considerable effort was required to remove the oil pan or crankcase cover from the crankshaft, then the crankshaft is probably bent and must be discarded. If moderate effort was required during removal, then the crankshaft should be checked for straightness by a shop with the necessary equipment. If oil pan or crankcase removal was relatively easy,

and the bearings are within specifications, then the crankshaft is probably straight.

3. Inspect all mating surfaces on the crankshaft for indications of scoring, scuffing and other damage.

4. Check any fastener threads, such as the flywheel nut and blade retaining screw, for damage and cross-threading and repair if possible.

5. Check the keyway (**Figure 122**) and remove burrs. The keyway must be straight and unworn.

6. Check the crankshaft gear for broken or pitted teeth.

7. Measure the crankpin journal (A, **Figure 123**). The crankshaft should be replaced if the crankpin journal diameter is not as specified in **Table 1**.

NOTE
On HM100 engines with an external ignition, record the crankpin diameter for future reference when measuring crankshaft end play.

8. If the engine is equipped with a ball bearing main bearing, refer to the *Ball Bearing Replacement* section.

9. Measure the main bearing journals (B, **Figure 123**). The crankshaft should be replaced if the journal diameter is not as specified in **Table 1**. Unless noted, the dimension applies to both main bearing journals.

NOTE
If not equipped with a ball bearing main bearing, the crankshaft rides directly in the aluminum of the crankcase and oil pan (on vertical crankshaft engines) or crankcase cover (on horizontal crankshaft engines), or the crankshaft rides in a renewable bushing. On some engines, a bushing can be installed if the aluminum is excessively worn or damaged. Most engines can be repaired using a bushing, but check parts availability on a specific engine. Main bearing bushing installation should be performed by a shop with the necessary experience and tools.

10. Inspect the main bearing (bushing) surfaces in the crankcase and oil pan (on vertical crankshaft engines) or crankcase cover (on horizontal crankshaft engines). The surface must be smooth with no sign of abrasion.

11. Measure the bearing (bushing) inside diameter in the crankcase (**Figure 124**) and oil pan or crankcase cover (**Figure 125**). If the diameter is not as specified in **Table 1**, either a bushing must be installed or the crankcase, oil pan or crankcase cover must be replaced.

12. Except on engines equipped with a ball bearing, the clearance between the main bearing journal on the crankshaft and the inside diameter of the main bearing in the crankcase, crankcase cover or oil pan should be 0.0015-0.0025 in. (0.038-0.064 mm).

13. Refer to the *Engine Assembly* section in this chapter for procedure to install the crankshaft.

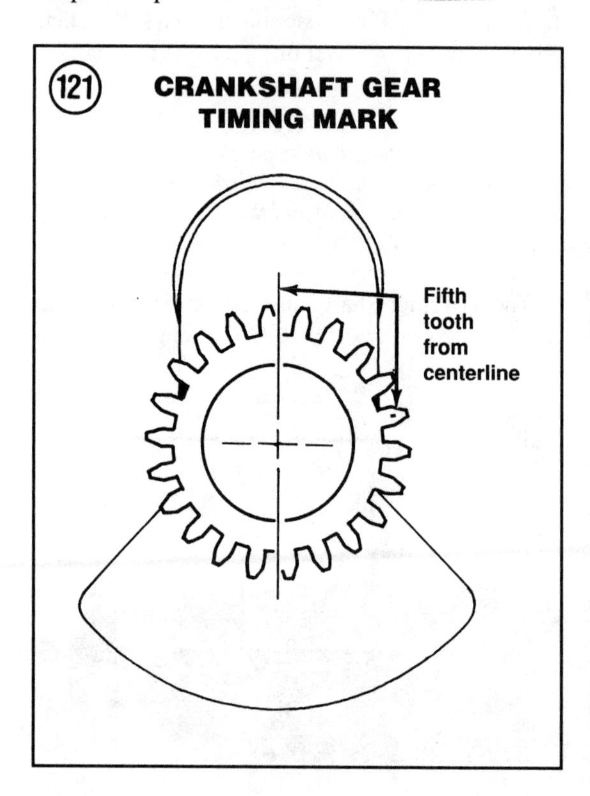

Ball Bearing Replacement

On engines so equipped, inspect the ball bearing for roughness, pitting, galling and play by rotating the bearing slowly by hand. If any roughness or play can be felt in the bearing, it must be replaced.

NOTE
*Refer to **Table 2** in Chapter Three for engine model number and displacement specifications.*

(123)

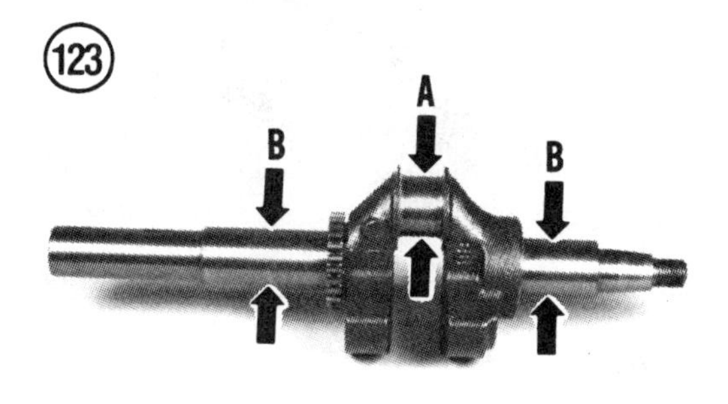

(124)

(125)

1. On horizontal crankshaft engines with a displacement of 12.04 cu. in. (197 cc) or less (except Model H40), the ball bearing is located in the crankcase cover.
   a. The bearing can be removed after unscrewing the retaining screw located inside the cover.
   b. Install a new bearing and secure with the retaining screw.
   c. On engines with less than 5 hp (3.7 kW), tighten the bearing retaining screw to 45-60 in.-lb. (5.1-6.8 N•m).
   d. On engines with 5 hp (3.7 kW) or greater, tighten the bearing retaining screw to 15-22 in.-lb. (1.7-2.5 N•m).
2. On Model H40 and horizontal crankshaft engines with a displacement greater than 12.04 cu. in. (197 cc), the ball bearing is a press fit on the crankshaft. The crankshaft should be replaced if the bearing is loose on the crankshaft.
   a. Ball bearing removal and installation can be performed by a shop equipped with a press.
   b. An alternate method for bearing installation is to warm the bearing in oil that is heated to no more than 250° F (121° C). The bearing must not contact the container. While hot, the bearing can be slipped down the crankshaft into place.

CYLINDER

Inspection/Reconditioning

The cylinder bore should be inspected for scratches, scoring, scuffing and other damage. If the bore is excessively worn or damaged, then an oversize piston can be installed.

NOTE
Machining (boring) the cylinder to accept an oversize piston should be performed by a shop with the required equipment and experience. Compare the cost of another engine assembly (short block) with the cost of installing an oversize piston.

1. Measure the cylinder bore to determine if the bore is excessively worn, tapered or out-of-round. Measure the cylinder bore at the top, middle and bottom at several points around the cylinder. Compare the measurements to determine the cylinder

bore size, taper and out-of-round. Refer also to the *Piston Clearance* section. The cylinder should be machined to the nearest oversize for which a piston and ring set are available if the cylinder is tapered or out-of-round more than 0.005 in. (0.13 mm).

NOTE
In some cases the engine may be built with an oversize cylinder. Look for an oversize marking on top of the cylinder and on the piston (see ***Piston and Piston Pin*** *section) to determine if the cylinder has been machined oversize.*

2. **Table 1** lists the standard cylinder bore diameter. Add the oversize dimension to the standard size listed to determine the oversize diameter (if the engine has external ignition, check the table for a different specification than engines not so equipped).

3. If new piston rings are being installed, the cylinder bore surface should be reconditioned with a hone or deglazing tool (**Figure 126**) to restore the crosshatch pattern (**Figure 127**). The crosshatch pattern retains oil in the grooves while also promoting piston ring seating. The tool is driven by an electric drill.

   a. With the hone rotating at approximately 600 rpm, move it in and out of the cylinder bore at 50 strokes per minute so that an angle of 45° between the intersecting lines of the crosshatch is produced.

   b. After deglazing, thoroughly wash the cylinder bore using a stiff brush with hot soapy water.

NOTE
Do not clean the bore with kerosene or solvent as a residue of abrasive grit will remain.

OIL SEALS

Removal/Installation

An oil seal is located adjacent to the main bearing in both the crankcase and the oil pan or crankcase cover. The oil seals should be replaced during any overhaul.

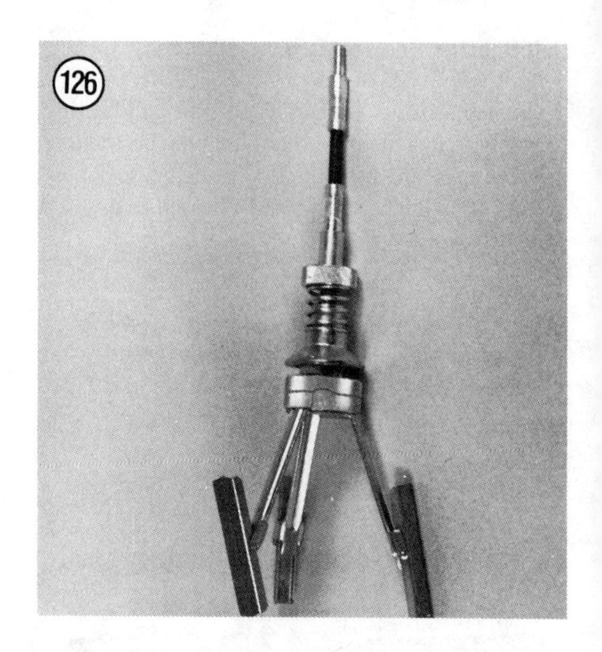

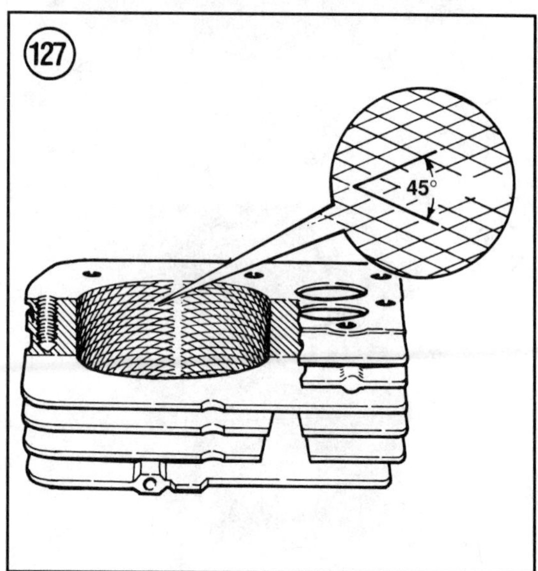

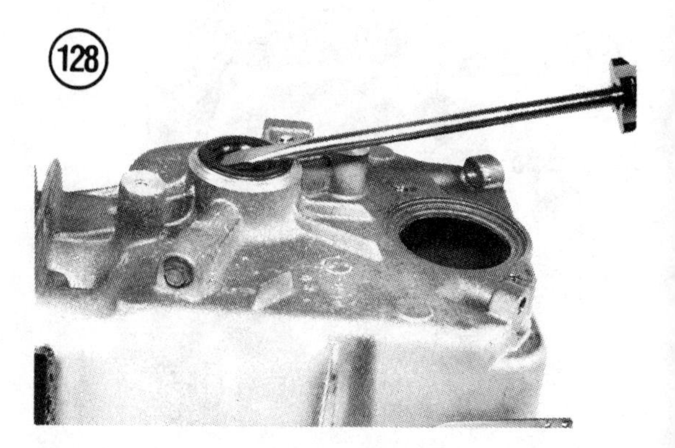

NOTE
Before removing an oil seal, note the position of the seal in the crankcase, oil pan or crankcase cover.

1. To remove an oil seal, pry the oil seal out using a screwdriver as shown in **Figure 128** or drive the seal out from the back side. Be careful so the metal of the crankcase, oil pan or crankcase cover surrounding the seal is not damaged.
2. Clean the seal seating area so the new seal will seat properly.

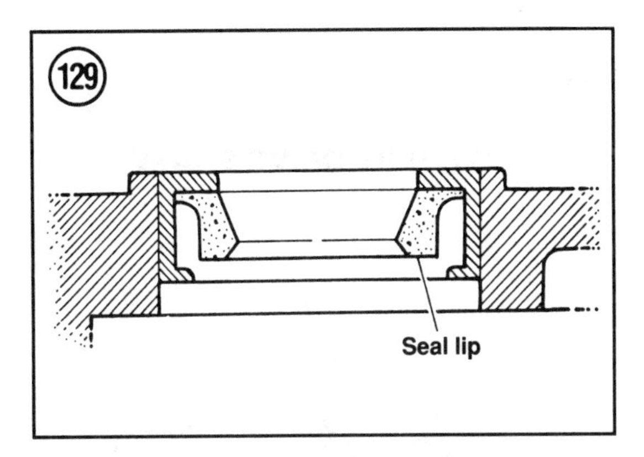

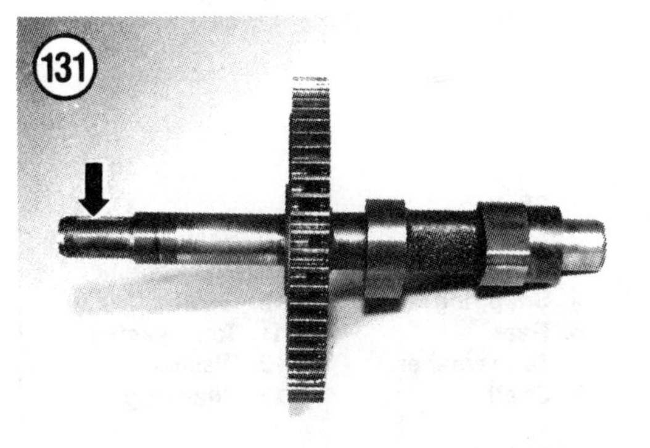

3. Before installing the new oil seal, apply a non-hardening sealer to the periphery of the seal.
4. To install the oil seal, position the seal so the seal lip is slanted toward the inside of the engine (**Figure 129**). Drive the seal into its original position using a seal driver or a wood block and a hammer.

NOTE
The oil seal lip must be lubricated with engine oil before inserting the crankshaft. Lack of lubrication when the engine is started may damage the seal lip.

BALANCER SHAFT

Inspection

Some larger engines with a vertical crankshaft are equipped with a balancer shaft (**Figure 130**). The weight on the balancer shaft offsets the counterweights on the crankshaft to reduce vibration.

No service is required other than checking for damage and excessive wear on the balancer shaft bearing journals and gear, as well as the crankshaft gear, crankcase bearing and oil pan bearing.

The balancer shaft gear must be positioned in time with the crankshaft gear during installation of the oil pan. See the *Engine Assembly* procedure.

AUXILIARY DRIVE SHAFT

Some engines are equipped with an auxiliary drive shaft that rotates slower than the crankshaft.

An extension shaft on the camshaft (**Figure 131**) serves as an auxiliary drive shaft on some engines. No service is required for this type other than for the oil seal around the shaft.

On some engines, an auxiliary drive shaft is located in the oil pan (**Figure 132**) and driven by a worm gear on the crankshaft (**Figure 133**). Two configurations have been used, depending on which direction the shaft is installed in the oil pan, as shown in **Figure 134**.

The shaft (1 or 7, **Figure 134**) is held in position by a snap ring (4 or 13). To disassemble the unit, detach the snap ring from the shaft and remove the components from the oil pan. When assembling the unit, note that the tang (6 or 11) is installed facing away from the gear.

ENGINE ASSEMBLY

Before assembling the engine, be sure all components are clean. Any residue or debris left in the engine will cause rapid wear and/or major damage when the engine runs.

The following components should be assembled or installed before proceeding with the assembly of the engine:

a. Piston, piston rings and connecting rod.
b. Oil seals.
c. Governor flyweight and on VLV50, VLXL50, VLV55 and VLXL55 engines, governor idler gear.
d. Bushings or bearings.
e. Camshaft and compression release.
f. Auxiliary drive unit.

Install Internal Engine Components

1. Lubricate the crankcase oil seal and crankshaft bearing with engine oil.

NOTE
*Thin tape or an oil seal protection sleeve should be used around the crankshaft keyway and threads (**Figure 135**) to protect the oil seal when the crankshaft is inserted.*

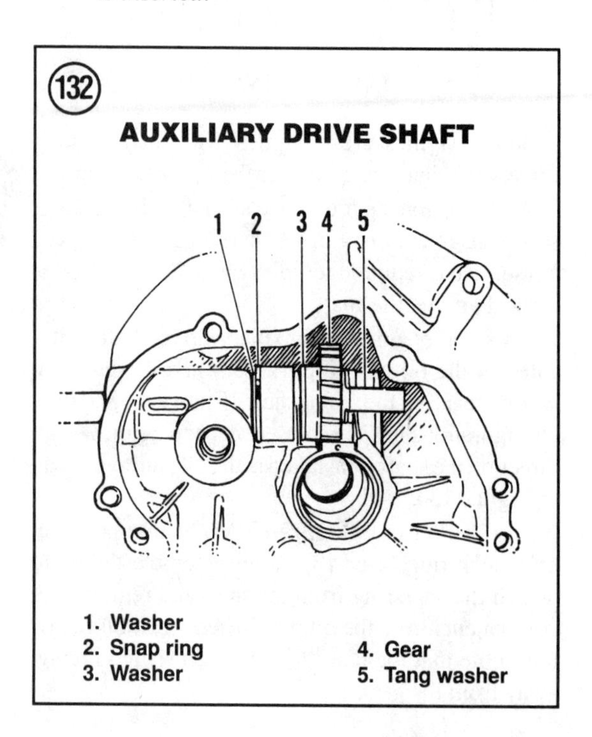

1. Washer
2. Snap ring
3. Washer
4. Gear
5. Tang washer

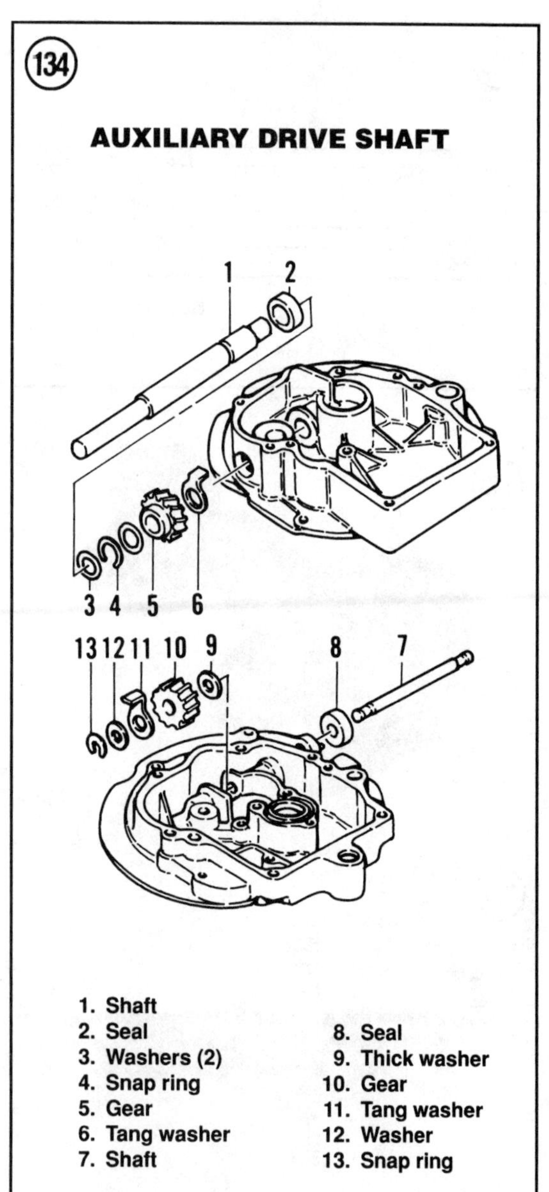

1. Shaft
2. Seal
3. Washers (2)
4. Snap ring
5. Gear
6. Tang washer
7. Shaft
8. Seal
9. Thick washer
10. Gear
11. Tang washer
12. Washer
13. Snap ring

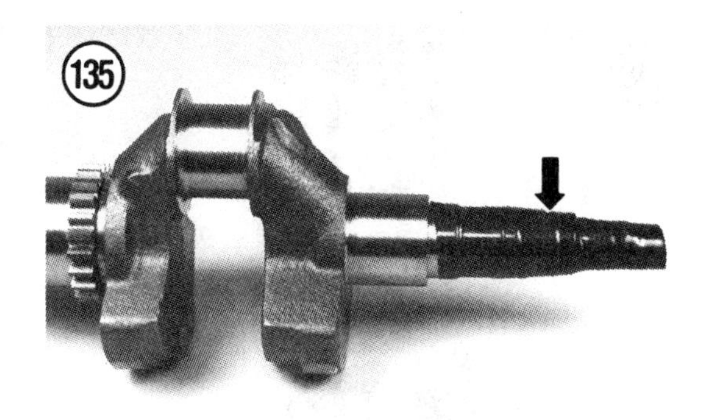

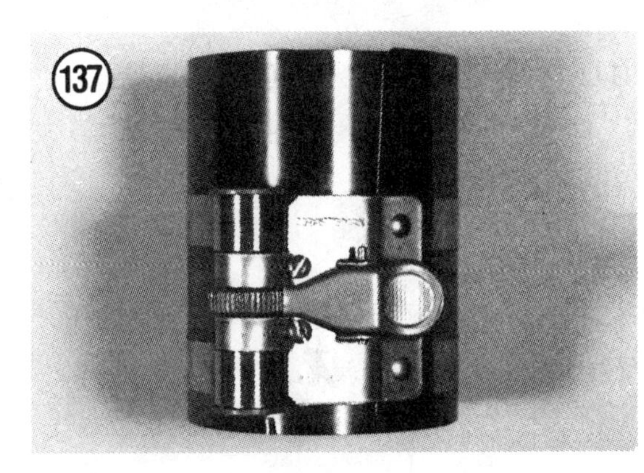

2. Insert the crankshaft in the crankcase.

3. Rotate the crankshaft so the crankpin is towards the cylinder bore.

4A. Look at the top of the cylinder bore and note if the area adjacent to the valves (**Figure 136**) has been machined so it is lower than the remaining bore circumference. If the bore area has been machined, position the piston rings on the piston so that the end gaps are staggered at 90° intervals, but none of the end gaps will coincide with the machined area. This prevents a piston ring end from catching the machined surface during installation.

NOTE
Keep in mind that the piston and rod will be installed so the match marks on the rod and cap will be towards the output end of the crankshaft.

4B. If the upper cylinder bore circumference has not been machined, position the piston ring end gaps so they are 120° apart.

5. Lubricate the piston, piston pin and piston rings with engine oil.

6. Compress the piston rings with a piston ring compressor (**Figure 137**). Turn the compressor tightener so the piston will just rotate in the compressor.

7. Lubricate the cylinder bore with engine oil.

8. Place the piston and connecting rod in the cylinder with the match mark on the connecting rod (**Figure 138**) towards the output end of the crankshaft (the marks will be visible after installation of the rod on the crankshaft if properly installed). Note that the arrow (if present) on the piston crown (**Figure 139**) should point towards the valve side of the engine.

NOTE
***Figure 138** illustrates one piston and rod configuration. The location of the match marks and long side of the rod may be opposite that shown.*

9. Insert the connecting rod and piston through the top of the engine (**Figure 140**) so the piston ring compressor rests against the engine.

10. Push the piston into the cylinder bore while guiding the connecting rod onto the crankshaft crankpin.

CAUTION
Do not use excessive force when installing the piston and rod. If binding occurs, remove the piston and rod and try again. Excessive force can damage or break the piston rings, piston ring lands, connecting rod or crankshaft.

11. Liberally lubricate the connecting rod bearing and the crankshaft crankpin with engine oil.
12. Mate the connecting rod with the crankshaft crankpin, then rotate the crankshaft so the cap can be installed.
13. Liberally lubricate the bearing surface of the connecting rod cap and install the cap on the connecting rod. Be sure the match marks on the rod and cap are on the same side.
    a. On Vector VLV50, VLXL50, VLV55 and VLXL55 engines, the marks are straight casting indentations on the rod and cap (**Figure 141**).
    b. On all other engines, the match marks are in the form of the casting projections shown in **Figure 142** and **Figure 143**.

14A. On horizontal crankshaft engines, install the oil dipper, if so equipped, or the lockplate, if so equipped, and the connecting rod screws (**Figure 144**). The dipper should point to the right. A new lockplate should be installed.
14B. On vertical crankshaft engines, install the lockplate, if so equipped, and connecting rod screws.
15. Tighten the connecting rod screws to the torque listed in **Table 2** after determining the type of fastener used. Older rods are equipped with standard type screws or bolts to secure the rod cap to the rod. Later rods are equipped with Durlock screws that can be identified by serrations on the underside of the screw head (**Figure 145**).

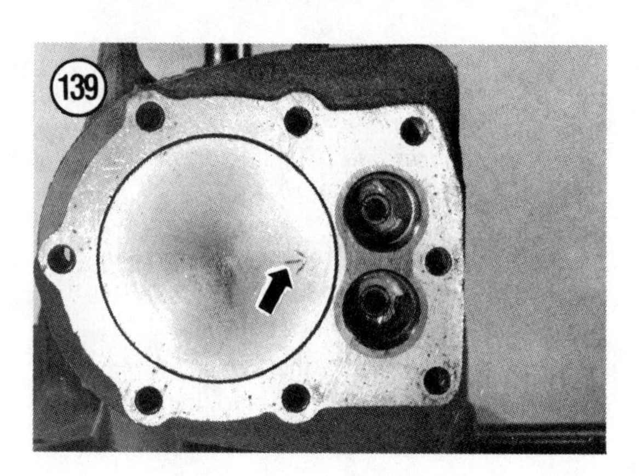
139

140

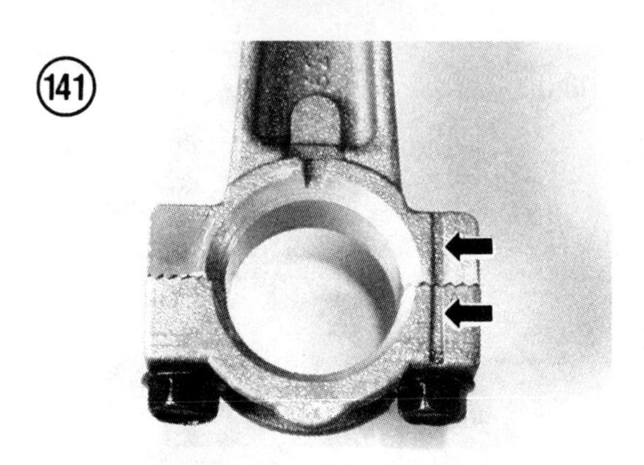
141

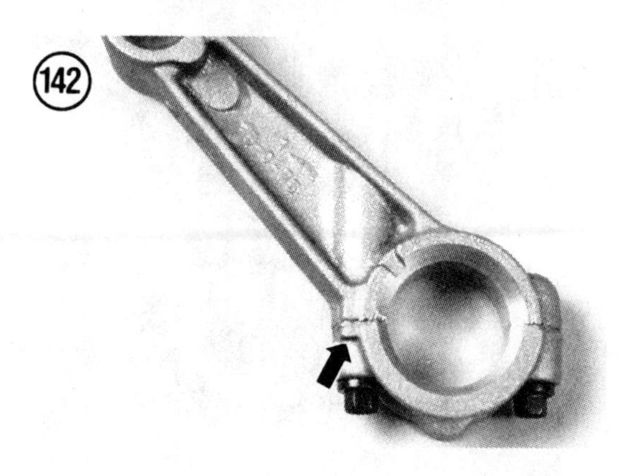
142

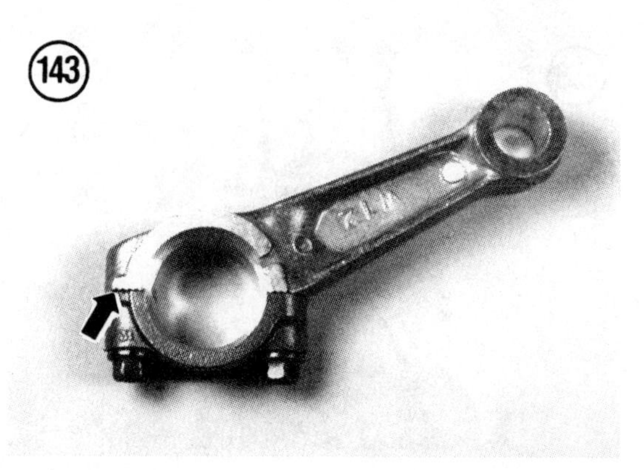
143

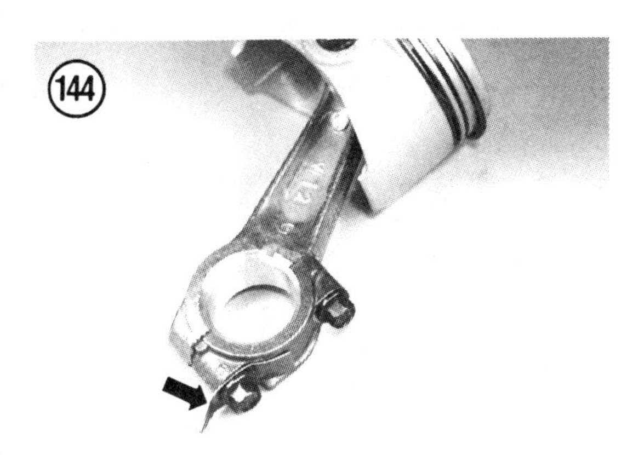

16. Position the engine so the tappets will not fall out and install the tappets (**Figure 146**) in their original positions.

17. Lubricate the camshaft and camshaft bearings with engine oil.

18. Install the camshaft so the timing marks on the crankshaft and camshaft gears are positioned correctly.

   a. On HM70, HMXL70, HM80, HM100, TVM170, TVM195, TVXL195, TVM220, TVXL220, VM70, VM80 and VM100 engines, the camshaft must be installed so the compression release on the camshaft will not contact the exhaust valve tappet. The crankshaft and camshaft gears must be mated so the crankshaft beveled tooth and camshaft timing mark are out of alignment by three teeth (**Figure 147**).

   b. On all engines except those listed in previous Step 18a, install the camshaft so the timing marks are aligned (**Figure 148**). The timing mark on the crankshaft gear may be either a round punch mark or a straight line mark.

NOTE

If there is no timing mark on the crankshaft gear, align the keyway in the crankshaft with the timing mark on the camshaft gear.

NOTE

*If the crankshaft is equipped with a ball bearing or the crankshaft gear is pressed on (no keyway), align the crankshaft gear tooth that is beveled or has a punch mark (**Figure 149**) with the mark on the camshaft.*

19. On engines equipped with a balancer shaft, install the balancer drive gear (A, **Figure 150**), then lubricate and install the balancer shaft (B). Be sure the timing marks (C) on the gears are aligned.
20. On engines with a vertical crankshaft, lubricate and install the oil pump (**Figure 151**) on the camshaft. When installing the pump, note that one side of the loop end of the pump is chamfered (**Figure 152**). The chamfered side must be towards the camshaft gear.

NOTE
*The position of the ball end of the pump may be other than as shown in **Figure 151**. Position the ball end so it corresponds to the position of the recess in the oil pan.*

21. Install the crankcase gasket.
22. Install the crankcase cover or oil pan while mating the gear teeth on the camshaft and governor gear (or idler gear on VLV50, VLXL50, VLV55 and VLXL55). On engines with a vertical crankshaft, be sure the oil pump plunger ball fits into the recess in the oil pan.

NOTE
On engines with an auxiliary shaft drive, it may be necessary to rotate the crankshaft or auxiliary shaft so the worm gear and auxiliary shaft gear don't bind during assembly.

NOTE
Do not force the oil pan or crankcase cover onto the crankcase. If binding occurs, remove the oil pan or crankcase cover and determine the cause.

23. Apply Loctite 242 (blue) to the crankcase cover or oil pan retaining screws and install the screws. Be sure any screws with different lengths are in the proper holes.
24. Tighten the oil pan or crankcase cover screws evenly in a crossing pattern to 110-130 in.-lb. (12.4-14.7 N•m).

NOTE
*Refer to **Table 2** in Chapter Three for engine model number and displacement specifications.*

25A. On horizontal crankshaft engines with a displacement of 12.04 cu. in. (197 cc) or less (except

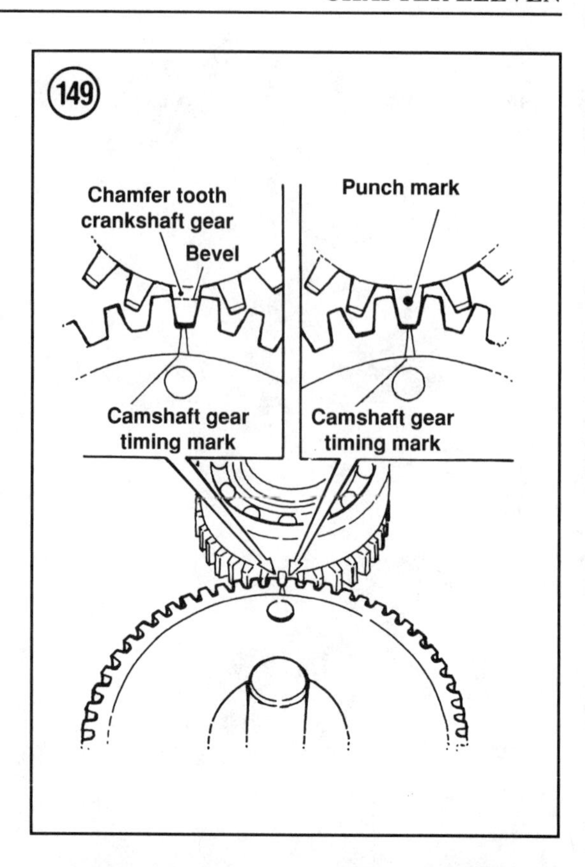

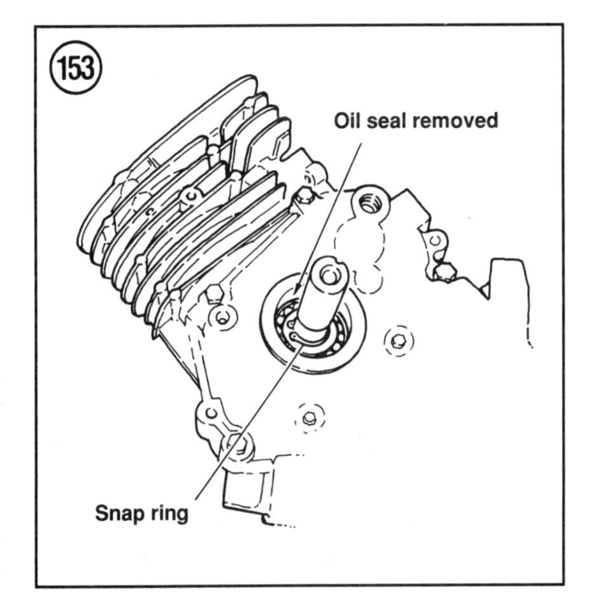

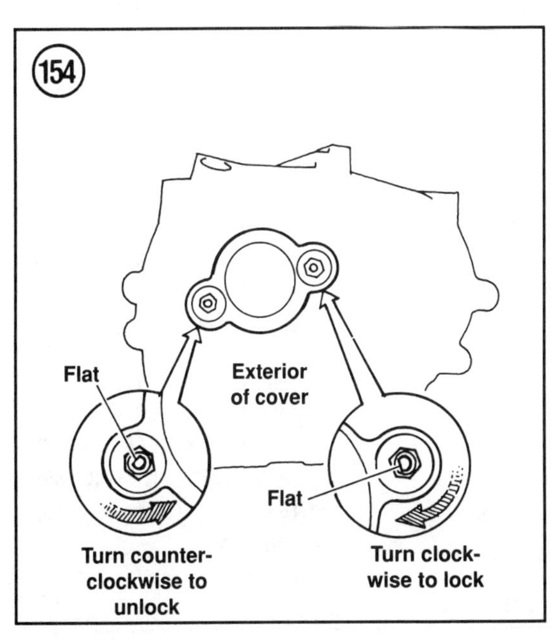

Model H40) that are equipped with a ball bearing on the output end of the crankshaft, install the snap ring (**Figure 153**) then install the oil seal.

25B. On Model H40 and horizontal crankshaft engines with a displacement greater than 12.04 cu. in. (197 cc) and equipped with a ball bearing in the crankcase cover, proceed as follows:

a. Rotate each retaining screw clockwise so the flat on the screw faces the crankshaft (**Figure 154**), then tighten the locknuts on the screws.
b. Install the oil seal.

26. Check crankshaft end play. End play is the distance the crankshaft can move along its axis.

a. Set up a dial indicator so that the in and out movement (end play) of the crankshaft can be measured. If the crankshaft is long, the dial indicator can be mounted on the crankshaft with the plunger against the oil pan or crankcase cover (**Figure 155**). If the crankshaft is short, the dial indicator plunger should rest against the end of the crankshaft (**Figure 156**).
b. Move the crankshaft in and out and measure the movement (end play). Refer to **Table 1** for the end play specification.

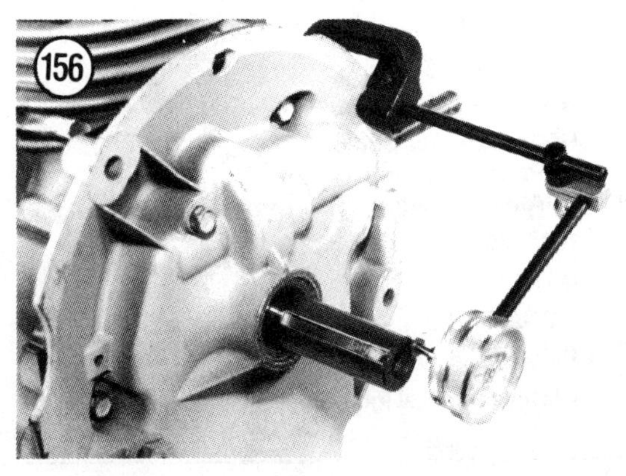

c. If end play is excessive, then there are worn parts which must be replaced or a required washer was not installed.

Install External Engine Components

Install the following components, if the engine is so equipped, as outlined in the appropriate sections of this manual and perform any needed adjustments.

1. Install the intake and exhaust valves, valve springs and retainers. Measure the clearance between the end of the valve stems and tappets as outlined in *Valve System* section in this chapter.
2. Install the crankcase breather.
3. Install the cylinder head with a new head gasket.
4. On models so equipped, install ignition system components that fit under the flywheel (if required, refer to Chapter Five and adjust the breaker point gap and ignition timing).
5. Install the alternator stator if so equipped.
6. Install the flywheel brake if so equipped.
7. Install the flywheel and key, and if so equipped, the plastic spacer on the crankshaft. Tighten the flywheel nut to the torque specified in **Table 2**.
8. On models so equipped, install external ignition components (refer to Chapter Five and adjust the ignition coil air gap).
9. Install the electric starter motor if so equipped.
10. Install the rewind starter and/or blower housing.
11. Install the carburetor and control linkage (refer to Chapter Six and adjust the governor linkage).
12. Install the muffler.
13. Install the air cleaner.
14. Fill the engine with oil.
15. Install the engine on the equipment.
16. Check and adjust the carburetor mixture settings, if applicable, and engine speed as outlined in Chapter Five.

Table 1 ENGINE SERVICE SPECIFICATIONS

	in. (mm)
Camshaft journal diameter	
H40, V40	0.6230-0.6235 (15.824-15.837)
V40 (external ignition)	0.4975-0.4980 (12.637-12.649)
VLV50, VLV55,VLXL50, VLXL55	0.4975-0.4980 (12.637-12.649)
Engines with less than 12.10 cu. in. (198 cc) displacement (except as noted above)	0.4975-0.4980 (12.637-12.649)
Engines with greater than 12.10 cu. in. (198 cc) displacement (except as noted above)	0.6230-0.6235 (15.824-15.837)
Connecting rod big end diameter	
ECV100, ECH90, H25, H30, H35 (prior to 1983), LAV25, LAV30, LAV35, LV35, TNT100, TVS75, TVS90, TVS100, TVS115	0.8620-0.8625 (21.895-21.908)
ECV105, ECV110, ECV120, H35 (after 1982), HS40, HS50, LAV40, LAV50, V40 (external ignition), TNT120, TVS105, TVS120	1.0005-1.0010 (25.413-25.425)
H40, H50, H60, TVM125, TVM140, V40, V50, V60	1.0630-1.0636 (27.000-27.013)
H70, HM70	1.1880-1.1885 (30.175-30.188)
H70 (external ignition)	1.0630-1.0636 (27.000-27.013)
HM70 (external ignition)	
Prior to type letter D	1.1880-1.1885 (30.175-30.188)
After type letter C	1.3760-1.3765 (34.950-34.963)
HMXL70	1.3760-1.3765 (34.950-34.963)
H80	1.1880-1.1885 (30.175-30.188)
HM80, HM100	1.1880-1.1885 (30.175-30.188)
HM80 (external ignition)	See note 1
HM100 (external ignition)	See note 1

(continued)

Table 1 ENGINE SERVICE SPECIFICATIONS (continued)

	in. (mm)
Connecting rod big end diameter (continued)	
TVM170	1.1880-1.1885 (30.175-30.188)
TVM170 (external ignition)	
Prior to type letter F	1.1880-1.1885 (30.175-30.188)
After type letter E	1.3760-1.3765 (34.950-34.963)
TVM195, TVM220	1.1880-1.1885 (30.175-30.188)
TVM195 (external ignition)	
Prior to type letter L	1.1880-1.1885 (30.175-30.188)
After type letter K	1.3760-1.3765 (34.950-34.963)
TVXL195 (external ignition)	See note 1
TVM220 (external ignition)	
Prior to type letter G	1.1880-1.1885 (30.175-30.188)
After type letter F	1.3760-1.3765 (34.950-34.963)
TVXL220 (external ignition)	See note 1
V70, V80, VM70, VM80, VM100	1.1880-1.1885 (30.175-30.188)
V70 (external ignition)	1.0630-1.0636 (27.000-27.013)
VLV50, VLV55,VLXL50, VLXL55	1.0240-1.0246 (25.088-25.103)
Crankpin diameter	
ECV100, ECH90, H25, H30, H35 (prior to 1983), LAV25, LAV30, LAV35,LV35, TNT100, TVS75, TVS90, TVS100, TVS115	0.8610-0.8615 (21.869-21.882)
ECV105, ECV110, ECV120, H35 (after 1982), HS40, HS50, LAV40, LAV50, V40 (external ignition), TNT120, TVS105, TVS120	0.9995-1.0000 (25.390-25.400)
H40, H50, H60, TVM125, TVM140, V40, V50, V60	1.0615-1.0620 (26.962-26.975)
H70, HM70	1.1870-1.1875 (30.150-30.162)
H70 (external ignition)	1.0615-1.0620 (26.962-26.975)
HM70 (external ignition)	
Prior to type letter D	1.1870-1.1875 (30.150-30.162)
After type letter C	1.3740-1.3745 (34.900-34.912)
HMXL70	1.3740-1.3745 (34.900-34.912)
H80	1.1870-1.1875 (30.150-30.162)
HM80, HM100	1.1870-1.1875 (30.150-30.162)
HM80 (external ignition)	See note 2
HM100 (external ignition)	See note 2
TVM170	1.1870-1.1875 (30.150-30.162)
TVM170 (external ignition)	
Prior to type letter F	1.0615-1.0620 (26.962-26.975)
After type letter E	1.3740-1.3745 (34.900-34.912)
TVM195, TVM220	1.1870-1.1875 (30.150-30.162)
TVM195 (external ignition)	
Prior to type letter L	1.1870-1.1875 (30.150-30.162)
After type letter K	1.3740-1.3745 (34.900-34.912)
TVXL195 (external ignition)	See note 2
TVM220 (external ignition)	
Prior to type letter G	1.1870-1.1875 (30.150-30.162)
After type letter F	1.3740-1.3745 (34.900-34.912)
TVXL220 (external ignition)	See note 2
V70, V80, VM70, VM80, VM100	1.1870-1.1875 (30.150-30.162)
V70 (external ignition)	1.0615-1.0620 (26.962-26.975)

(continued)

Table 1 ENGINE SERVICE SPECIFICATIONS (continued)

	in. (mm)
Crankpin diameter (continued)	
VLV50, VLV55, VLXL50, VLXL55	1.0230-1.0235 (25.088-25.103)
Crankshaft end play	
TVM220 (after type letter F) & TVXL220	0.005-0.035 (0.13-0.89)
TVM220 & TVXL220 with balancer	0.000-0.040 (0.00-1.02)
HM100 with external ignition	
1.186 in. crankpin OD	0.005-0.027 (0.13-0.69)
1.374 in. crankpin OD	0.005-0.035 (0.13-0.89)
All other models	0.005-0.027 (0.13-0.69)
Cylinder bore diameter (standard)	
ECH90, H30 (after 1982), H35, H40, LAV35, LV35, TVS90, V40	2.5000-2.5010 (63.500-63.525)
ECV100, ECV105, HS40, LAV40, TNT100, TVS100, TVS105, V40 (external ignition)	2.6250-2.6260 (66.675-66.700)
ECV110	2.7500-2.7510 (69.850-69.875)
ECV120, HS50, LAV50, TNT120, TVS115, TVS120	2.8120-2.8130 (71.425-71.450)
H25, H30 (prior to 1983), LAV25, LAV30, TVS75	2.3125-2.3135 (58.738-58.763)
H50, H60, TVM125, TVM140, V50, V60	2.6250-2.6260 (66.675-66.700)
H70, V70, VM70	2.750-2.751 (69.85-69.88)
H80, V80	3.062-3.063 (77.77-77.80)
HM70	2.9375-2.9385 (74.612-74.638)
HM70 (external ignition)	
Prior to type letter E	2.9375-2.9385 (74.612-74.638)
After type letter D	3.125-3.126 (79.38-79.40)
HMXL70	3.125-3.126 (79.38-79.40)
HM80, VM80	
Prior to type letter E	3.0620-3.0630 (77.775-77.800)
After type letter D	3.1250-3.1260 (79.375-79.400)
HM100	
Cyl. bore 3.187 in.	3.1870-3.1880 (80.950-80.975)
Cyl. bore 3.312 in.	3.3120-3.3130 (84.125-84.150)
TVM170	
Prior to type letter F	2.9375-2.9385 (74.612-74.638)
After type letter E	3.125-3.126 (79.38-79.40)
TVM195, TVXL195	3.125-3.126 (79.38-79.40)
TVM220, TVXL220	3.312-3.313 (84.12-84.15)
VLV50, VLV55, VLXL50, VLXL55	2.7950-2.7960 (70.993-71.018)
VM100	3.1870-3.1880 (80.950-80.975)
Cylinder bore taper or out-of-round (max.)	0.005 (0.13)
Flywheel brake pad (min.)	0.060 (1.52)
Governor shaft height	
H25, H30, H35, H40, HS40, HS50, ECH90, ECV100, ECV105, ECV110, ECV120, LAV25, LAV30, LAV35, LAV40, LAV50, LV35, TNT100, TNT120, TVS75, TVS90, TVS100, TVS105, TVS115, TVS120, V40	1-21/64 (34.92)
H50, H60, H70, H80, HM70, HM80, HM100, HMXL70	1-7/16 (36.51)
TVM125, TVM140, TVM170,TVM195, TVM220, TVXL195, TVXL220, V50, V60, V70, V80, VM70, VM80, VM100	1-19/32 (40.48)

(continued)

Table 1 ENGINE SERVICE SPECIFICATIONS (continued)

	in. (mm)
Ignition coil armature air gap	0.0125 (0.32)
Main bearing clearance	0.0015-0.0025 (0.038-0.064)
Main bearing diameter	
ECH90, ECV100, H25, H30 (prior to 1983), LAV25, LAV30, LAV35, LV35, TVS75, TVS90	0.8755-0.8760 (22.238-22.250)
ECV105, ECV110, ECV120, H35 (after 1982), H40, H50, H60, HS40, HS50, LAV40, LAV50, TNT120, TVM125, TVM140, TVS105, TVS115, TVS120, V40, V50, V60	1.0005-1.0010 (25.413-25.425)
ECV100 (external ignition), H30 (after 1982), TNT100, T VS75 (external ignition), TVS90 (external ignition), TVS100	
Crankcase	1.0005-1.0010 (25.413-25.425)
Crankcase cover/oil pan	0.8755-0.8760 (22.238-22.250)
H70, V70, VM70	1.0005-1.0010 (25.413-25.425)
HM70, HM80, HM100	
Early models	
Crankcase	1.0005-1.0010 (25.413-25.425)
Crankcase cover	1.1890-1.1895 (30.201-30.213)
Later models	1.3765-1.3770 (34.963-34.976)
HMXL70	1.3765-1.3770 (34.963-34.976)
H80	
Crankcase	1.0005-1.0010 (25.413-25.425)
Crankcase cover	1.1890-1.1895 (30.201-30.213)
V80, VM80, VM100	
Crankcase	1.0005-1.0010 (25.413-25.425)
Oil pan	1.1890-1.1895 (30.201-30.213)
TVM170, TVM195, TVM220	
Early models	
Crankcase	1.0005-1.0010 (25.413-25.425)
Oil pan	1.1890-1.1895 (30.201-30.213)
Later models	1.3765-1.3770 (34.963-34.976)
TVXL195, TVXL220	1.3765-1.3770 (34.963-34.976)
VLV50, VLV55, VLXL50, VLXL55	1.0257-1.0262 (26.053-26.066)
Main bearing journal diameter	
ECH90, ECV100, H25, H30 (prior to 1983) LAV25, LAV30, LAV35, LV35, TVS75, TVS90	0.8735-0.8740 (22.187-22.200)
ECV105, ECV110, ECV120, H35 (after 1982), H40, H50, H60, HS40, HS50, LAV40, LAV50, TNT120, TVM125, TVM140, TVS105, TVS115, TVS120, V40, V50, V60	0.9985-0.9990 (25.362-25.375)
ECV100 (external ignition), H30 (after 1982), TNT100, TVS75 (external ignition), TVS90 (external ignition), TVS100	
Flywheel end	0.9985-0.9990 (25.362-25.375)
Output end	0.8735-0.8740 (22.187-22.200)
H70, V70, VM70	0.9985-0.9990 (25.362-25.375)
HM70, HM80, HM100	
Early models	
Flywheel end	0.9985-0.9990 (25.362-25.375)
Output end	1.1870-1.1875 (30.150-30.162)
Later models	1.3745-1.3750 (34.912-34.925)
HMXL70	1.3745-1.3750 (34.912-34.925)

(continued)

Table 1 ENGINE SERVICE SPECIFICATIONS (continued)

	in. (mm)
Main bearing journal diameter (continued)	
H80	
Flywheel end	0.9985-0.9990 (25.362-25.375)
Output end	1.1870-1.1875 (30.150-30.162)
TVM170, TVM195, TVM220	
Early models	
Flywheel end	0.9985-0.9990 (25.362-25.375)
Output end	1.1870-1.1875 (30.150-30.162)
Later models	1.3745-1.3750 (34.912-34.925)
TVXL195, TVXL220	1.3745-1.3750 (34.912-34.925)
V80, VM80, VM100	
Flywheel end	0.9985-0.9990 (25.362-25.375)
Output end	1.1870-1.1875 (30.150-30.162)
VLV50, VLV55, VLXL50, VLXL55	1.0237-1.0242 (26.002-26.015)
Piston clearance	
ECH90, ECV100, ECV105, ECV120, H30 (after 1982), H35 (prior to 1983), HS40, HS50, LAV35, LAV40, LAV50, LV35, TNT100, TNT120, TVS90, TVS100, TVS105, TVS115, TVS120, V40 (external ignition)	0.0040-0.0058 (0.102-0.147)
ECV110	0.0045-0.0060 (0.114-0.152)
H25, H30 (prior to 1983), LAV25, LAV30, TVS75	0.0025-0.0043 (0.064-0.110)
H40, V40	0.0055-0.0070 (0.140-0.178)
H50, H60, TVM125, TVM140, V50, V60	0.0035-0.0050 (0.089-0.127)
H50 (external ignition), H60 (external ignition), TVM125 (external ignition), TVM14O (external ignition)	0.0030-0.0048 (0.076-0.123)
H70, V70, VM70	0.0045-0.0060 (0.114-0.152)
H80, V80, VM80	0.0035-0.0055 (0.089-0.140)
HM80	
Cyl. bore 3.062 in.	0.0035-0.0055 (0.089-0.140)
Cyl. bore 3.125 in.	0.0045-0.0065 (0.114-0.165)
HM80 (external ignition)	0.0045-0.0065 (0.114-0.165)
HM70, HMXL70, TVM170	0.004-0.006 (0.10-0.15)
HM100	
Cyl. bore 3.187 in.	0.0045-0.0065 (0.114-0.165)
Cyl. bore 3.313 in.	0.0015-0.0040 (0.038-0.102)
TVM170 (external ignition)	
Prior to type letter F	0.0030-0.0048 (0.076-0.123)
After type letter E	0.0045-0.0065 (0.114-0.165)
TVM195, TVXL195, VM100	0.0045-0.0065 (0.114-0.165)
TVM220	0.0015-0.0040 (0.038-0.102)
TVM220 (external ignition)	
Prior to type letter G	0.0015-0.0040 (0.038-0.102)
After type letter F	0.0012-0.0032 (0.030-0.081)
TVXL220	0.0012-0.0032 (0.030-0.081)
V70 (external ignition)	0.0030-0.0048 (0.076-0.123)
VLV50, VLXL50, VLV55, VLXL55	0.004-0.006 (0.10-0.15)
Piston diameter (standard)	
ECH90, H30 (after 1982), H35, LAV35, LV35, TVS90	2.4952-2.4960 (63.378-63.398)
ECV100, ECV105, HS40, LAV40, TNT100, TVS100, TVS105, V40 (external ignition)	2.6202-2.6210 (66.553-66.573)

(continued)

Table 1 ENGINE SERVICE SPECIFICATIONS (continued)

	in. (mm)
Piston diameter (standard) (continued)	
ECV110	2.7450-2.7455 (69.723-69.736)
ECV120, HS50, LAV50, TNT120, TVS115, TVS120	2.8072-2.8080 (71.303-71.323)
H25, H30 (prior to 1983), LAV25, LAV30, TVS75	2.3092-2.3100 (58.654-58.674)
H40, V40	2.4945-2.4950 (63.360-63.373)
H50, H60, TVM125, TVM140, V50, V60	2.6210-2.6215 (66.573-66.586)
TVM125 (external ignition), TVM140 (external ignition)	2.6212-2.6220 (66.578-66.599)
H70, V70, VM70	2.7450-2.7455 (69.723-69.736)
H80, V80	3.0575-3.0585 (77.660-77.686)
HM70	2.9325-2.9335 (74.486-74.511)
HM70 (external ignition)	
Prior to type letter E	2.9325-2.9335 (74.486-74.511)
After type letter D	3.1195-3.1205 (79.235-79.261)
HMXL70	3.1195-3.1205 (79.235-79.261)
HM80, VM80	
Prior to type letter E	3.0575-3.0585 (77.660-77.686)
After type letter D	3.1195-3.1205 (79.235-79.261)
HM100	
Cyl. bore 3.187 in.	3.1817-3.1842 (80.815-80.879)
Cyl. bore 3.313 in.	3.308-3.310 (84.023-84.074)
TVM170	
Prior to type letter F	2.9325-2.9335 (74.486-74.511)
After type letter E	3.1195-3.1205 (79.235-79.261)
TVM195, TVXL195	3.1195-3.1205 (79.235-79.261)
TVM220, TVXL220	3.308-3.310 (84.023-84.074)
VLV50, VLV55, VLXL50, VLXL55	2.7900-2.7910 (70.866-70.891)
VM100	3.1817-3.1842 (80.815-80.879)
Piston ring end gap	
Engine displacement 12.04 cu. in. (197 cc) or less (without external ignition)	0.007-0.017 (0.18-0.43)
Engine displacement exceeds 12.04 cu. in. (197 cc) or with external ignition	0.010-0.020 (0.25-0.50)
Piston ring side clearance	
ECH90, H25, H30, H35, LAV25, LAV30, LAV35, LV35, TVS75, TVS90	
Compression rings	0.002-0.005 (0.05-0.13)
Oil control ring	0.0005-0.0035 (0.013-0.089)
ECV100, ECV105, ECV120, H40, H50 (external ignition), H60 (external ignition), HS40, HS50, LAV40, LAV50, TNT100, TNT120, TVM125 (external ignition), TVM140 (external ignition), TVS100, TVS105, TVS115, TVS120, V40	
Compression rings	0.002-0.005 (0.05-0.13)
Oil control ring	0.001-0.004 (0.03-0.10)
ECV110	
Compression rings	0.002-0.004 (0.05-0.10)
Oil control ring	0.001-0.002 (0.03-0.05)
H50, H60, TVM125, TVM140, V50, V60	
Compression rings	0.002-0.004 (0.05-0.10)
Oil control ring	0.002-0.004 (0.05-0.10)

(continued)

11

Table 1 ENGINE SERVICE SPECIFICATIONS (continued)

	in. (mm)
Piston ring side clearance (continued)	
H70, V70, VM70	
Compression rings	0.002-0.003 (0.05-0.08)
Oil control ring	0.001-0.003 (0.03-0.08)
H70 (external ignition), H80, HM70, HMXL70, HM80, HM100, TVM170, TVM195, TVXL195, TVM220, V70 (external ignition), V80, VM80, VM100	
Compression rings	0.002-0.005 (0.05-0.13)
Oil control ring	0.001-0.004 (0.03-0.10)
HM100 (external ignition), TVM220 (external ignition), TVXL220	
Compression rings	0.0015-0.0035 (0.038-0.089)
Oil control ring	0.001-0.004 (0.03-0.10)
VLV50, VLXL50, VLV55, VLXL55	
Compression rings	0.002-0.005 (0.05-0.13)
Oil control ring	0.0005-0.0035 (0.013-0.089)
Spark plug electrode gap	0.030 (0.76)
Tecumseh float carburetor float level	0.162-0.215 (4.11-5.46)
Valve head margin (min.)	1/32 (0.8)
Valve seat width	
H40, H50, H60, H70, HM70, H80, HM80, HM100, HMXL70, TVM125, TVM140, TVM170, TVM195, TVXL195, TVM220, TVXL220, V40, V50, V60, V70, VM70, V80, VM80, VM100	0.042-0.052 (1.07-1.32)
All other models	0.035-0.045 (0.89-1.14)
Valve tappet gap (both valves)	
H50 (ignition under flywheel), H60, H70, HM70, H80, HM80, HM100, TVM125 (ignition under flywheel), TVM140, TVM170, TVM195, TVXL195, TVM220, TVXL220, V50, V60, V70, VM70, V80, VM100	0.010 (0.25)
All other models	0.008 (0.20)
Walbro LME carburetor float level	
8 horsepower engines	0.070-0.110 (1.78-2.79)
All other engines	0.110-0.130 (2.79-3.30)

1. HM80, HM100, TVXL195 and TVXL220 engines with an external ignition may have a connecting rod big end diameter of either 1.1880-1.1885 in. (30.175-30.188 mm) or 1.3760-1.3765 in. (34.950-34.963 mm). Check the unworn portion of the crankshaft crankpin to determine the original specification.
2. HM80, HM100, TVXL195 and TVXL220 engines with an external ignition may have a crankpin diameter of either 1.1870-1.1875 in. (30.150-30.162 mm) or 1.3740-1.3745 in. (34.900-34.912 mm). Check the unworn portion of the crankshaft crankpin to determine the original specification.

Table 2 ENGINE TIGHTENING TORQUES

	ft.-lb.	in.-lb.	N·m
Carburetor to intake pipe		48-72	5.4-8.1
Connecting rod			
Standard fastener			
2.5 hp, 3.0 hp, ECH90, ECV100, TNT100		65-75	7-9
4 hp & 5 hp light frame models, ECV105, ECV110, ECV120, TNT120		80-95	9-11
5 hp medium frame, 6 hp		86-110	10-12
7 hp and up		106-130	12-14.7

Table 2 ENGINE TIGHTENING TORQUES (continued)

	ft.-lb.	in.-lb.	N•m
Durlock Screw			
5 hp medium frame, 6 hp		160-180	18-20.3
7 hp and up		200-220	22.6-24.8
All other models		95-110	11-12
Crankcase cover		100-130	11.3-14.7
Cylinder head bolts			
Vector VLV50, VLXL50, VLV55, VLXL55		180-220	20.3-24.9
All other engines		160-200	18.1-22.6
Flywheel nut			
Light frame & VLV50, VLXL50, VLV55 and VLXL55	33-36		45-49
Medium frame (ignition coil behind flywheel)	35-42		48-57
Medium frame (ignition coil outside flywheel)	50-55		68-75
Intake pipe to cylinder		72-96	8.2-10.8
Muffler (small frame)		20-35	2.3-3.9
Muffler (medium frame)		90-150	10.2-16.9
Oil pan		100-130	11.3-14.7
Spark plug	15		20
Starter, electric		50-80	5.7-9.0
Starter, rewind			
Top mount		40-60	4.5-6.7
Side mount		50-70	5.7-7.9

Table 3 INCH SERIES TORQUE CHART*

SAE grade	Head markings	SAE grade	Nut markings
SAE grade 1 SAE grade 2	 No mark	2	 No mark
SAE grade 5		5	
SAE grade 5.1			
SAE grade 5.2			
SAE grade 8		8	
SAE grade 8.2			

11

Table 3 INCH SERIES TORQUE CHART* (continued)

Diameter	Wrench size	SAE grade 1 Oil in.-lb. (N•m)	SAE grade 1 Dry in.-lb. (N•m)	SAE grade 2 Oil in.-lb. (N•m)	SAE grade 2 Dry in.-lb. (N•m)
#6	—	4.5 (0.5)	6 (0.7)	7 (0.8)	10 (1)
#8	—	8 (0.9)	11 (1.2)	13 (1.5)	18 (2)
#10	—	12 (1.4)	16 (1.8)	19 (2)	25 (2.8)
#12	—	19 (2)	25 (2.8)	30 (3.4)	40 (4.5)
1/4	7/16	2.5 (3.5)	3.0 (4)	4.0 (5)	5.0 (7)
5/16	1/2	5.0 (7)	6.5 (9)	7.5 (10)	10.0 (14)
3/8	9/16	8.5 (12)	12.0 (16)	14.0 (19)	18.0 (24)
7/16	5/8	14.0 (19)	19.0 (26)	22.0 (30)	30 (41)
1/2	3/4	21.0 (24)	30 (41)	35 (47)	45 (61)
9/16	13/16	30 (41)	40 (54)	50 (68)	65 (88)
5/8	15/16	40 (54)	55 (75)	65 (88)	90 (122)
3/4	1-1/8	75 (102)	100 (136)	120 (163)	160 (217)
7/8	1-5/16	120 (163)	165 (224)	120 (163)	165 (224)
1	1-1/2	180 (244)	245 (332)	180 (244)	245 (332)

Diameter	Wrench size	SAE grade 5 Oil in.-lb. (N•m)	SAE grade 5 Dry in.-lb. (N•m)	SAE grade 8 Oil in.-lb. (N•m)	SAE grade 8 Dry in.-lb. (N•m)
#6		12 (1.4)	15 (1.7)	—	—
#8		21 (2.4)	28 (3.2)	—	—
#10		30 (3.4)	41 (4.6)	—	—
#12		48 (5.4)	65 (7.3)	—	—
1/4	7/16	6.0 (8)	8.0 (11)	8.5 (12)	12 (16)
5/16	1/2	12.0 (16)	17.0 (23)	18.0 (24)	24 (33)
3/8	9/16	22.0 (30)	30 (41)	30 (41)	40 (54)
7/16	5/8	35 (47)	50 (68)	50 (68)	70 (95)
1/2	3/4	55 (75)	75 (102)	75 (102)	105 (142)
9/16	13/16	80 (108)	105 (142)	110 (149)	150 (203)
5/8	15/16	110 (149)	145 (197)	150 (203)	205 (278)
3/4	1-1/8	190 (258)	260 (353)	270 (366)	365 (495)
7/8	1-5/16	305 (414)	415 (563)	435 (590)	590 (800)
1	1-1/2	460 (624)	625 (848)	650 (881)	880 (1193)

* Tighten cap screws having lock nuts to approximately 50 percent of amount shown in chart.

Table 4 METRIC SERIES TORQUE CHART

Property class	Head markings	Property class	Nut markings
4.6	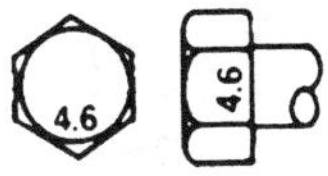	5	
4.8	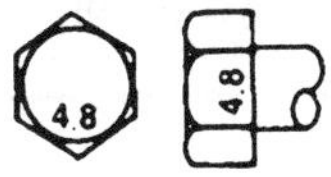		
8.8	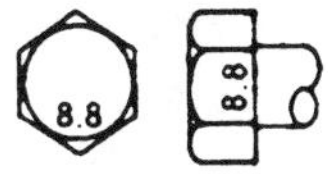	8	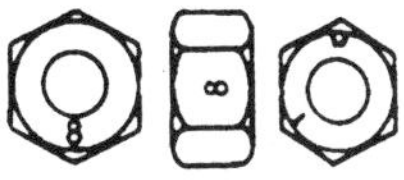
9.8	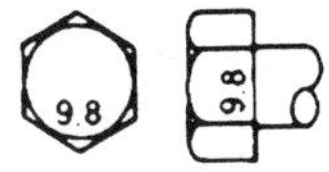	10	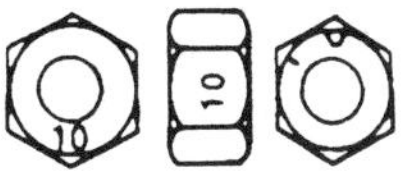
10.9	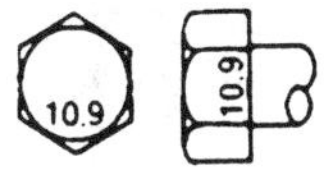		
12.9	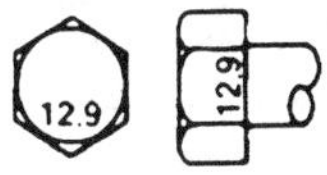	12	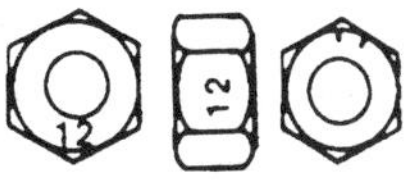

		4.6		4.8	
Diameter	**Wrench size**	**Oil ft.-lb.(N•m)**	**Dry ft.-lb.(N•m)**	**Oil ft.-lb.(N•m**	**Dry ft.-lb.(N•m)**
M5	8 mm	1 (1.5)	1.5 (2.5)	1.5 (2.5)	2 (3.0)
M6	10 mm	2 (3.0)	3 (4.0)	3 (4.0)	4 (5.5)
M8	13 mm	5 (7.0)	7 (9.5)	7.5 (10.0)	10 (13.0)
M10	16 mm	10 (14.0)	14 (19.0)	15 (20.0)	18 (25)
M12	18 mm	18 (25)	26 (35)	26 (35)	33 (45)
M14	21 mm	30 (40)	37 (50)	41 (55)	55 (75)
M16	24 mm	44 (60)	59 (80)	52 (85)	85 (115)
M18	27 mm	59 (80)	81 (110)	74 (115)	118 (160)
M20	30 mm	85 (115)	118 (160)	122 (165)	166 (225)
M22	33 mm	118 (160)	159 (215)	167 (225)	225 (305)
M24	36 mm	148 (200)	203 (275)	210 (285)	288 (390)
M27	41 mm	218 (295)	295 (400)	306 (415)	417 (565)
M30	46 mm	295 (400)	402 (545)	417 (565)	568 (770)

		8.8		9.8	
Diameter	**Wrench size**	**Oil ft.-lb. (N•m)**	**Dry ft.-lb. (N•m)**	**Oil ft.-lb. (N•m)**	**Dry ft.-lb. (N•m)**
M5	8 mm	3.5 (4.5)	4.5 (6.0)	3.5 (5.0)	5 (7.0)
M6	10 mm	5.5 (7.5)	7.5 (10.0)	6 (8.5)	9 (12.0)
M8	13 mm	13 (18.0)	18 (25)	15 (21.0)	22 (30)
M10	16 mm	26 (35)	37 (50)	30 (40)	41 (55)

(continued)

11

Table 4 METRIC SERIES TORQUE CHART (continued)

Diameter	Wrench size	8.8 Oil ft.-lb. (N•m)	8.8 Dry ft.-lb. (N•m)	9.8 Oil ft.-lb. (N•m)	9.8 Dry ft.-lb. (N•m)
M12	18 mm	48 (65)	63 (85)	52 (70)	74 (100)
M14	21 mm	74 (100)	103 (140)	85 (115)	114 (155)
M16	24 mm	118 (160)	159 (215)	133 (180)	180 (245)
M18	27 mm	166 (225)	225 (305)	—	—
M20	30 mm	236 (320)	321 (435)	—	—
M22	33 mm	321 (435)	435 (590)	—	—
M24	36 mm	409 (555)	553 (750)	—	—
M27	41 mm	597 (810)	811 (1100)	—	—
M30	46 mm	811 (1100)	1103 (1495)	—	—

Diameter	Wrench size	10.9 Oil ft.-lb. (N•m)	10.9 Dry 6.5ft.-lb. (N•m)	12.9 Oil ft.-lb. (N•m)	12.9 Dry ft.-lb. (N•m)
M5	8 mm	4.5 (6.5)	6.5 (9.0)	5.5 (7.5)	7.5 (10.0)
M6	10 mm	8 (11.0)	11 (15.0)	9.5 (13.0)	13 (18.0)
M8	13 mm	18 (25)	26 (35)	22 (30)	33 (45)
M10	16 mm	41 (55)	55 (75)	48 (65)	63 (85)
M12	18 mm	70 (95)	97 (130)	81 (110)	111 (150)
M14	21 mm	111 (150)	151 (205)	129 (175)	177 (240)
M16	24 mm	173 (235)	232 (315)	203 (275)	273 (370)
M18	27 mm	236 (320)	321 (435)	277 (375)	376 (510)
M20	30 mm	356 (455)	457 (620)	395 (535)	535 (725)
M22	33 mm	457 (620)	620 (840)	535 (725)	726 (985)
M24	36 mm	583 (790)	789 (1070)	682 (925)	926 (1255)
M27	41 mm	852 (1155)	1154 (1565)	996 (1350)	1353 (1835)
M30	46 mm	1158 (1570)	1571 (2130)	1353 (1835)	1837 (2490)

CHAPTER TWELVE

TECUMSEH TOOLS

Tecumseh offers a wide range of tools and products to service and maintain their engines. Refer to Tecumseh Specialty Tools manual 694862 for a complete list. Tecumseh tools can be obtained through a dealer or distributor. See Chapter Three for a list of Tecumseh distributors who can provide the name of a local dealer.

Some special Tecumseh tools as well as more commonly used engine tools are found in **Table 1**.

Table 1 TECUMSEH TOOLS

Description	Tool No.	Applicable model series
Carburetor float measuring tool	670253A	Tecumseh float carburetors
Dial indicator	670241	Determine piston position
Flywheel puller	670306	Flywheels with puller holes
Flywheel knockoff nut	670103	7/16 in. diameter crankshaft
Flywheel knockoff nut	670169	1/2 in. diameter crankshaft
Flywheel knockoff nut	670314	5/8 in. diameter crankshaft
Ignition coil air gap gauge (0.0125 in.)	670297	All engines
Oil seal remover	670287	Crankshaft output end of ECV90-100, H & LAV30-35, TNT100, TVS75-90
Oil seal remover	670288	Crankshaft flywheel end of most H25-35, LAV30-35, and ECH, ECV, TNT, TVS without external ignition
Oil seal remover	670289	Crankshaft flywheel end of most H50-60, HS40-50, LAV40-50, V50-60 and TNT, TVS with external ignition
Oil seal remover	670290	Crankshaft flywheel end of most H70, HM70-100, V70, VM70-100 Crankshaft output end of ECV120, H50-70, HS40-50, LAV40-50, TNT120, TVS105-120, V50-70
Oil seal driver/protector	670260	Crankshaft output end of HM70-100, VM70-100
Oil seal driver/protector	670261	Crankshaft flywheel end of H50-60, HS40-50, LAV40-50, V50-60 and ECV, TNT, TVS with external ignition

(continued)

Table 1 TECUMSEH TOOLS (continued)

Description	Tool No.	Applicable model series
Oil seal driver/protector	670262	Crankshaft flywheel end of H25-35, LAV30-35 and ECH, ECV, TNT, TVS without external ignition
Oil seal driver/protector	670263	Seal on extended camshaft of medium frame engines
Oil seal driver/protector	670264	Seal on extended camshaft of light frame engines
Oil seal driver/protector	670265	Crankshaft flywheel end of H50-70, HM70-100, V50-70, VM70-100 Crankshaft output end of H50-70, HS40-50, LAV40-50, TNT120, TVS105-120, V50-70
Oil seal driver/protector	670268	Crankshaft output end of ECV90-100, H30-35, LAV30-35, TNT100, TVS75-90
Oil seal driver/protector	670277	Seal on auxiliary drive shaft
Oil seal driver/protector	670292	Crankshaft flywheel end of H70, HM70-100, V70, VM70-100 Crankshaft output end of ECV120, H50-70, HS40-50, LAV40-50, V50-70, TNT120, TVS105-120
Oil seal driver/protector	670293	Crankshaft output end of ECV100, H30-35, LAV30-35, TNT100, TVS75-90
Oil seal driver	670308	Crankshaft flywheel end of HM70-100, TVM170-220
Oil seal protector	670309	Crankshaft flywheel end of HM70-100, TVM170-220
Oil seal driver/protector	670310	Crankshaft output end of HM70-100, TVM170-220
Oil seal driver/protector	670327	Vector engines
Strap wrench	670305	All engines
Tachometer	670156	All engines
Valve guide reamer	670283	Ream 1/4 in. valve guide for oversize valve
Valve guide reamer	670284	Ream 5/16 in. valve guide for oversize valve
Valve lapping tool	670154	All engines

CHAPTER THIRTEEN

CRAFTSMAN/TECUMSEH MODEL NUMBERS

The following Craftsman engines were manufactured by Tecumseh for Sears Roebuck & Co. Locate the Craftsman model in **Table 1** and note the respective Tecumseh model. Follow the service procedures outlined in this manual for the Tecumseh model when servicing the Craftsman engine.

Table 1 CRAFTSMAN/TECUMSEH MODEL NUMBER CROSS-REFERENCE

Craftsman model number	Tecumseh model number	Craftsman model number	Tecumseh model number	Craftsman model number	Tecumseh model number
143.304012	TVS90	143.314092	TVS90	143.314332	TVS90
143.304032	ECV100	143.314102	TVS90	143.314342	TVS90
143.304042	ECV100	143.314112	TVS90	143.314372	ECV100
143.304052	TVS90	143.314122	ECV100	143.314382	TVS90
143.304072	ECV100	143.314132	ECV100	143.314392	ECV100
143.304092	TVS90	143.314142	ECV100	143.314402	TVS90
143.304102	ECV100	143.314152	ECV100	143.314412	TVS90
143.304112	ECV100	143.314162	ECV100	143.314422	ECV100
143.304122	ECV100	143.314172	ECV100	143.314432	LAV35
143.305012	ECV120	143.314182	TVS90	143.314442	ECV100
143.305022	ECV120	143.314192	ECV100	143.314452	ECV100
143.305032	ECV120	143.314202	ECV100	143.314462	ECV100
143.305042	LAV50	143.314212	ECV100	143.314472	ECV100
143.305052	ECV120	143.314222	ECV100	143.314512	ECV100
143.305062	LAV50	143.314232	ECV100	143.314522	ECV100
143.313012	TVS75	143.314242	ECV100	143.314532	LAV35
143.314012	ECV100	143.314252	ECV100	143.314542	TVS90
143.314022	ECV100	143.314262	TVS90	143.314552	TVS90
143.314032	TVS90	143.314272	TVS90	143.314562	TVS900
143.314042	TVS90	143.314282	TVS90	143.314572	TVS90
143.314052	TVS90	143.314292	TVS90	143.314692	ECV100
143.314062	TVS90	143.314302	TVS90	143.314702	LAV35
143.314072	TVS90	143.314312	ECV100	143.315012	ECV120
143.314082	TVS90	143.314322	TVS90	143.315022	LAV50

(continued)

Table 1 CRAFTSMAN/TECUMSEH MODEL NUMBER CROSS-REFERENCE (continued)

Craftsman model number	Tecumseh model number	Craftsman model number	Tecumseh model number	Craftsman model number	Tecumseh model number
143.315032	TVS105	143.334162	TVS90	143.344232	ECV100
143.315042	TVS105	143.334172	ECV100	143.344242	ECV100
143.314642	ECV100	143.334182	ECV100	143.344252	ECV100
143.314652	ECV100	143.334192	LAV35	143.344262	ECV100
143.314662	ECV100	143.334202	ECV100	143.344272	ECV100
143.314672	ECV100	143.334212	ECV100	143.344282	ECV100
143.314682	ECV100	143.334222	ECV100	143.344292	ECV100
143.314582	ECV100	143.334232	ECV100	143.344302	ECV100
143.314592	ECV100	143.334242	ECV100	143.344312	ECV100
143.314612	ECV100	143.334252	ECV100	143.344322	ECV100
143.314622	ECV100	143.334262	TVS90	143.344332	ECV100
143.314632	ECV100	143.334272	TVS90	143.344342	ECV100
143.315062	LAV50	143.334282	TVS90	143.344352	ECV100
143.315072	TVS105	143.334292	TVS90	143.344362	ECV100
143.315082	ECV120	143.334302	TVS90	143.344372	ECV100
143.315092	LAV50	143.334312	TVS90	143.344382	ECV100
143.315102	LAV50	143.334322	ECV100	143.344392	ECV100
143.315112	LAV50	143.334332	TVS90	143.344402	TVXL105
143.315122	LAV50	143.334342	ECV100	143.344412	TVXL105
143.321012	TVS75	143.334352	TVS90	143.344422	TVS90
143.321022	TVS75	143.334362	TVS90	143.344432	TVS90
143.324012	ECV100	143.334372	TVS90	143.344442	TVS105
143.324022	ECV100	143.334382	TVS90	143.344452	ECV100
143.324042	ECV100	143.335012	ECV120	143.344462	TVS105
143.324052	TVS90	143.335022	ECV120	143.344472	ECV100
143.324062	ECV100	143.335032	LAV50	143.345012	ECV120
143.324072	ECV100	143.335042	LAV50	143.345022	ECV120
143.324082	ECV100	143.335052	TVS120	143.345032	TVS120
143.324102	ECV100	143.335062	LAV50	143.345042	LAV50
143.324112	TVS90	143.335072	TVS120	143.345052	ECV120
143.324132	ECV100	143.336012	TVM140	143.345062	ECV120
143.324142	TVS90	143.336022	TVM220	143.346012	TVM220
143.324152	TVS90	143.336032	TVM220	143.346022	TVM220
143.324162	TVS90	143.336042	TVM220	143.346032	TVM170
143.324172	TVS90	143.341012	TVS75	143.346042	TVM195
143.324182	TVXL105	143.344022	TVS90	143.346052	TVM195
143.324192	TVS90	143.344032	TVS90	143.346062	TVM220
143.324202	ECV100	143.344042	TVS90	143.346072	TVM220
143.324212	ECV100	143.344052	ECV100	143.346082	TVM170
143.324222	ECV100	143.344062	ECV100	143.346092	TVM195
143.324232	ECV100	143.344072	TVS90	143.346102	TVM195
143.331012	TVS75	143.344082	ECV100	143.346112	TVM195
143.331022	TVS75	143.344092	ECV100	143.346122	TVM195
143.334022	TVS75	143.344102	TVS90	143.346132	TVM195
143.334032	TVS90	143.344112	TVXL105	143.346142	TVM220
143.334042	ECV100	143.344122	ECV100	143.346152	TVM220
143.334052	TVXL105	143.344132	ECV100	143.346162	TVM220
143.334062	TVS90	143.344142	TVS90	143.346172	TVM220
143.334072	TVS90	143.344152	ECV100	143.346182	TVM220
143.334082	ECV100	143.344162	TVS90	143.346192	TVM220
143.334102	ECV100	143.344172	ECV100	143.346202	TVM125
143.334112	TVS90	143.344182	TVS90	143.351012	TVS75
143.334122	TVS900	143.344192	TVS90	143.351022	TVS75
143.334132	ECV100	143.344202	TVS90	143.354012	TVS90
143.334142	TVS90	143.344212	TVS90	143.354022	ECV100
143.334152	TVS90	143.344222	TVS90	143.354032	ECV100

(continued)

Table 1 CRAFTSMAN/TECUMSEH MODEL NUMBER CROSS-REFERENCE (continued)

Craftsman model number	Tecumseh model number	Craftsman model number	Tecumseh model number	Craftsman model number	Tecumseh model number
143.354042	ECV100	143.356062	TVM125	143.364372	TVS90
143.354052	ECV100	143.356072	TVM195	143.364382	ECV100
143.354062	TVS90	143.356082	TVM220	143.364392	TVS90
143.354072	ECV100	143.356092	TVM220	143.364402	TVS105
143.354082	ECV100	143.356102	TVM170	143.365012	ECV120
143.354092	TVS90	143.356122	TVM195	143.365022	ECV120
143.354102	TVS90	143.356132	TVM195	143.366022	TVM195
143.354112	ECV100	143.356142	TVM195	143.366032	TVM220
143.354122	TVS90	143.356152	TVM195	143.366042	TVM195
143.354132	TVXL105	143.356162	TVM220	143.366052	TVM220
143.354142	TVS90	143.356172	TVM220	143.366062	TVM220
143.354152	ECV100	143.356182	TVM220	143.366082	TVM125
143.354162	TVS90	143.356192	TVM220	143.366102	TVM195
143.354172	TVS90	143.356202	TVM220	143.366112	TVM220
143.354182	TVS90	143.356212	TVM220	143.366122	TVM220
143.354192	TVS90	143.356222	TVM220	143.366132	TVM220
143.354202	TVS90	143.356232	TVM220	143.366152	TVM195
143.354212	TVS90	143.356252	TVM220	143.366182	TVM125
143.354222	ECV100	143.356362	TVM125	143.366192	TVM220
143.354232	TVS90	143.361012	TVS75	143.366222	TVM220
143.354242	ECV100	143.364012	TVS90	143.371012	TVS75
143.354252	ECV100	143.364022	ECV100	143.371022	TVS75
143.354262	ECV100	143.364032	ECV100	143.371032	TVS75
143.354272	ECV100	143.364042	ECV100	143.374012	TVS90
143.354282	LAV35	143.364052	ECV100	143.374022	TVS90
143.354292	TVS90	143.364062	ECV100	143.374032	TVS90
143.354302	ECV100	143.364072	ECV100	143.374052	TVS90
143.354312	TVS90	143.364082	TVS90	143.374062	TVS90
143.354312	TVS90	143.364092	ECV100	143.374072	TVS90
143.354322	TVS90	143.364102	TVS90	143.374082	TVS90
143.354332	TVS90	143.364112	TVS90	143.374092	ECV100
143.354342	TVS90	143.364122	TVS90	143.374102	ECV100
143.354352	TVS90	143.364132	TVS90	143.374112	ECV100
143.354362	ECV100	143.364142	TVS90	143.374122	ECV100
143.354372	ECV100	143.364152	TVXL105	143.374132	ECV100
143.354382	ECV100	143.364162	ECV100	143.374142	ECV100
143.354392	ECV100	143.364172	ECV100	143.374152	ECV100
143.354402	ECV100	143.364182	ECV100	143.374162	ECV100
143.354412	ECV100	143.364192	ECV100	143.374172	ECV100
143.354422	ECV100	143.364202	TVS90	143.374182	ECV100
143.354432	ECV100	143.364212	ECV100	143.374192	ECV100
143.354442	ECV100	143.364222	TVS90	143.374202	ECV100
143.354452	ECV100	143.364232	ECV100	143.374212	TVS90
143.354462	ECV100	143.364242	ECV100	143.374222	TVS90
143.354482	TVS105	143.364252	ECV100	143.374232	TVS90
143.354492	TVS105	143.364262	TVS105	143.374282	TVS90
143.354502	TVS105	143.364272	ECV100	143.374292	TVS105
143.355012	ECV120	143.364282	ECV100	143.374302	TVS90
143.355022	ECV120	143.364292	ECV100	143.374312	TVS105
143.355032	LAV50	143.364302	ECV100	143.374322	TVS90
143.355032	LAV50	143.364312	ECV100	143.374332	TVS90
143.356012	TVM220	143.364322	ECV100	143.374342	ECV100
143.356022	TVM125	143.364332	ECV100	143.374362	TVS90
143.356032	TVM195	143.364342	ECV100	143.374372	TVS105
143.356042	TVM220	143.364352	TVS90	143.374382	TVS90
143.356052	TVM195	143.364362	TVS90	143.374402	ECV100

(continued)

Table 1 CRAFTSMAN/TECUMSEH MODEL NUMBER CROSS-REFERENCE (continued)

Craftsman model number	Tecumseh model number	Craftsman model number	Tecumseh model number	Craftsman model number	Tecumseh model number
143.374412	ECV100	143.384472	ECV100	143.395012	ECV120
143.374422	TVS105	143.384482	ECV100	143.395022	ECV120
143.374432	TVS90	143.384492	ECV100	143.396022	TVXL220
143.374452	ECV100	143.384502	ECV100	143.396042	TVXL220
143.375012	ECV120	143.384512	ECV100	143.396052	TVXL220
143.375022	ECV120	143.384522	ECV100	143.396082	TVXL220
143.375032	LAV50	143.384532	ECV100	143.396102	TVM125
143.375042	LAV50	143.384542	ECV100	143.396122	TVXL220
143.375052	ECV120	143.384552	TVS90	143.401012	TVS75
143.376022	TVM220	143.384562	ECV100	143.404022	TVS75
143.376042	TVM195	143.384572	TVS90	143.404032	TVS90
143.376052	TVM220	143.385012	ECV120	143.404042	TVS105
143.376062	TVM195	143.385022	ECV120	143.404082	TVS105
143.376092	TVM220	143.385032	ECV120	143.404092	TVS105
143.381012	TVS75	143.385042	LAV50	143.404122	TVS120
143.381022	TVS75	143.385052	LAV50	143.404132	TVS105
143.384012	TVS90	143.386022	TVM220	143.404142	TVS105
143.384022	TVS90	143.386042	TVM220	143.404152	TVS120
143.384032	TVS90	143.386052	TVM195	143.404162	TVS105
143.384042	TVS90	143.386062	TVM220	143.404172	TVS105
143.384052	TVS90	143.386072	TVM220	143.404182	TVS120
143.384062	TVS90	143.386082	TVM220	143.404202	TVS105
143.384072	TVS90	143.386122	TVM195	143.404222	TVS105
143.384082	TVS90	143.386132	TVM195	143.404232	TVS105
143.384092	ECV100	143.386142	TVM220	143.404242	TVS105
143.384102	ECV100	143.386172	TVM220	143.404252	TVS105
143.384112	ECV100	143.386182	TVM220	143.404282	TVS105
143.384122	ECV100	143.391012	TVS75	143.404292	TVS120
143.384172	ECV100	143.391022	TVS75	143.404312	TVS105
143.384202	ECV100	143.394012	ECV100	143.404322	TVS105
143.384212	ECV100	143.394022	TVS90	143.404332	TVS105
143.384222	ECV100	143.394032	TVS90	143.404342	TVS90
143.384232	ECV100	143.394042	TVS90	143.404352	TVS90
143.384242	ECV100	143.394052	TVS90	143.404362	TVS105
143.384252	ECV100	143.394062	TVS90	143.404372	TVS105
143.384262	ECV100	143.394072	TVS90	143.404382	TVS105
143.384272	TVS90	143.394082	ECV100	143.404392	TVS105
143.384282	TVS90	143.394122	TVS90	143.404402	TVS120
143.384292	TVS90	143.394132	TVS90	143.404412	TVS105
143.384302	TVS90	143.394142	TVS90	143.404422	TVS105
143.384312	TVS90	143.394152	TVS90	143.404432	TVS105
143.384322	ECV100	143.394162	ECV100	143.404442	TVS105
143.384332	ECV100	143.394172	ECV100	143.404452	TVS105
143.384342	TVS90	143.394182	ECV100	143.404462	TVS105
143.384352	ECV100	143.394222	ECV100	143.404472	TVS120
143.384362	ECV100	143.394232	ECV100	143.404482	TVS120
143.384372	ECV100	143.394242	TVS90	143.404502	TVS90
143.384382	TVS90	143.394252	ECV100	143.404532	TVS90
143.384392	TVS90	143.394262	ECV100	143.406022	TVXL220
143.384402	TVS105	143.394272	ECV100	143.406032	TVXL220
143.384412	TVS105	143.394282	ECV100	143.406042	TVXL220
143.384422	TVS105	143.394302	TVS90	143.406082	TVM125
143.384432	TVS100	143.394322	TVS90	143.406092	TVM195
143.384442	TVS90	143.394372	ECV100	143.406102	TVXL220
143.384452	TVS90	143.394492	TVS90	143.406122	TVXL220
143.384462	ECV100	143.394502	LAV35	143.406172	TVXL195

(continued)

Table 1 CRAFTSMAN/TECUMSEH MODEL NUMBER CROSS-REFERENCE (continued)

Craftsman model number	Tecumseh model number	Craftsman model number	Tecumseh model number	Craftsman model number	Tecumseh model number
143.414012	TVS90	143.414592	TVS105	143.706172	H70
143.414022	TVS105	143.414602	TVS105	143.706182	H70
143.414032	TVS90	143.414612	TVS90	143.706222	HM80
143.414042	TVS90	143.414622	TVS120	143.706232	HM100
143.414052	TVS90	143.414632	TVS105	143.716012	HM70
143.414062	TVS105	143.414642	TVS120	143.716022	HM80
143.414072	TVS105	143.414652	TVS105	143.716052	HM70
143.414082	TVS90	143.414662	TVS105	143.716062	HM70
143.414092	ECV100	143.414672	TVS105	143.716072	HM80
143.414102	ECV100	143.414682	ECV100	143.716082	HM80
143.414112	ECV100	143.414692	TVS100	143.716092	HM80
143.414122	ECV100	143.416022	TVXL195	143.716202	H70
143.414132	ECV100	143.416032	TVXL220	143.716212	HM100
143.414142	ECV100	143.416052	TVM125	143.716222	HM80
143.414152	ECV100	143.416062	TVM125	143.716252	H70
143.414162	ECV100	143.416072	TVXL220	143.716312	H70
143.414172	TVS90	143.424012	TVS90	143.716322	H70
143.414182	TVS90	143.424022	TVS105	143.716332	HM100
143.414192	ECV100	143.424032	TVS90	143.716342	HM70
143.414202	ECV100	143.424042	TVS105	143.716372	HM100
143.414212	TVS90	143.424052	TVS90	143.716382	HM100
143.414232	TVS90	143.424062	TVS120	143.716392	HM80
143.414242	TVS90	143.424072	TVS120	143.716402	H60
143.414252	TVS90	143.424082	TVS105	143.716412	H60
143.414262	ECV100	143.424102	TVS120	143.716422	H50
143.414272	ECV100	143.424112	TVS100	143.716432	HM100
143.414282	TVS90	143.424122	TVS100	143.717012	HS40
143.414292	TVS105	143.424132	TVS100	143.717022	HS40
143.414302	TVS120	143.424142	TVS105	143.717032	HS40
143.414312	TVS105	143.424152	TVS120	143.717042	HS50
143.414322	TVS105	143.424162	TVS105	143.717052	HS40
143.414332	TVS90	143.424172	TVS120	143.717062	HS50
143.414342	TVS105	143.424182	TVS100	143.717072	HS50
143.414352	TVS120	143.424202	TVS90	143.717082	HS50
143.414362	TVS105	143.424312	TVS105	143.717092	HS50
143.414372	TVS105	143.424322	TVS105	143.717102	HS40
143.414382	TVS105	143.424332	TVS120	143.717112	HS50
143.414392	TVS120	143.424342	TVS120	143.721012	H30
143.414402	TVS105	143.424352	TVS105	143.721022	H30
143.414412	TVS105	143.424362	TVS90	143.721032	H30
143.414422	TVS120	143.424372	TVS90	143.724012	H35
143.414442	TVS105	143.424382	TVS105	143.724022	H35
143.414452	TVS105	143.424392	TVS105	143.724032	H35
143.414462	TVS105	143.424402	TVS120	143.724042	H35
143.414472	TVS120	143.426072	TVXL220	143.724052	HS40
143.414482	TVS105	143.706032	HM80	143.725012	HS50
143.414492	TVS120	143.706042	HM80	143.726012	H60
143.414502	TVS90	143.706052	HM80	143.726022	HM100
143.414512	TVS90	143.706072	HM80	143.726032	HM100
143.414522	TVS90	143.706082	HM80	143.726042	H70
143.414532	TVS90	143.706092	H70	143.726052	H70
143.414542	TVS105	143.706112	H70	143.726082	H60
143.414552	TVS120	143.706122	HM80	143.726092	H50
143.414562	TVS105	143.706132	HM100	143.726102	HM100
143.414572	TVS120	143.706152	HM100	143.726112	H60
143.414582	TVS105	143.706162	HM80	143.726132	H60

(continued)

Table 1 CRAFTSMAN/TECUMSEH MODEL NUMBER CROSS-REFERENCE (continued)

Craftsman model number	Tecumseh model number	Craftsman model number	Tecumseh model number	Craftsman model number	Tecumseh model number
143.726182	H70	143.746012	H60	143.766112	HM80
143.726192	H70	143.746022	HM80	143.766122	HM100
143.726202	H70	143.746072	HM100	143.766132	H70
143.726212	H70	143.746082	HM80	143.766142	HM100
143.726222	H60	143.746092	HM80	143.766152	HM80
143.726232	HM100	143.746102	HM100	143.774012	H35
143.726242	H60	143.751012	H30	143.774102	H30
143.726252	H70	143.751022	H30	143.774112	HS40
143.726262	H60	143.751032	H30	143.774122	H35
143.726272	HM80	143.751042	H30	143.775132	HS50
143.726282	H70	143.751052	H30	143.776012	HM80
143.726302	H50	143.751062	H30	143.776022	H70
143.726312	HM100	143.754012	H35	143.776042	HM80
143.726322	HM100	143.754022	H35	143.776052	HM100
143.731012	H30	143.754042	H35	143.776062	HM100
143.734012	H35	143.754052	H35	143.776072	H70
143.734022	H35	143.754062	HS50	143.780012	HH120
143.734032	H35	143.754072	H35	143.784012	HS40
143.734042	HS40	143.754082	HS50	143.784022	HS50
143.735012	HS50	143.754092	HS50	143.784032	HS50
143.735022	HS50	143.754102	H35	143.784042	HS40
143.736032	H60	143.754112	HS50	143.784052	H30
143.736042	HM80	143.754122	HS40	143.784062	H30
143.736052	H60	143.754132	HS50	143.784072	HS50
143.736062	H60	143.754142	HS50	143.784082	H30
143.736072	H50	143.754152	H35	143.784092	H30
143.736082	HM80	143.756012	H60	143.784102	H30
143.736092	H70	143.756022	H60	143.784112	H35
143.736102	HM100	143.756042	H70	143.784122	H35
143.736112	H50	143.756052	H70	143.784132	HS50
143.736122	HM80	143.756062	HM100	143.784142	H35
143.736132	HM80	143.756072	H60	143.784152	HS50
143.736142	HM100	143.756082	HM80	143.784162	H30
143.741012	H30	143.756092	H60	143.784172	H30
143.741022	H30	143.756102	HM80	143.784182	H35
143.741032	H30	143.756132	H70	143.784192	H35
143.741042	H30	143.756152	HM100	143.784202	HS40
143.741052	H30	143.756162	H70	143.786012	HM80
143.741062	H30	143.756172	HM100	143.786022	HM80
143.741072	H30	143.756182	HM80	143.786032	HM80
143.741082	H30	143.756192	HM80	143.786042	HM100
143.741092	H30	143.756202	HM100	143.786052	HM80
143.742032	HM80	143.756212	HM80	143.786062	HM100
143.742042	H50	143.756222	H70	143.786072	HM100
143.742052	H50	143.764012	HS50	143.786082	H60
143.744012	H35	143.764022	HS50	143.786092	HM80
143.744022	H35	143.764032	HS40	143.786102	H70
143.744032	H35	143.764042	HS50	143.786112	HM80
143.744052	H35	143.764052	H35	143.786122	HM100
143.744062	H35	143.764062	HS50	143.786132	H70
143.744072	H35	143.764072	HS50	143.786142	H70
143.744082	H35	143.766012	HM80	143.786152	H70
143.744092	HS50	143.766072	HM80	143.786162	H60
143.744102	HS40	143.766082	HM80	143.786172	HM80
143.744112	HS50	143.766092	HM100	143.786182	HM100
143.744122	H35	143.766102	HM80	143.786192	HM100

(continued)

Table 1 CRAFTSMAN/TECUMSEH MODEL NUMBER CROSS-REFERENCE (continued)

Craftsman model number	Tecumseh model number	Craftsman model number	Tecumseh model number	Craftsman model number	Tecumseh model number
143.786202	HM100	143.796162	HM100	143.806152	HM80
143.794012	HS40	143.796172	HM100	143.806162	HM100
143.794022	HS40	143.796182	H70	143.806172	HM100
143.794032	HS50	143.796192	HM80	143.806182	HM80
143.794042	HS50	143.804062	HS40	143.814012	HS40
143.794052	HS40	143.804072	HS50	143.814022	HS50
143.794053	HS50	143.804082	H30	143.814032	H30
143.794072	HS40	143.804092	H30	143.814042	H30
143.794082	HS50	143.804102	H35	143.814052	H35
143.796012	HM80	143.804112	H30	143.814062	H35
143.796022	HM80	143.806012	HM80	143.814072	H30
143.796032	HM80	143.806022	HM80	143.816012	HM100
143.796042	HM80	143.806032	HM80	143.816022	HM80
143.796052	HM100	143.806042	HM100	143.816032	HM100
143.796062	HM100	143.806052	HM100	143.816052	HM80
143.796072	HM100	143.806072	HM80	143.816072	H70
143.796082	HM80	143.806082	HM80	143.824012	H30
143.796092	HM80	143.806092	HM100	143.824022	H30
143.796102	HM100	143.806102	HM80	143.826012	HM80
143.796122	H70	143.806112	H70	143.826022	HM100
143.796132	HM80	143.806122	H70	143.826032	HM80
143.796142	H70	143.806132	HM100	143.826042	HM80
143.796152	HM80	143.806142	H70		

INDEX

A

Air cleaner 59-60
Auxiliary drive shaft 195

B

Balancer shaft 195
Battery 136
Breaker point cam 109

C

Camshaft and tappets 186-188
Carburetor adjustment
  except Vector VLV50, VLX150, VLV55 and VLX155 68-73
  Vector VLV50, VLX150, VLV55 and VLX155 73-74
Carburetors
  diaphragm 76-82
  Tecumseh float 82-97
  Walbro LME 97-100
Charging/lighting system 136-142
Combustion chamber 75
Compression test 61
Condenser, and ignition breaker points 62-65
Connecting rod 185-186
Controls 49-50
Cooling system 60
Craftsman/Tecumseh model numbers 215-221
Crankcase breather 161-164
Crankshaft and main bearings 191-193
Cylinder 193-194
Cylinder head 166-168

D

Diaphragm carburetors 76-82

E

Electrical precautions 135
Electrical system 39-40
  battery 136
  charging/lighting system 136-142
  precautions 135
  testing 135-136
  troubleshooting 142
Electric starter motors 142-150
Electric starter system 142
Engine
  basic specifications 44
  components, internal 176
  disassembly 176-180
  electrical system 39-40
  fuel system 32-36
  identification 43-44
  ignition system 36-39
  major components 22-32
  operating principles 21-22
Engine overhaul
  assembly 196-202
  auxiliary drive shaft 195
  balancer shaft 195
  camshaft and tappets 186-188
  connecting rod 185-186
  crankcase breather 161-164
  crankshaft and main bearings 191-193
  cylinder 193-194
  cylinder head 166-168
  disassembly 176-180

flywheel 164-166
governor 188-191
internal engine components 176
muffler 161
oil pump 188
oil seals 194-195
oil seals, service with
crankshaft installed 174-176
piston and piston pin 182-185
piston rings 180-182
valve system 168-174

F

Failure analysis 151-154
Fasteners 15-17
Float carburetors, Tecumseh 82-97
Flywheel 164-166
brake 110-114
flywheel and key 108-109
Fuel 42, 49
Fuel system 32-36

G

Governor 188-191
Governor linkage 74-75
Governor system 100-105

H

Hoses and wires 60

I

Ignition breaker points and condenser 62-65
Ignition coil armature air gap (models
with external ignition coil) 65
Ignition coil (breaker-point ignition) 106-107
Ignition module/coil 107-108
Ignition system 36-39
Ignition timing 65-68
Internal engine components 176

L

Lubricants 19-20
Lubrication 57

M

Maintenance 57-58
Measurements, precision 154-157
Model numbers, Craftsman/Tecumseh . . . 215-221
Muffler 61, 161

O

Oil 42-43
Oil change 58-59
Oil pump 188
Oil seals 194-195
service with crankshaft installed 174-176
Operating cable 60-61
Operating linkage 76

P

Piston and piston pin 182-185
Piston rings 180-182
Purchasing parts 44

R

Repair techniques 157-159
Rewind starters
mounted on blower housing 115
stylized 124-128
vertical-pull 128-134

S

Safety devices 58
Sealants, cements and cleaners 17-19
Service hints 3-4
Spark plug 61-62
Starters
early "teardrop" shaped 115-121
stylized rewind 124-128

Starters (continued)
used on HM and VM series engines . . . 121-124
vertical-pull rewind 128-134
Stop switch 109-110
Stylized rewind starter 124-128

T

Tecumseh/craftsman model numbers 215-221
Tecumseh float carburetors 82-97
Test equipment 11-12
Tools
basic hand 5-11
precision measuring 12-14
special 4
Tecumseh 213-214
Troubleshooting 50-55

V

Valve system 168-174
Vertical-pull rewind starters 128-134

W

Walbro LME carburetor 97-100

NOTES

NOTES